The New Nuclear Forensics

Analysis of Nuclear Materials for Security Purposes

STOCKHOLM INTERNATIONAL PEACE RESEARCH INSTITUTE

SIPRI is an independent international institute dedicated to research into conflict, armaments, arms control and disarmament. Established in 1966, SIPRI provides data, analysis and recommendations, based on open sources, to policymakers, researchers, media and the interested public.

STOCKHOLM INTERNATIONAL PEACE RESEARCH INSTITUTE

Signalistgatan 9
SE-169 70 Solna, Sweden
Telephone: +46 8 655 97 00
Fax: +46 8 655 97 33
Email: sipri@sipri.org
Internet: www.sipri.org

The New Nuclear Forensics

Analysis of Nuclear Materials for Security Purposes

EDITED BY VITALY FEDCHENKO

sipri

OXFORD UNIVERSITY PRESS
2015

OXFORD
UNIVERSITY PRESS

Great Clarendon Street, Oxford OX2 6DP,
United Kingdom

Oxford University Press is a department of the University of Oxford.
It furthers the University's objective of excellence in research, scholarship,
and education by publishing worldwide. Oxford is a registered trade mark of
Oxford University Press in the UK and in certain other countries

British Library Cataloguing in Publication Data
Data available

Library of Congress Cataloging in Publication Data
Data available

ISBN 978–0–19–873664–6

Typeset and originated by SIPRI

Contents

Part II. Nuclear forensics in practice

Preface

This book provides a survey and an analysis of the scientific discipline of nuclear forensic analysis, and the way it is applied to specific issues of international peace and security, from the 1940s to the present day. The book uses a language and methodology that opens the issue of nuclear forensics, and its potential applications, to a non-specialist readership.

The world's first nuclear explosion took place on 16 July 1945 in New Mexico, United States. The second and third followed soon after—over two heavily populated Japanese cities—and the military applications of nuclear power became part and parcel of national security policies, and a subject of central concern to international peace and security, where they have remained for 70 years. After 1945, many states realized the need to understand the impact of nuclear weapons on war and politics, and more states began to develop their national nuclear capabilities, sometimes resulting in the production and testing of nuclear weapons.

The starting point for an understanding of the role and impact of nuclear weapons is an understanding of nuclear materials. It is nearly impossible to produce nuclear materials, and especially test nuclear weapons, without introducing some nuclear or other radioactive material into the environment, even if in small quantities. After release, this material can be collected and analysed. Intelligence services were exploring ways and means to learn more about nuclear material since the 1940s, and spent considerable effort developing detection and monitoring capacities and highly sensitive analytical techniques that are applied to nuclear explosions and nuclear material production around the world.

After the cold war, international cooperation was essential, and, for the first time, possible. After the collapse of the Soviet system of governance a number of cases of illicit trafficking of nuclear materials came to light, including materials directly usable in nuclear weapons. The discovery of the undeclared nuclear weapons programme in Iraq after the 1990–91 Gulf War reinforced the need for more and better knowledge about nuclear developments around the world.

International cooperation allowed nuclear intelligence techniques to be applied in new, often less restricted, contexts. Innovative frameworks like the cooperative threat reduction programme, initiatives to strengthen the International Atomic Energy Agency (IAEA) safeguards system, and the extensive verification mechanism of the 1996 Comprehensive Nuclear-Test-Ban Treaty all expanded international efforts to analyse nuclear and other radioactive materials.

The analysis of radioactive material for security purposes has migrated from the mostly classified area of intelligence into open scientific settings and peer-reviewed journals. The number of laboratories engaged in it has increased worldwide, boosted by political support at the highest level at three nuclear security summits, and the maturity and popularity of these methods have reached the point where nuclear forensic analysis (nuclear forensics) should be treated as a separate and unified scientific discipline.

International cooperation in nuclear forensic analysis has found ways to manage concerns that sensitive information may be revealed, and in this way it builds trust and reassurance across a community of states—including those that possess nuclear weapons and those that do not. Moving forward, deeper and wider cooperation in nuclear forensic analysis can be a platform for the verification of future nuclear arms control commitments.

As the editor of this volume, Vitaly Fedchenko should be commended on numerous counts for his achievement in bringing this project into harbour: the seas have not always been smooth.

Dr Ian Anthony
SIPRI Director
Stockholm, June 2015

Acknowledgements

This book would not have been written and published without two organizations that provided necessary funding. I appreciate very much the generosity of the Swedish Radiation Safety Authority (SSM) whose grant made possible the research and writing of this book; and well as crucial assistance of the European Commission's Joint Research Centre's Institute for Transuranium Elements (ITU), which provided the funding necessary for its editing and publication.

I am especially grateful to Klaus Mayer, who provided help and support every step of the way, including at the time when it was particularly needed. David Smith has also provided kind encouragement and taught me a lot about nuclear forensic analysis.

It was a privilege to participate in the sessions of the Nuclear Forensics International Technical Working Group (ITWG), as well as the International Atomic Energy Agency (IAEA) conferences and consultancy meetings on various aspects of nuclear forensics. I greatly appreciate an opportunity to learn from so many inspired individuals there, including Klaus Mayer, Benjamin Garrett, David Smith, Michael Curry, Frank Wong, Uri Admon, Carey Larsson, Emily Kröger, Tamás Biró, Zsolt Varga, Diane Fischer, Gene Cheney, Tanya Hinton, Tegan Bull, David Hill, Stephen La Mont, Maria Wallenius, Henrik Ramebäck, Bjorn Sandström, Victor Erastov, Yuri Panteleev, Vladimir Stebelkov, Jon Schwantes, John Wacker and Jim Blankenship.

James Acton, Lars van Dassen, Ian Anthony, David Cruickshank and Page Stoutland were kind enough to discuss with me an idea for this book at a rather early stage. I am very appreciative of the efforts of Maxim Peńkin and Vladimir Stebelkov, who reviewed an earlier draft of this book and provided comments.

It would have been impossible to even imagine this book without Lars-Erik de Geer, Nicole Erdmann, Sophie Grape, Magnus Hedberg, Robert Kelley, Maria Wallenius, Klaus Mayer and Zsolt Varga, who wrote large parts of the text. I am indebted to them all.

Finally, I would like to express my gratitude to SIPRI editors David Cruickshank, Jetta Gilligan-Borg and, last but not least, Joey Fox, who all went above and beyond their duties while preparing this book for publication.

Vitaly Fedchenko
Stockholm, June 2015

Abbreviations

3F	Fission–fusion–fission (bombs)
ADC	Analogue-to-digital converter
AMS	Accelerator mass spectrometry
BGCS	Beta–gamma coincidence spectrometry
BWR	Boiling water reactor
D	Deuterium (hydrogen-2)
DA	Destructive assay (destructive analysis)
CD	Committee on Disarmament (later the Conference on Disarmament)
CSA	Comprehensive safeguards agreement
CTBT	Comprehensive Nuclear-Test-Ban Treaty
CTBTO	Preparatory Commission for the Comprehensive Nuclear-Test-Ban Treaty Organization
CZT	Cadmium zinc telluride
DIQ	Design information questionnaire
DOE	Department of Energy
EMIS	Electromagnetic isotope separation
ENCD	Eighteen Nation Committee on Disarmament
ES	Environmental Sampling (database)
EU	European Union
FGPu	Fuel-grade plutonium
FMCT	Fissile material cut-off treaty
FOA	Försvarets forskningsanstalt (Swedish National Defence Research Establishment)
FSB	Federal'naya sluzhba bezopasnosti (Russian Federal Security Service)
FT-IR	Fourier-transform infrared spectroscopy
FT-TIMS	Fission-track thermal ionization mass spectrometry
FWHM	Full width at half maximum
GIRM	Graphite isotope ratio method
GSE	Group of Scientific Experts
HASS	High-activity sealed sources
HEU	Highly enriched uranium
HPGe	High-purity germanium
HRGS	High-resolution gamma spectrometry
HWR	Heavy-water reactor

IAEA	International Atomic Energy Agency
IC	Ion chromatography
ICP-MS	Inductively coupled plasma mass spectrometry
ICP-SFMS	Inductively coupled plasma sector field mass spectrometry
IDC	International Data Centre
ID-TIMS	Isotope dilution thermal ionization mass spectrometry
IMS	International Monitoring System
INF	Incident notification form
IR	Infrared
ISO	International Organization for Standardization
ITDB	Incident and Trafficking Database
ITU	Institute for Transuranium Elements
ITWG	Nuclear Forensics International Technical Working Group
LA-ICP-MS	Laser ablation inductively coupled plasma mass spectrometry
LEU	Low-enriched uranium
LWR	Light-water reactor
MCA	Multichannel analyser
MC-ICP-MS	Multi-collector inductively couple plasma mass spectrometry
MOX	Mixed oxide
NDA	Non-destructive assay (non-destructive analysis)
NMIP	Nuclear Materials Information Program
NNFL	National nuclear forensic libraries
NPT	Treaty on the Non-Proliferation of Nuclear Weapons (Non-Proliferation Treaty)
NRL	Naval Research Laboratory
NTM	National technical means of verification
NWAL	Network of Analytical Laboratories
PMT	Photo multiplier tube
PWR	Pressurized-water reactor
RDD	Radiological dispersal device
RGPu	Reactor-grade plutonium
RIAN	Radievyi Institut Akademii Nauk (Radium Institute of the Soviet Academy of Sciences)
RTG	Radioisotope thermoelectric generator
SAL	Safeguards Analytical Laboratories
SAUNA	Swedish Automatic Unit for Noble Gas Acquisition
SEM	Scanning electron microscopy
SGAB	Swedish Geological AB
SIMS	Secondary ion mass spectrometry
SMHI	Swedish Meteorological and Hydrological Institute

T	Tritium (hydrogen-3)
$T_{1/2}$	Half-life
TIMS	Thermal ionization mass spectrometry
TOF	Time-of-flight
TsAGI	Tsentralniy Aerogidrodinamicheskiy Institut (Central Aerohydrodynamic Institute)
UNSCEAR	United Nations Scientific Committee on the Effects of Atomic Radiation
UTC	Universal Coordinated Time
UOC	Uranium ore concentrate
VNIINM	High-Technology Scientific Research Institute for Inorganic Materials (Bochvar Institute)
WGPu	Weapon-grade plutonium
XRF	X-ray fluorescence analysis

Units

°C	degree Celsius
Bq	becquerel
eV	electonvolt
g	gram
kt	kiloton (explosive power equivalent to 1000 tonnes of TNT, or 4.184 terajoules)
m	metre
Mt	megaton (explosive power equivalent to 1 000 000 tonnes of TNT, or 4.184 petajoules)
ppm	parts per million
Sv	Sievert

Metric prefixes

P	peta (10^{15})
T	tera (10^{12})
G	giga (10^{9})
M	mega (10^{6})
k	kilo (10^{3})
h	hecto (10^{2})
da	deca (10^{1})
d	deci (10^{-1})
c	centi (10^{-2})
m	milli (10^{-3})
μ	micro (10^{-6})
n	nano (10^{-9})
p	pico (10^{-12})
f	femto (10^{-15})

Chemical elements, by atomic number

1	H	Hydrogen
2	He	Helium
3	Li	Lithium
4	Be	Beryllium
5	B	Boron
6	C	Carbon
7	N	Nitrogen
8	O	Oxygen
9	F	Fluorine
10	Ne	Neon
11	Na	Sodium
12	Mg	Magnesium
13	Al	Aluminium
14	Si	Silicon
15	P	Phosphorus
16	S	Sulphur

17	Cl	Chlorine
18	Ar	Argon
19	K	Potassium
20	Ca	Calcium
21	Sc	Scandium
22	Ti	Titanium
23	V	Vanadium
24	Cr	Chromium
25	Mn	Manganese
26	Fe	Iron
27	Co	Cobalt
28	Ni	Nickel
29	Cu	Copper
30	Zn	Zinc
31	Ga	Gallium
32	Ge	Germanium
33	As	Arsenic
34	Se	Selenium
35	Br	Bromine
36	Kr	Krypton
37	Rb	Rubidium
38	Sr	Strontium
39	Y	Yttrium
40	Zr	Zirconium
41	Nb	Niobium
42	Mo	Molybdenum
43	Tc	Technetium
44	Ru	Ruthenium
45	Rh	Rhodium
46	Pd	Palladium
47	Ag	Silver
48	Cd	Cadmium
49	In	Indium
50	Sn	Tin
51	Sb	Antimony
52	Te	Tellurium
53	I	Iodine
54	Xe	Xenon
55	Cs	Caesium
56	Ba	Barium
57	La	Lanthanum
58	Ce	Cerium
59	Pr	Praseodymium
60	Nd	Neodymium
61	Pm	Promethium
62	Sm	Samarium
63	Eu	Europium
64	Gd	Gadolinium
65	Tb	Terbium
66	Dy	Dysprosium
67	Ho	Holmium
68	Er	Erbium
69	Tm	Thulium
70	Yb	Ytterbium
71	Lu	Lutetium
72	Hf	Hafnium
73	Ta	Tantalum
74	W	Tungsten
75	Re	Rhenium
76	Os	Osmium
77	Ir	Iridium
78	Pt	Platinum
79	Au	Gold
80	Hg	Mercury
81	Tl	Thallium
82	Pb	Lead
83	Bi	Bismuth
84	Po	Polonium
85	At	Astatine
86	Rn	Radon
87	Fr	Francium
88	Ra	Radium
89	Ac	Actinium
90	Th	Thorium
91	Pa	Protactinium
92	U	Uranium
93	Np	Neptunium
94	Pu	Plutonium

95	Am	Americium	107	Bh	Bohrium
96	Cm	Curium	108	Hs	Hassium
97	Bk	Berkelium	109	Mt	Meitnerium
98	Cf	Californium	110	Ds	Darmstadtium
99	Es	Einsteinium	111	Rg	Roentgenium
100	Fm	Fermium	112	Uub	Ununbium
101	Md	Mendelevium	113	Uut	Ununtrium
102	No	Nobelium	114	Uuq	Ununquadium
103	Lr	Lawrencium	115	Uup	Ununpentium
104	Rf	Rutherfordium	116	Uuh	Ununhexium
105	Db	Dubnium	117	Uus	Ununseptium
106	Sg	Seaborgium	118	Uuo	Ununoctium

Glossary

Additional protocol	A legal agreement signed by states individually with the IAEA—based on a Model Protocol Additional to the Safeguards Agreements that was approved in 1997—granting the IAEA inspection authority that complements that provided in underlying safeguards agreements.
Accuracy	The ability to obtain a value that is close to the true value (see Appendix 2A).
Age (of radioactive material)	The time elapsed since the most recent separation or chemical purification.
Allotropes	Structural modifications of a chemical element, occurring when atoms of the element are bonded together in a different manner. For example, diamond and graphite are allotropes of carbon.
Atom bomb (A-bomb)	A bomb which derives its destructive power from the rapid release of energy by fission of heavy atomic nuclei.
Attribution	The process of *reconstruction* in the narrower context of investigations of nuclear trafficking and nuclear terrorism events.
Barn	A unit of area, equal to 10^{-28} square metres.
Bulk material	Material in loose form, such as liquid, gas or powder, or in a large number of small units (e.g. pellets or pebbles) that are not each individually identified.
Burn-up	Burn-up is a measure of how much energy was extracted from the nuclear fuel. It could be measured in percents of heavy metal (U or Pu) atoms that underwent fission, for example. More fission means higher burn-up and more fission products, lower quality of plutonium. So higher burn-up results in reactor-grade Pu, lower burn-up - in weapons-grade Pu.
Categorization	The quick assignment of material of interest to a predefined group, thus determining its further handling.

Characterization	The determination (i.e. measurement) of a sample's characteristics.
Collimate	Make rays of light or particles accurately parallel.
Comprehensive safeguards agreement	An agreement between a state and the IAEA with a view to prevent diversion of nuclear energy from peaceful uses to nuclear weapons or other nuclear explosive devices; comprehensive safeguards are based on nuclear material accountancy, complemented by containment and surveillance techniques.
Cross-contamination	An unintended introduction of material to the sample, which could lead to false results. Possible sources of cross-contamination are the sampling medium itself, the sampling kit, another sample, the sampling team and/or post-sampling handling, including analysis.
Cross section	A concept used to express the likelihood of interaction between incident and target particles (e.g. a neutron and a nucleus)—the likelihood is represented by the size of an area (cross section, often measured in *barns*) that has to be hit by an incident particle for reaction to occur: the larger the cross section, the more probable the interaction.
Daughter nuclide	See *decay chain*.
Debris	Residue of nuclear explosive device and any other matter that surrounded it and got disintegrated and dispersed by the explosion. The debris that was transferred into the atmosphere will gradually descend to earth, constituting the phenomenon referred to as 'fallout'.
Decay chain	A series of nuclides in which each member (parent) transforms into the next (daughter) through nuclear decay until a stable nuclide has been formed. Synonymous with radioactive chain and radioactive series.
Decoupled explosion	An underground explosion that takes place in a chamber large enough for only elastic motion to be produced in the walls; nearly all energy then goes into increasing the gas pressure within the chamber, which results in a significantly increased risk of dynamic gas venting.

Delta (or δ) phase of plutonium	One of the *allotropes* of plutonium, particularly appropriate for nuclear weapons manufacturing.
Depleted uranium	Uranium with a lower proportion of the isotope uranium-235 than occurs in natural uranium (i.e. less than 0.7 per cent), for example uranium in spent fuel from natural uranium fuelled reactors and tails from uranium enrichment processes.
Dose	An amount of ionizing radiation received or absorbed at one time.
Dose commitment	The infinite integral of the average (per capita) dose rate to the population due to a specified event like a nuclear test in the atmosphere; in line with this it is common to say that a specified event or aggregate of events (e.g. the full practice of nuclear testing in the atmosphere) has committed a certain average dose to the population.
Dose rate	A measure of the energy deposited by radiation in a target per unit of time.
Dual-use	Having both military and non-military applications.
Electronvolt	A unit of energy often used in nuclear physics: $1\ \text{eV} = 1.602 \cdot 10^{-19}$ joules, a value that equals the kinetic energy gained by an electron accelerating through an electric field of 1 volt.
Elimination	See *Method of exclusion*.
Environmental sampling	Defined by the IAEA as 'the collection of samples from the environment with a view to analysing them for traces of materials that can reveal information about nuclear material handled or activities conducted. The media sampled include various surfaces (e.g. of equipment and building structures), air, water, sediments, vegetation, soil and biota' International Atomic Energy Agency (IAEA), *IAEA Safeguards Glossary: 2001 Edition*, International Nuclear Verification Series no. 3 (IAEA: Vienna, 2001), p. 72.
Fissile material	Material composed of atoms which can be split by either fast or slow (thermal) neutrons. Uranium-235 and plutonium-239 are the most common fissile materials; see also *highly enriched uranium*, *low-enriched uranium*, *separated plutonium* and *weapon-grade uranium*.

Fresh nuclear fuel	Nuclear fuel that has not yet been irradiated is a reactor; see also *spent nuclear fuel*.
Forensic	Relating to or denoting the application of scientific methods and techniques to matters of law and regulations.
Fractionation	Enrichment of one component of a mixture relative to another in a chemical or physical process. In the context of nuclear weapon testing, this term covers a number of processes, other than radioactive decay, leading to different condensation rates of chemical elements after a nuclear explosion. In geochemistry, phase transitions of water between vapour, liquid and ice result in isotopic fractionation of hydrogen and oxygen.
Ground zero (or hypocentre)	The point on the earth's surface directly above or below an exploding nuclear bomb.
Growing-in (or in growth)	The generation of a nuclide through radioactive decay of its parent nuclide.
Highly enriched uranium	Uranium that has been enriched to 20 per cent or more uranium-235.
Hydrogen bomb (H-bomb)	A nuclear bomb whose destructive power comes from the fusion of isotopes of hydrogen (deuterium and tritium).
Hypocentre	See *ground zero*.
Isotopics	Isotopic composition.
Item form	Material that consists of individually identifiable units (e.g. fuel assembly, bundle, pin, plate or coupon) that are kept intact. IAEA, *IAEA Safeguards Glossary: 2001 Edition*, International Nuclear Verification Series no. 3 (2001), p. 26.
Low-enriched uranium	Uranium that has been enriched to between 0.72 and 20 per cent uranium-235 (typically, 3–5 per cent) and it thus suitable for use in reactors.
Method of exclusion	Determination of the true hypothesis by successive elimination as false of all the possible or plausible hypotheses except one; also known as elimination.

National security	Favourable condition in a state's national defence and foreign relations that allow it to resist hostile or destructive actions from within or without, whether overt or covert. Broader definitions that include other facets, such as environmental security, also exist. If such definitions are used, then other applications of *nuclear forensic analysis* might also be discussed, such as the analysis of debris from the 1986 accident at the Chernobyl nuclear power plant.
Neutron fluence	The total number of neutrons that traverse a unit area in a certain point in space over a defined period of time, measured in neutrons per cm^2; it is time-integrated *neutron flux*.
Neutron flux	Essentially, the intensity of neutron flow: the number of neutrons per unit volume multiplied by their speed, measured in neutrons per cm^2 per second.
Neutron reflector	A layer of material immediately surrounding the fissile material in a nuclear weapon that reflects neutrons back to the core; it thus reduces the critical mass and increases the weapon's efficiency; see also *tamper*.
Nuclear forensic analysis	The analysis of a sample of nuclear or radioactive material and any associated information, done in the *national security* context, in order to provide evidence for determining the history of the material.
Nuclear forensic interpretation	The process of correlating the characteristics of a sample with information on known methods of material production, handling and use.
Nuclear fuel cycle	A system of nuclear installations and activities involved in the production of nuclear power or nuclear materials, which are interconnected by streams of nuclear material.
Nuclear materials accounting	Activities carried out to establish the quantities of nuclear material present within defined areas and the changes in those quantities within defined periods.
Nuclear security	Defined by the IAEA as 'the prevention and detection of, and response to, theft, sabotage, unauthorized access, illegal transfer or other malicious acts involving nuclear material, other radioactive substances or their associated facilities'.

Nuclear terrorism	According to the 2005 UN International Convention for the Suppression of Acts of Nuclear Terrorism, an offense committed if a person unlawfully and intentionally uses in any way radioactive material with the intent to cause death or serious bodily injury, or substantial damage to property or to the environment; or with the intent to compel others to do or refrain from doing an act.
Nuclear smuggling (trafficking)	Illicit trafficking of nuclear materials.
Nuclide	A distinct kind of atom or nucleus characterized by a specific number of protons and neutrons.
Parent nuclide	See *decay chain*.
Precision	The ability to reproduce approximately the same measurement result given the same input (see Appendix 2A).
Radiation	Energy emitted as electromagnetic waves or subatomic particles.
Radioactive	Emitting or relating to the emission of ionizing radiation or particles.
Radioassay	An assay of a radioactive sample to determine the intensity of its radiation.
Radiological dispersal device	Any device used to deliberately disperse radioactive material to create terror or harm.
Radionuclide	A radioactive *nuclide*.
Reactor-grade plutonium	Plutonium containing over 18 per cent plutonium-240.
Reconstruction	The process of combining the information produced by *nuclear forensic interpretation* with all other available information to determine as full a history as possible of a sample of nuclear or radioactive material or a nuclear or radioactive event.
Safeguards	See *comprehensive safeguards agreement* and *additional protocol*.

Secular equilibrium	If the half-life of the parent radionuclide is so much longer than the half-life of *daughter radionuclide* that the change of its activity can be ignored during the period of interest, then after a period of ingrowth the daughter radionuclide's activity becomes equal to that of a parent radionuclide. This state is called *secular equilibrium*.
Separated plutonium	Plutonium chemically extracted from irradiated nuclear fuel (i.e. separated from other components of nuclear fuel).
Signature	Characteristics that enable a material to be identified.
Sinter	Coalesced from powder to solid by heating, and usually also by compression.
Source material	Uranium containing the mixture of isotopes occurring in nature; uranium depleted in the isotope 235; thorium; any of the foregoing in the form of metal, alloy, chemical compound, or concentrate; any other material containing one or more of the foregoing in such concentration as the IAEA Board of Governors shall from time to time determine; and such other material as the Board of Governors shall from time to time determine. This term does not apply to ore or ore residue, but does apply to the *uranium ore concentrate*.
Species	A particular kind of atom, molecule, ion, or particle.
Spent nuclear fuel	Nuclear fuel that has been irradiated in a nuclear reactor. See also *fresh nuclear fuel*.
Sputtering	A process of ejecting atoms from a solid target by bombarding the target surface with energetic particles. Ejected atoms can then be used, for example, for mass spectrometry purposes.
Standard deviation	The standard deviation is a numerical value used to indicate how widely individual observations vary from other observations in the same group. A low standard deviation indicates that the observation results tend to be very close to the mean value. A high standard deviation indicates that individual observations vary greatly from the group mean.

Tamper	A layer of a dense material that surrounds the fissile material in a nuclear weapon. It lengthens the short time during which the material holds together under the extreme pressures of the explosion, thereby increasing the efficiency of the weapon by increasing the proportion of the fissile material that undergoes fission. See also *neutron reflector*.
Thermonuclear	Using or denoting nuclear fusion reactions that occur at very high temperatures.
Total activity	The total number of decays in the whole of a radioactive object or material per unit of time.
Trace element	Any element having an average concentration of less than about 100 parts per million atoms (ppma) or less than 100 μg/g.
Uranium ore concentrate (UOC)	UOC, sometimes known as the *yellow cake*, is a generic name for a product of a uranium mill. In most cases, it is an oxide, U3O8, although alternative forms of UOC exist and include sodium diuranate, magnesium diuranate and ammonium diuranate, among others. All UOCs would also contain significant impurities. Despite the name, the colour of the UOC depends on a number of factors including the production process specifics, and the U3O8-based UOC may be green, khaki, brown or black. Magnesium diuranate and ammonium diuranate are indeed yellow.
Weapon debris	The highly radioactive material consisting of fission products, various products of neutron interactions, and uranium and plutonium that have escaped fission that remain after a nuclear explosion.
Weapon-grade plutonium	Plutonium containing less than 7 per cent plutonium-240.
Weapon-grade uranium	Generally considered to be uranium enriched to more than 90 per cent uranium-235.
Yellowcake	Impure uranium oxide obtained during processing of uranium ore; see also *uranium ore concentrate*.

Yield	The energy released in a nuclear explosion, measured in kilotons or megatons; a 1-kiloton nuclear explosion releases 4.184 terajoules of energy, which is equivalent to the energy released by the detonation of 1000 tonnes of trinitrotoluene (TNT).

Sources: SIPRI Yearbook, various editions; *Concise Oxford English Dictionary*, 12th edn (Oxford University Press: Oxford, 2011); International Atomic Energy Agency (IAEA), *IAEA Safeguards Glossary: 2001 Edition*, International Nuclear Verification Series no. 3 (2001); International Atomic Energy Agency (IAEA), *IAEA Safety Glossary: Terminology Used in Nuclear Safety and Radiation Protection* (IAEA: Vienna, 2007); McNaught, A. D. and Wilkinson, A., International Union of Pure and Applied Chemistry (IUPAC), *Compendium of Chemical Terminology*, 2nd edn (Blackwell Scientific Publications: Oxford, 1997); Argonne National Laboratory, 'Radiological dispersal device (RDD)', Human Health Fact Sheet, Aug. 2005, <http://www.evs.anl.gov/pub/doc/rdd.pdf>; and Glasstone, S., Dolan, P. J. (eds.), *The Effects of Nuclear Weapons* (US Government Printing Office: Washington, DC, 1977); IAEA, *Physical Model, Volume 1, Mining and Milling*, STR-314, 1999; Albright, D., Berkhout, F. and Walker, W., *Plutonium and Highly Enriched Uranium 1996: World Inventories, Capabilities and Policies* (Oxford University Press: Oxford, 1997).

1. Introduction

VITALY FEDCHENKO

The idea that nuclear energy could be used in weapons of an unprecedented power as well as in production of electricity and heat was present from the earliest days of nuclear science. The earliest estimates of the energy of the atomic nucleus were made in 1903, and in the same year Ernest Rutherford suggested that 'could a proper detonator be found, it was just conceivable that a wave of atomic disintegration might be started through matter, which would indeed make this old world vanish in smoke'.[1] The term 'atomic bomb' was probably first used not by the scientist but by the British fiction writer H. G. Wells, in his 1913 novel *The World Set Free*, which describes both the use of 'atomic energy' for transportation and industrial purposes and the development of 'atomic bombs', used to destroy major cities in a global war.[2] As the historian Thomas Powers has put it, 'the bomb was given a name thirty years before the first research dollar was spent to build one'.[3] Indeed, the first nuclear weapon was detonated in 1945, almost a decade before the first nuclear power plant was connected to the electricity grid, in 1954.[4]

Although the great military and industrial potential of nuclear energy remained unconfirmed for many years, the sheer magnitude of the problem of the dual uses of nuclear energy—both military and civilian—made sure that it received the close attention of national governments, which quickly filed it under the 'national security' category.[5]

I. National security policies and the analysis of nuclear materials

National security policies on the applications of nuclear materials were first developed by the United Kingdom and the United States in the 1940s and then by other countries. They included three mutually influencing

[1] Eve, A. S., *Rutherford: Being the Life and Letters of the Rt. Hon. Lord Rutherford, O. M.* (Macmillan: New York, 1939), p. 102. For background see Rhodes, R., *The Making of the Atomic Bomb* (Simon & Schuster: New York, 1986), pp. 24, 43–44.

[2] Wells, H. G., *The World Set Free* (Macmillan: London, 1914), p. 96.

[3] Powers, T., *Heisenberg's War: The Secret History of the German Bomb* (Da Capo Press: Cambridge, MA, 2000), p. 51.

[4] See e.g. Kramish, A., 'Atomic energy in the USSR', *Bulletin of the Atomic Scientists*, vol. 15, no. 8 (Oct. 1959), p. 326.

[5] On the narrow definition of 'national security' used here see the glossary in this volume.

goals: the development and control of military applications of nuclear energy; finding ways to curb the proliferation of states, or even non-state actors, with nuclear capabilities; and, failing that, receiving the best possible information on the nuclear programme or weapon arsenal of other states. Achieving the first goal principally meant the domestic production of nuclear weapon arsenals. In some cases, however, one state provided assistance in manufacturing nuclear weapons to another—for example, the UK helped the USA in the 1940s and the Soviet Union helped China in the 1950s.[6] The most developed approach to achieving the second goal has been the introduction of national and international legal and regulatory barriers to transfers of nuclear materials and nuclear technologies. Meanwhile, achieving the third goal was initially entrusted to national intelligence agencies.[7]

The application of these national security policies developed into national laws and international treaties. For example, the growth of national arsenals led to the conclusion of international nuclear arms control treaties and created the need to verify them. Similarly, preventing the transfer of nuclear materials and nuclear technologies has shaped the non-proliferation regime as it exists today: although nuclear facilities are owned and operated nationally, most are subject to regulations and restrictions imposed by international treaties and agreements.[8] The 1968 Non-Proliferation Treaty (NPT) is the legal and political foundation of this regime.[9] It mandates most states to apply the 'safeguards' verification mechanism of the International Atomic Energy Agency (IAEA). The extension of this approach to non-state actors, known as 'nuclear security', deals

[6] Reed, T. C. and Stillman, D. B., *The Nuclear Express: A Political History of the Bomb and Its Proliferation* (Zenith Press: Minneapolis, MN, 2009), pp. 92–101.

[7] Information on the intelligence efforts of governmental agencies to assess the nuclear fuel cycle and weapon complex of other states is available in published sources in the cases of the USA and, to a lesser extent, the UK and the Soviet Union. See e.g. Richelson, J. T., *Spying on the Bomb: American Nuclear Intelligence from Nazi Germany to Iran and North Korea* (W. W. Norton: New York, 2006); Ziegler, C. A. and Jacobson, D., *Spying without Spies: Origins of America's Secret Nuclear Surveillance System* (Praeger: Westport, CT, 1995); Goodman, M. S., *Spying on the Nuclear Bear: Anglo-American Intelligence and the Soviet Bomb* (Stanford University Press: Stanford, CA, 2007); Vasil'ev, A. P., [Created by the nuclear age], vols 1–3 (Self published: Moscow, 2002) (in Russian); and Vasil'ev, A. P., [The long-range system to detect nuclear explosions and the Soviet atomic programme], ed. V. P. Vizgin, [History of the Soviet atomic program: documents, memoirs, research], 2nd edn (Russian Christian Humanitarian Institute: St Petersburg, 2002) (in Russian).

[8] For a detailed discussion on this and other approaches to non-proliferation see Fedchenko, V., 'Multilateral control of the nuclear fuel cycle', *SIPRI Yearbook 2006: Armaments, Disarmament and International Security* (Oxford University Press: Oxford, 2006).

[9] Treaty on the Non-Proliferation of Nuclear Weapons (Non-Proliferation Treaty, NPT), opened for signature 1 July 1968, entered into force 5 Mar. 1970, IAEA Information Circular INFCIRC/140, 22 Apr. 1970, <http://www.iaea.org/Publications/Documents/Treaties/npt.html>. An example of a national law aimed at achieving this goal is the Nuclear Non-Proliferation Act of 1978, US Public Law 95-242, approved 10 Mar. 1978, <http://www.gpo.gov/fdsys/pkg/STATUTE-92/pdf/STATUTE-92-Pg120.pdf>.

with nuclear trafficking (i.e. nuclear smuggling) and, ultimately, the threat of nuclear terrorism.[10]

As the development of international nuclear arms control and non-proliferation treaties progressed, intelligence agencies, methods and facilities began to contribute to the first two policy goals in a new way. International treaties began to include provisions explicitly foreseeing the use of national technical means of verification (NTMs), and even prohibiting interference with them or otherwise impeding their use.[11] Relevant multilateral treaties include the 1963 Partial Test-Ban Treaty (PTBT) and the 1996 Comprehensive Nuclear-Test-Ban Treaty (CTBT).[12] While the use of NTMs is not specifically discussed in the text of the NPT, the states parties are able to provide the IAEA with information concerning other states that they believe could be relevant to safeguards purposes.[13] Some bilateral Soviet–US or Russian–US treaties also permit the use of NTMs in their implementation, including the 1974 Threshold Test-Ban Treaty (TTBT), the 1991 Treaty on the Reduction and Limitation of Strategic Offensive Arms (START I) and the 2010 Treaty on Measures for the Further Reduction and Limitation of Strategic Offensive Arms (New START).[14]

In most cases, in order to advance towards the goals they were designed to achieve, the policies and laws in question have to focus on nuclear materials and, to a lesser extent, radioactive materials, rather than on the related equipment. One reason for this is trivial: nuclear material is the source of the nuclear energy. Nuclear materials, by definition, are present at all stages of the nuclear fuel cycle, as well as in weapons, often accompanied by other radioactive materials. In addition, there is a consensus that the single most difficult and expensive step in the manufacturing of a nuclear explosive device is the production of sufficient nuclear material of the necessary quality. It is thus not surprising that the analysis of nuclear and radioactive materials has always been a crucial component of many

[10] On the definitions of 'nuclear security' and 'nuclear terrorism' see the glossary in this volume.

[11] For a partial list of such treaties and a discussion on NTMs see Krass, A. S., SIPRI, *Verification: How Much is Enough?* (Taylor and Francis: London, 1985), pp. 4–7.

[12] Treaty Banning Nuclear Weapon Tests in the Atmosphere, in Outer Space and Under Water (Partial Test-Ban Treaty, PTBT), opened for signature 8 Aug. 1963, entered into force 10 Oct. 1963, *United Nations Treaty Series*, vol. 480 (1963); and Comprehensive Nuclear-Test-Ban Treaty (CTBT), opened for signature 24 Sep. 1996, not in force, <http://treaties.un.org/Pages/CTCTreaties.aspx?id=26>.

[13] On the example of the discovery of an underground enrichment plant in Iran see Kile, S. N., 'Nuclear arms control and non-proliferation', *SIPRI Yearbook 2010: Armaments, Disarmament and International Security* (Oxford University Press: Oxford, 2010), p. 385.

[14] Soviet–US Treaty on the Limitation of Underground Nuclear Weapon Tests (Threshold Test-Ban Treaty, TTBT), signed 3 July 1974, entered into force 11 Dec. 1990, *United Nations Treaty Series*, vol. 1714 (1993); Soviet–US Treaty on the Reduction and Limitation of Strategic Offensive Arms (START I), signed 31 July 1991, entered into force 5 Dec. 1994, expired 5 Dec. 2009, <http://www.state.gov/t/avc/trty/146007.htm>; and Russian–US Treaty on Measures for the Further Reduction and Limitation of Strategic Offensive Arms (New START), signed 8 Apr. 2010, entered into force 5 Feb. 2011, <http://www.state.gov/t/avc/newstart/c44126.htm>.

technical and scientific methods related to the military use of nuclear energy.

The nuclear fuel cycle is a system of nuclear installations and activities involved in the production of nuclear power or nuclear materials, which are interconnected by streams of nuclear material.[15] Nuclear material can be imagined as 'moving' through a web of conduits from one facility to another, changing its chemical and physical properties, from ore to nuclear fuel to waste.[16] Each step of the fuel cycle or each use of the nuclear material will inevitably leave its mark in the material. In other words, nuclear materials retain some information about what happened to them in the past, or what they were like before that. This is possible because in reality there is only a limited number of physical and chemical processes that can be applied to a limited number of existing nuclear material types, and researchers in most cases know, at least approximately, what those processes and materials are. Thus the analysis of the nuclear (or radioactive) material after an event can, in principle, yield information on the event in question. (See chapters 5 and 6 for more information on these nuclear forensic signatures.) This is a trivial fact that makes measurement and analysis of nuclear and other radioactive materials indispensable for all three of the national security policy goals identified above.

II. Nuclear forensic analysis as a collective term

This book describes how the measurements and analysis of nuclear and relevant radioactive materials have been used in applications connected to the potential military uses of nuclear energy, and specifically in applications advancing the three goals outlined above: weapon development and arms control; non-proliferation and nuclear security; and verification and intelligence. Certain techniques of measurement and analysis have been used for many years in these applications in an isolated manner, without being explicitly linked. However, the maturity and popularity of the technologies involved have recently increased to the point where they should all be treated as a separate and unified scientific discipline.

The proposal presented here is that this new discipline be called 'nuclear forensic analysis' or 'nuclear forensics'. This is an expansion of an existing term to include all applications dealing with the analysis of nuclear materials for national security purposes. The terms 'nuclear forensic analysis' and 'nuclear forensics' were probably first coined in the context of

[15] International Atomic Energy Agency (IAEA), *IAEA Safeguards Glossary: 2001 Edition*, International Nuclear Verification Series no. 3 (IAEA: Vienna, 2001), p. 37.

[16] See e.g. Wilson, P. D. (ed.), *The Nuclear Fuel Cycle: From Ore to Waste* (Oxford University Press: Oxford, 1996).

combating nuclear smuggling, a problem that emerged in the early 1990s.[17] The investigations and prosecutions of the first such cases called for the development and application of techniques to analyse the nuclear materials involved in order to produce evidence for use in courts of law—hence the term 'forensics'.

The *Oxford English Dictionary* defines 'forensic' as 'Pertaining to, connected with, or used in courts of law; suitable or analogous to pleadings in court'.[18] More broadly, 'forensics' is understood in the specialized literature as 'the application of science to law'.[19] Although such definitions refer mostly to national laws, they could be interpreted as including international laws, regulations and, in particular, treaties. It could further be argued that the term can be expanded even more to include the implementation of policies. Policies and laws are normally designed to achieve the same goals and differ only in the degree to which they are codified and enforced.

Between 2006 and 2015 the IAEA defined 'nuclear forensics' as 'the analysis of intercepted illicit nuclear or radioactive material and any associated material to provide evidence for nuclear attribution', where 'attribution' refers to 'the process of identifying the source of nuclear or radioactive material used in illegal activities, to determine the point of origin and routes of transit involving such material, and ultimately to contribute to the prosecution of those responsible'.[20] The IAEA adopted a refined definition in 2015, where 'nuclear forensics' is 'the examination of nuclear or other radioactive materials, or of evidence that is contaminated with radionuclides, in the context of legal proceedings under international or national law related to nuclear security'.[21] These definitions are used in the specialized context of the IAEA's work on nuclear security, which is separate from its safeguards activities.[22] However, the analytical techniques used in the combating of nuclear trafficking have much greater potential and, in fact, have been extensively used for many years in other fields.

Expansion of the term 'nuclear forensic analysis' to encompass all aspects of national security, not just breaches of national law, is sometimes criti-

[17] Moody, K. J., Hutcheon, I. D. and Grant, P. M., *Nuclear Forensic Analysis* (CRC Press: Boca Raton, FL, 2005), pp. vi–vii.

[18] *The Oxford English Dictionary*, vol. IV (Oxford University Press: Oxford, 1978), p. F-438.

[19] Saferstein, R., *Criminalistics: An Introduction to Forensic Science*, 4th edn (Prentice Hall: Englewood Cliffs, NJ, 1990), p. 1, quoted in Moody et al. (note 17), p. vi.

[20] International Atomic Energy Agency (IAEA), *Nuclear Forensics Support: Reference Manual*, IAEA Nuclear Security Series no. 2, Technical Guidance (IAEA: Vienna, 2006), pp. 2–3. These IAEA definitions are based on the work of the US Department of Energy's National Laboratories community involved in combating nuclear smuggling. Kristo, M. J. et al., *Model Action Plan for Nuclear Forensics and Nuclear Attribution*, UCRL-TR-202675 (US Department of Energy, Lawrence Livermore National Laboratory: Livermore, CA 5 Mar. 2004).

[21] IAEA, *Nuclear Forensics in Support of Investigations: Implementing Guide*, IAEA Nuclear Security Series (IAEA: Vienna, forthcoming 2015). This publication is a revision of IAEA (note 20).

[22] IAEA, 'Nuclear security', <http://www-ns.iaea.org/security/>.

Box 1.1. Definitions[a]

Nuclear forensic analysis (or *nuclear forensics*) is the analysis of a sample of nuclear or radioactive material and any associated information to provide evidence for determining the history of the material. Nuclear forensic analysis includes categorization, characterization, nuclear forensic interpretation and reconstruction.

Categorization is the quick assignment of the material of interest to a predefined group, thus determining its further handling.

Characterization is the determination (i.e. measurement) of a sample's characteristics. It typically involves an elemental analysis of the sample, most often including isotopic analysis of nuclear materials (i.e. uranium or plutonium) and selected minor constituents (e.g. lead). It also includes physical characterization, for example, measuring the key dimensions of solid samples or, in the case of powders, determining particle size and shape distribution.

Nuclear forensic interpretation is the process of correlating the characteristics of the sample with information on known methods of material production, handling and use. The information obtained as a result of this process is the end product of nuclear forensic laboratories.[b]

Reconstruction is the process of combining the information produced by nuclear forensic interpretation with all other available information (e.g. from forensic analysis of non-nuclear evidence associated with the sample or from intelligence sources) to determine as full a history as possible of the nuclear or radioactive material or an event. This phase is called *attribution* in the narrower contexts of investigations of nuclear trafficking and nuclear terrorism events.

[a] The definitions of 'nuclear forensic analysis' and 'reconstruction' were developed in cooperation with Dr James Acton on the basis of the definitions in International Atomic Energy Agency (IAEA), *Nuclear Forensics Support: Reference Manual,* IAEA Nuclear Security Series no. 2, Technical Guidance (IAEA: Vienna, 2006).

[b] IAEA (note a), pp. 3–4; and Mayer, K., Wallenius, M. and Ray, I., 'Tracing the origin of diverted or stolen nuclear material through nuclear forensic investigations', eds R. Avenhaus et al., *Verifying Treaty Compliance: Limiting Weapons of Mass Destruction and Monitoring Kyoto Protocol Provisions* (Springer: Heidelberg, 2006), p. 402.

cized as invoking too many connotations with criminalistics. Alternative terms, such as 'nuclear material analysis', have been proposed. In Russian, for example, the equivalent of the term 'nuclear forensic analysis' is used in case of a crime or illicit activity, and a term equivalent to 'expert-technical nuclear analysis' is suggested as an alternative in other cases.[23]

The term 'nuclear forensic analysis' is adopted here because it is already being used in an extended sense more often than any other phrase. For example, the best-known textbook on nuclear forensics (in the narrow sense) discusses the use of the discipline to determine the source of a nuclear explosive device used in a terrorist attack, but in doing so it unavoidably lists the indicators of the device performance, which is clearly information pertinent to the arms control, disarmament and intelligence

[23] IAEA, Division of Conference and Document Services, English and Russian Translation Sections, Personal communication with author, 14 Sep. 2010.

domains.[24] This is probably not coincidental, since properties of nuclear materials are independent of the political context in which they are being used.

In order to capture all possible applications of the techniques in question, this book uses the broad definitions given in box 1.1. These and other definitions also appear in the glossary.

III. The applications and purposes of nuclear forensic analysis

Usually there are specific features that interest a researcher in the history of a sample of material, such as its origin and producer, the processes that led to its production, the point of its diversion from legitimate uses, its age, its routes of transit and its planned end-use. The goal of nuclear forensics—to reconstruct the history of the material or an event—makes it a technique of choice in a number of applications. The specific application determines what is required to be found out from the sample material—in particular, there are a number of legal frameworks in which nuclear forensics has been used or could be applied in the future (see table 1.1). In line with the definitions given above, the following criteria are chosen to guide the selection of these frameworks: a sample or item containing nuclear or other radioactive material is analysed; the purpose of the analysis is to determine the history of the material in the sample or a related event; and the analysis serves a purpose pertinent to national security.

For example, investigators of a nuclear smuggling case would want to determine the source of the material, at which point it was diverted from legitimate uses, what its possible illegitimate use could be, and so on. Investigators of a nuclear or radiological terrorism incident would look for the material's origin in order to ensure a correctly targeted response. IAEA safeguards inspectors may want to know if the isotopic composition and production date of sample material gathered from a state's nuclear facilities correspond to the state's declared inventory. The Comprehensive Nuclear-Test-Ban Treaty Organization (CTBTO) collects air samples to detect suspected explosions, as well as to verify their nuclear nature. The verification procedures of a fissile materials cut-off treaty (FMCT), if and when such a treaty is negotiated, might include determination of the age of nuclear materials and some kind of environmental sampling to ensure that production of new nuclear materials subject to the treaty does not continue.

This book is divided into two parts. Part I describes the nuclear forensics as an analysis of nuclear materials for various security purposes, and explains its process and scientific techniques. Chapter 2 enumerates the

[24] Moody et al. (note 17), pp. 203–205.

Table 1.1. Applications of nuclear forensic analysis

Framework	Information to be inferred concerning the history of the material or item
Development and control of nuclear armaments and stockpiles, disarmament	
National weapon development programmes	The device's explosion yield, efficiency and other performance characteristics
Partial Test-Ban Treaty	Nuclear explosive origin of debris, and its age and location, especially if leaked from an underground test by a treaty party
Comprehensive Nuclear-Test-Ban Treaty	Nuclear explosive origin of debris
Fissile material cut-off treaty	Origin and age of nuclear materials and fuel cycle effluents; lifetime material output of a nuclear facility
Non-proliferation and nuclear security	
Non-Proliferation Treaty (environmental sampling for IAEA safeguards)	Age and production process (consistency with declaration)
Attribution in a trafficking case	Age, production process and manufacturer
Attribution of a nuclear terrorism event	Design features of explosive device, the nuclear material used, the explosion yield, and the origin of the device and material
Nuclear intelligence	
Monitoring of foreign explosions	Weapon performance characteristics
Monitoring of foreign facilities and materials	Nuclear material stockpiles and production rates

stages of the nuclear forensic process and discusses differences in their application in various contexts. Appendix 2A briefly introduces the essential terms and techniques of measurement. Chapters 3 and 4 describe the most important measurement techniques used for nuclear forensic analysis: mass spectrometry and gamma spectrometry. Chapter 5 presents an extensive and systematic compendium of nuclear forensic signatures that can be found in materials related to the nuclear fuel cycle. Chapter 6—which is distinctive of the broader definition of nuclear forensics—provides information on nuclear forensic signatures that can be used after a nuclear explosion.

Part II illustrates how nuclear forensic techniques have been used. In chapters 7 and 8, a history of the origins of nuclear forensics describes the problems that motivated nuclear scientists to develop the techniques in the USA, the Soviet Union and Sweden. Chapter 9 uses contemporary examples of how these techniques have or could be used to demonstrate the current and future relevance of the new nuclear forensics.

Part I

Nuclear forensic methods

2. The process of nuclear forensic analysis[1]

VITALY FEDCHENKO

As implied in the definitions given in chapter 1, nuclear forensic investigation or analysis would normally comprise four stages: sample collection and categorization; characterization of the material in the sample; interpretation of the results; and reconstruction of the history of the material or an event to which the material relates. A nuclear forensic investigation is normally launched in anticipation of or as a response to some event or action. It could be the theft or diversion of nuclear material, a nuclear explosion or the clandestine production of nuclear material. The purpose of nuclear forensics is to help produce knowledge that decision makers can use to formulate their response to the event or action.

The first step in this process—sample collection and categorization—provides a usually small quantity of material or a physical object (an item), which is assumed to contain clues concerning an event or action of interest. The second step—characterization—provides a description of the sample in the form of raw data. This raw data—known as the 'sample characteristics'—has no meaning by itself, it is simply an unprocessed collection of numbers, images or other outputs of measurement devices, which represent physical quantities as symbols.

The third step—interpretation—converts the raw data into the meaningful information. For example, characterization of the uranium sample might reveal that it contains 0.37 per cent of the isotope ^{236}U. Interpretation of this data would yield the conclusion that this material was most likely irradiated in a nuclear reactor. This conclusion would be corroborated if trace amounts of plutonium are also found. Expert knowledge and judgement are often required to interpret the raw data.

The final step is reconstruction of the event or the history of the material. At this stage data and information from nuclear forensics are combined with all data and information available from all other sources. The results of the reconstruction represent the knowledge that will be passed on to

[1] This chapter expands on International Atomic Energy Agency (IAEA), *Nuclear Forensics Support: Reference Manual*, IAEA Nuclear Security Series no. 2, Technical Guidance (IAEA: Vienna, 2006), pp. 21–37; and Mayer, K., Wallenius, M. and Ray, I., 'Tracing the origin of diverted or stolen nuclear material through nuclear forensic investigations', eds R. Avenhaus et al., *Verifying Treaty Compliance: Limiting Weapons of Mass Destruction and Monitoring Kyoto Protocol Provisions* (Springer: Heidelberg, 2006), pp. 390–400.

Table 2.1. Materials and their forms sampled for nuclear forensic purposes

Framework	Typical sample form, content (and source)
Development and control of nuclear armaments and stockpiles, disarmament	
National weapon development programmes	Bulk material, particles and gases (from weapon debris)
Partial Test-Ban Treaty	Particles and gases (from weapon debris)
Comprehensive Nuclear-Test-Ban Treaty:	
IMS radionuclide component	Particles, xenon gas (from weapon debris)
On-site inspection	Particles and gases in air (from weapon debris)
Fissile material cut-off treaty	Noble gases (from nuclear reactors, fuel reprocessing plants or isotope-production facilities); bulk graphite samples (from shut-down plutonium-producing reactors)
Non-proliferation and nuclear security	
Non-Proliferation Treaty (environmental sampling for IAEA safeguards)	Particles (on swipes from nuclear material production and handling areas); air, water, sediments, vegetation, soil, biota
Attribution in a trafficking case	Nuclear or radioactive materials, items or bulk form (from nuclear fuel cycle facilities)
Attribution of a nuclear terrorism event	Particles and gases (from weapon debris); particles (from a radiological dispersal device); human tissue, clothing and excretions
Nuclear intelligence	
Monitoring of foreign explosions	Particles and gases (weapon debris)
Monitoring of foreign facilities and materials	Noble gases (nuclear reactors, fuel-reprocessing plants or isotope-production facilities); particles in available man-made media (e.g. wine or clothing), in the air, water, sediments, vegetation, soil and biota

IAEA = International Atomic Energy Agency; IMS = International Monitoring System.

decision-making circles. Thus, the level of abstraction increases with every step of the nuclear forensic analysis process.

The first three of the steps are described in turn in the three sections of this chapter. In order to identify the origin of the material, nuclear forensics often employs the process of elimination (also known as the 'method of exclusion'). Interpretation of the first results of characterization will rule out some potential origins of the material in the sample. However, it is likely that many candidate origins will still remain, and further exclusion would demand additional characterization or interpretation. Thus, the process of nuclear forensic analysis is most often not linear, but iterative—that is, it requires the repetition of all or some of the steps described above. Nuclear forensic analysis may be unable to produce any conclusive evidence, which cannot be disputed or dismissed. Instead, it might return results consistent with a certain theory but perhaps not clear enough to satisfy the standard of proof required by the application for which it is employed. For example, the legal standard of proof in criminal trials in the USA—'beyond reasonable doubt'—is quite rigorous, and so are the requirements applied internationally to scientific evidence. However, policy-makers and intelligence officials often have to act on the basis of incomplete information and under tight time constraints, which may lead to acceptance of less rigorous standards of proof.[2]

I. Sample collection and categorization

Sample collection

The process of obtaining a sample for nuclear forensic purposes may differ in some respects from a similar process undertaken for purely scientific applications. On the one hand, nuclear forensic sample collection might have to be quite versatile in order to be able to process often diverse and unexpected samples. The sample itself might look very different in various frameworks (see table 2.1). Unlike most scientific disciplines, which normally deal with clearly defined sample types, nuclear forensic investigation may demand an analysis of almost anything—from gases and dust particles to wines and human bodily excretions. Although the material of interest in any sample can only have one of four forms—item, bulk, particles or gas—it can come attached to a vast number of carriers or be incorporated in some matrix, which will inevitably have a bearing on the sample-collection procedure. On the other hand, the nuclear forensic sample-collection process is often subject to many restrictions and limitations that would be unusual for general scientific sampling procedures. There are two reasons for this, one legal and one practical.

Legal or regulatory restrictions

The nature of nuclear forensics as a discipline ('application of science to law') means that some kind of legal or regulatory restriction will most often apply to the way in which a sample is collected and handled. Collection must yield a sample that is eligible—in legal or regulatory terms—for further analysis.

For example, if a sample is collected during investigation of a trafficking case, the chain of custody has to be maintained in order to be able to prove

[2] For discussion of standards of proof used to determine if North Korea used plutonium in its 2006 nuclear test see Kang, J., von Hippel, F. N. and Zhang, H., 'Letter to the editor: the North Korean test and the limits of nuclear forensics', and Smith, H., 'Harold Smith Responds', *Arms Control Today*, vol. 37, no. 1 (Jan./Feb. 2007). On the North Korean nuclear test see chapter 9 in this volume.

the sample's legal integrity. Otherwise, the sample will not be admissible in court as evidence and so will fail to serve its purpose for prosecution.[3] Sample-collection guidelines exist for nuclear forensic investigators in such cases.[4]

An example of regulatory restrictions on sample collection are the rules under which the International Atomic Energy Agency (IAEA) safeguards inspectors are allowed to conduct environmental sampling while verifying the absence of undeclared nuclear material and activities. The idea underlying environmental sampling is that every nuclear process, no matter how well sealed, emits some nuclear or radioactive material. While these emissions are small enough to pose no environmental or health concern, they can be collected and used as a source of information on the process which produced them.[5] The comprehensive safeguards agreements (CSAs) that are concluded by the IAEA with individual states only allow environmental sampling (which most often means swiping equipment or buildings with special cloths that collect dust particles) to be conducted at the locations to which the IAEA inspectors already have access when conducting routine inspections or design information verification.[6] The optional Model Additional Protocol to the CSAs relaxes that restriction and introduces another collection technique, wide-area environmental sampling, which can be used only with the approval of the IAEA Board of Governors.[7] Moreover, IAEA inspectors are allowed to use only preapproved sample-collection procedures and 'swipe sampling kits'.[8]

Practical limitations

Practical limitations stem from the investigative nature of nuclear forensics. A nuclear forensic investigation is expected to use the available sample to produce information about the processes or events that caused the

[3] Champion, D. J., *The American Dictionary of Criminal Justice: Key Terms and Major Court Cases*, 3rd edn (Scarecrow Press: Lanham, MD, 2005), p. 40.

[4] IAEA (note 1), p. 17.

[5] On the definition of 'environmental sampling' see the glossary in this volume.

[6] IAEA, The structure and content of Agreements between the Agency and States required in connection with the Treaty on the Non-Proliferation of Nuclear Weapons, INFCIRC/153 (Corrected), June 1972, <http://www.iaea.org/Publications/Documents/Infcircs/Others/ infcirc153.pdf>, p. 18. See also Kuhn, E., Fischer, D. and Ryjinski, M., 'Environmental sampling for IAEA safeguards: a five year review', IAEA-SM-367/10/01, *Symposium on International Safeguards, Verification and Nuclear Material*, Vienna, 29 Oct.–2 Nov. 2001 (IAEA: Vienna, 2001). For a brief description of CSAs see the glossary in this volume.

[7] International Atomic Energy Agency (IAEA), 'Model protocol additional to the agreement(s) between state(s) and the International Atomic Energy Agency for the application of safeguards', INFCIRC/540 (Corrected), Sep. 1997, <http://www.iaea.org/Publications/Documents/Infcircs/1997/infcircnr197.shtml>, articles 5, 9; and International Atomic Energy Agency (IAEA), *IAEA Safeguards Glossary: 2001 Edition*, International Nuclear Verification Series no. 3 (IAEA: Vienna, 2001), p. 72. For a brief description of additional safeguards protocols see the glossary in this volume.

[8] International Atomic Energy Agency (IAEA), *Safeguards Techniques and Equipment: 2003 Edition*, International Nuclear Verification Series no. 1 (IAEA: Vienna, 2003), p. 77.

material in the sample to appear at the time and place of sampling. The difficulty common to almost all nuclear forensic applications is that the investigator normally has no control over those processes or events. Moreover, there may be no opportunity for additional sample collection. Regardless of the sampling procedures applied to collect the sample for forensic investigation, investigators cannot assume that the material in the sample is representative of the larger piece of material—the batch or group—from which it originated.

For example, in case of nuclear materials accounting for IAEA safeguards, sample collection is conducted in accordance with predefined, elaborate sampling plans.[9] In this case the IAEA inspector gets access to the whole batch of the material and is free to conduct necessary manipulations. In contrast, sampling of material intercepted during trafficking is limited to the intercepted material only. Environmental sampling for safeguards purposes returns only the material leaked from some (not necessarily known) equipment over some period of time. The monitoring stations of the Comprehensive Nuclear-Test-Ban Treaty Organization (CTBTO) and the national technical means used for verification of bilateral treaties or intelligence are designed to collect noble gases and particulate material that carry information on an event at a location that is probably inaccessible, lying continental or global distances away.

By the time the investigator collects the sample, some information in the material may have been lost or altered. In some cases, such as collection of short-lived isotopes of noble gases, radioactive decay may cause the sample to deteriorate so much that it becomes useless. Similarly, fractionation of radionuclides in nuclear weapon fallout changes the composition of nuclear weapon debris collected after an explosion.[10] This process has to be accounted for since much of the analysis of the properties of a nuclear weapon relies on knowledge of ratios of fission fragments before the fractionation occurred.[11]

Since the material in the forensic sample is not necessarily representative of the batch or group it came from, special care is taken later in the nuclear forensic analysis process to account for this possibility. At the sample-collection stage it might be necessary to collect as many samples as possible and use procedures for their expedited handling.

[9] IAEA, *IAEA Safeguards Glossary* (note 7), pp. 59, 76.

[10] For the definition of 'fractionation' see the glossary in this volume. Glasstone, S. and Dolan, P. J. (eds), *The Effects of Nuclear Weapons*, 3rd edn (Department of Defense/Energy Research and Development Administration: Washington, DC, 1977), pp. 634, 389; and Chamberlain, A. C., *Radioactive Aerosols* (Cambridge University Press: Cambridge, 1991), p. 65.

[11] May, M. et al., Joint Working Group of the American Physical Society (APS) and the American Association for the Advancement of Science (AAAS), *Nuclear Forensics: Role, State of the Art, Program Needs* (AAAS: Washington, DC, 2008), p. 22.

Table 2.2. Categories of nuclear materials

Category	Type of material	Radioactive components
Unirradiated direct-use nuclear material	Highly enriched uranium	>20% U-235
	Plutonium	<80% Pu-238
	U-233	Separated isotope
Irradiated direct-use nuclear material	The three materials above in irradiated nuclear fuel	Material in irradiated fuel elements or in spent fuel reprocessing solutions
Alternative nuclear material	Americium (Am-241)	Separated element or present in irradiated nuclear material, in separated plutonium or in their mixtures
	Neptunium (Np-237)	
Indirect-use nuclear material	Depleted uranium	<0.7% U-235
	Natural uranium	~0.7% U-235
	Low-enriched uranium	More than ~0.7% and less than 20% of U-235
	Plutonium (Pu-238)	>80% Pu-238
	Thorium	Th-232

Source: International Atomic Energy Agency (IAEA), *IAEA Safeguards Glossary: 2001 Edition*, International Nuclear Verification Series no. 3 (IAEA: Vienna, 2001), pp. 30–33; and International Atomic Energy Agency (IAEA), *Nuclear Forensics Support: Reference Manual*, IAEA Nuclear Security Series no. 2, Technical Guidance (IAEA: Vienna, 2006), p. 5.

Sample categorization

The sample-collection process is sometimes accompanied by categorization, which is a quick and rough on-site measurement of the material of interest by non-destructive analysis in order to determine further ways of dealing with it. Categorization is especially important in cases of trafficking or a nuclear terrorism event, where a first responder (i.e. a firefighter, police constable or paramedic) might unexpectedly encounter a range of different items or materials, and every sample-collection environment will be unique.

Categorization has two goals: 'to identify the risk to the safety of first responders, law enforcement personnel and the public' and 'to determine if there is criminal activity or a threat to national security'.[12] The first goal, the identification of safety risks posed by the nuclear or radioactive material at the scene of the event, is normally understood as determining the risks associated with the radiation which it might be emitting. This requires the application of techniques and methods of radiological analysis, such as estimates of total activity and dose rates from alpha, beta, gamma or neutron radiation.[13]

[12] IAEA (note 1), p. 3.

[13] For the definition of 'total activity' and 'dose rate' see the glossary in this volume.

Table 2.3. Categories of radioactive materials other than nuclear materials

Category of radioactive source	Type of device (example)	Typical radioactive components
Category 1	Radioisotope thermoelectric generators	Pu-238, Cm-244, Sr-90
	Irradiators/sterilizers	Co-60, Cs-137
	Teletherapy sources	Co-60, Cs-137
Category 2	Industrial gamma radiography sources	Co-60, Ir-192, Se-75, Yb-169, Tm-170
	High- and medium-dose-rate brachyotherapy sources	Co-60, Cs-137, Ir-192
	Calibration sources	Co-60, Cs-137
Category 3	Fixed industrial gauges	Co-60, Cs-137, Cf-252, Am-241
	Well-logging sources	Am-241/Be, Cs-137, Cf-252
	Pacemakers	Pu-238
Category 4	Low-dose-rate brachyotherapy sources	Cs-137, Ra-226, I-125, Ir-92, Au-198, Cf-252
	Thickness gauges	Kr-85, Sr-90, Am-241, Pm-147
	Fill level gauges	Am-241, Cs-137, Co-60
	Portable gauges	Am-241/Be, Cs-137
	Bone densitometers	Cd-109, Gd-153, I-125, Am-241
	Static eliminators	Am-241, Po-210
Category 5	X-ray fluorescence devices	Fe-55, Cd-109, Co-57
	Lightning preventers	Am-241, Ra-226, H-3
	Mössbauer spectrometers	Co-57
	Medical diagnostic sources	Short-lived isotopes, e.g. I-131
	Fire detectors	Am-241 and Pu-238

Sources: International Atomic Energy Agency (IAEA), *Categorization of Radioactive Sources: Safety Guide*, IAEA Safety Standards Series no. RS-G-1.9 (IAEA: Vienna, 2005), pp. 4–6, 15–29; and International Atomic Energy Agency (IAEA), *Nuclear Forensics Support: Reference Manual*, IAEA Nuclear Security Series no. 2, Technical Guidance (IAEA: Vienna, 2006), p. 5.

The second goal is achieved when on-site measurements by non-destructive analysis techniques, such as gamma spectrometry, reveal the quantity and main constituents of the found materials or items. This allows the materials or items to be assigned to a convenient category. Categorization will help to determine the magnitude of the threat posed by a specific incident, which 'may range from environmental contamination, through risk to public health and safety, to proliferation concerns, each requiring a different response. Further analysis will be guided by the initial categorization.'[14] Each category should have predefined procedures associated with it, so the personnel on site would know exactly what to do next.

[14] IAEA (note 1), p. 3.

Table 2.4. Categories of characteristics of materials or items subject to measurement

Category of characteristic	Characteristics
Physical	Sizes, shapes and textures of solid objects
	Characteristics of powders: particle size distribution; morphology
	Characteristics of liquids
Chemical (molecular)	Chemical form of nuclear or radioactive material
	Non-radioactive chemicals used in the nuclear fuel cycle
Elemental	Major elements constituting the sample (if not clear from determination of chemical composition)
	Minor elements[a]
	Trace elements[b]
Isotopic	Isotopic composition of nuclear materials
	Isotopic composition of non-nuclear materials

[a] Minor elements are usually those intentionally added to the material in small quantities to improve or change its physical or chemical characteristics.

[b] Trace elements are usually added unintentionally during material production or manufacturing. E.g. machining of uranium metal with steel instruments will leave traces of iron and chromium behind. For the definition see the glossary in this volume.

Source: International Atomic Energy Agency (IAEA), *Nuclear Forensics Support: Reference Manual*, IAEA Nuclear Security Series no. 2, Technical Guidance (IAEA: Vienna, 2006), pp. 29–30.

The IAEA has recommended categories for nuclear forensic categorization (see tables 2.2 and 2.3). They are formulated separately for nuclear and radioactive material, because of the fundamentally different natures of the threat represented by the misuse of these materials. The IAEA categorizes nuclear material according to the time and effort required to manufacture it into its most destructive form—the components of a nuclear explosive device. In contrast, radioactive materials are assigned to one of five numbered categories according to their potential to affect human health, with Category 1 being the most dangerous and Category 5 the least.

In some applications, especially where sample collection is happening in a much more controlled or codified environment, categorization may be skipped on the assumption that the composition of the material in the sample is known to some degree, or on the realization that the amount of the material in the sample is too small for on-site measurements. IAEA safeguards inspector collecting swipe samples at a uranium-enrichment plant, for example, will package and label a swipe and send it to the laboratory without attempting to categorize the material on site.

II. Sample characterization

Once a sample has been obtained, it is sent to a laboratory to determine the characteristics of the material or the item it contains. The laboratory measures the material in a sample and compiles as precise a description of it as practicable or necessary for investigation purposes.

Material characteristics

The list of material characteristics subject to measurement, and thus the list of measurement techniques that should be applied to a sample, is not necessarily known in advance. In some cases, an obvious set of basic measurements can satisfy the purpose of a nuclear forensic investigation. In others, the characterization process will continue after it has been informed by the interpretation of initially obtained results, and new, perhaps unexpected, measurements will have to be ordered. Since nuclear forensics deals with various nuclear or radioactive materials from the nuclear fuel cycle, there is a large but limited number of them, and it is possible to group their characteristics into a straightforward list of categories. The IAEA has adopted a list of categories of characteristics for the purposes of its nuclear security work (see table 2.4).

Measurement techniques and equipment

Various analytical techniques (i.e. measurement methods) and the appropriate equipment are used to extract characteristics from a sample (see table 2.5). According to an established classification, three categories of analytical tool are used: bulk analysis, imaging and microanalysis (see table 2.6).[15] These categories and techniques are explained below.

The analytical techniques can be destructive or non-destructive. Destructive analysis (destructive assay, DA) is measurement of the content and elemental or isotopic concentration of the nuclear material in a sample that involves a change in the material's physical and chemical form (i.e. destruction).[16] It is normally conducted after the sample has been processed in some way into a state suitable for the specific analysis (e.g. by dissolution in acid). The prepared material is then used in, and destroyed by, the analysis. In contrast, non-destructive analysis (non-destructive assay, NDA) is measurement of the content and elemental or isotopic concentration of the nuclear material in a sample without causing significant physical or chemical changes to the material.[17] It is normally carried out by

[15] IAEA (note 1), pp. 29–30.
[16] IAEA (note 7), p. 60.
[17] IAEA (note 7), p. 62.

Table 2.5. Typical material characteristics categories and measurement techniques

Framework	Characteristics of particular interest	Measurement techniques
Development and control of nuclear armaments and stockpiles, disarmament		
National weapon development programmes[a]	All	HRGS, mass spectrometry
Partial Test-Ban Treaty[a]	Isotopic	HRGS, mass spectrometry
Comprehensive Nuclear-Test-Ban Treaty[b]:		
IMS radionuclide component	Isotopic	HRGS, BGCS
On-site inspection	Isotopic	HRGS, BGCS
Fissile material cut-off treaty[b, c]	Isotopic	Mass spectrometry, HRGS
Non-proliferation and nuclear security		
Non-Proliferation Treaty (environmental sampling for IAEA safeguards)[a]	Elemental, isotopic	HRGS, XRF, SEM, mass spectrometry
Attribution in a trafficking case[a]	All	HRGS, SEM, mass spectrometry
Attribution of a nuclear terrorism event[a]	All	HRGS, mass spectrometry
Nuclear intelligence		
Monitoring of foreign explosions	All	
Monitoring of foreign facilities and materials	All	

BGCS = beta–gamma coincidence spectrometry; HRGS = high-resolution gamma spectrometry; IAEA = International Atomic Energy Agency; IMS = International Monitoring System; SEM = scanning electron microscopy; XRF = X-ray fluorescence analysis.

[a] This framework does not apply restrictions on characteristics to be measured or the measurement techniques that can be used. The table gives typical examples.

[b] This framework restricts both characteristics that can be measured and the measurement techniques that can be employed.

[c] The measurement techniques permitted by a fissile material cut-off treaty (FMCT) and characteristics of interest will be determined if and when such a treaty is negotiated and will depend on the specific details of the FMCT verification regime. The characteristics chosen for measurement may allow determination of the nuclear material's age, the fact of its ongoing production (or absence thereof) and the total amount of plutonium produced in a reactor.

Sources: For measurement techniques used in IAEA safeguards see International Atomic Energy Agency (IAEA), *Safeguards Techniques and Equipment: 2003 Edition,* International Nuclear Verification Series no. 1 (IAEA: Vienna, 2003), pp. 5–34. For techniques used by the Comprehensive Nuclear-Test-Ban Treaty Organization (CTBTO) see Kalinowski, M. B. et al., 'The complexity of CTBT verification: taking noble gas monitoring as an example', *Complexity,* vol. 14, no. 1 (Sep./Oct. 2008), p. 93; and Takano, M. and Krioutchenkov, V., 'Technical methods employed for the on-site inspection', *Kerntechnik,* vol. 66, no. 3 (2001), p. 144.

measuring the radiation or heat emitted from the sample and comparing that emission to a calibration based on essentially similar material whose contents have been determined through destructive analysis. NDA can be passive or active. Passive NDA measures radiation or heat emitted spontaneously. For example, gamma spectrometers register gamma radiation emitted from a sample and measure its energy, which

Table 2.6. Measurement techniques typical for sample characterization in nuclear forensic analysis

Type of tool	Destructive assay	Non-destructive assay
Bulk analysis	Thermal ionization mass spectrometry (TIMS), inductively coupled plasma mass spectrometry (ICP-MS) Radiochemical separation	High-resolution gamma spectrometry (HRGS), beta–gamma coincidence spectrometry (BGCS), beta-gated gamma spectrometry (BGGS), X-ray fluorescence analysis (XRF), alpha particle spectroscopy (APS)
Imaging	–	Direct visual inspection, photography, optical microscopy Scanning electron microscopy (SEM), transmission electron microscopy (TEM)
Microanalysis	Fission-track TIMS (FT-TIMS), ICP-MS Resonance ionization mass spectrometry (RIMS), accelerator mass spectrometry (AMS)	Surface ionization mass spectrometry (SIMS) Scanning electron microanalysis (SEM) with energy dispersive sensor (SEM/EDS), SEM with wavelength dispersive sensor (SEM/WDS)

Sources: International Atomic Energy Agency (IAEA), *Nuclear Forensics Support: Reference Manual*, IAEA Nuclear Security Series no. 2, Technical Guidance (IAEA: Vienna, 2006), pp. 46–54; and International Atomic Energy Agency (IAEA), Department of Safeguards, *Research and Development Programme for Nuclear Verification 2010–2011* (IAEA: Vienna, 2010), p. 147.

allows identification of the isotope whose decay has led to the emission of the photon with this particular energy. In active NDA the emission to be measured is stimulated by the measurement instrument, for example, by a beam of neutrons.

Bulk analysis

Bulk analysis tools are designed to characterize the measured material as a whole and therefore provide information about the average composition of the sample. This may be necessary for adequate detection and identification of trace constituents.[18] If the sample is homogenous (i.e. if the composition and state of the material are uniform, such as with a piece of uranium metal), then bulk analysis might be sufficient for material characterization. Bulk analysis can be either destructive or non-destructive.

Imaging

Imaging tools produce high-magnification images or maps of the material. Such imagery can serve two interrelated goals. First, it determines if a

[18] IAEA (note 1), pp. 29–30.

sample is heterogeneous (i.e. consists of different components or diverse parts) or homogenous. In the case of a heterogeneous sample, bulk analysis alone may be insufficient, because it can mask important characteristics by averaging them over the whole sample. The microanalysis of individual components would then be required. For example, a sample of a powder seized at Munich Airport in 1994 was analysed by scanning electron microscopy (SEM), which revealed the presence of three very different types of particle—uranium oxide (U_3O_8) and two forms of plutonium oxide (PuO_2). This called for a separate microanalysis of each type (see below).[19]

Second, imaging tools produce information on the shapes and sizes of particles, grains, platelets and other components of the material. Such information can be characteristic of the mode of production and history of the material in a sample. For example, enrichment plants use uranium in a gaseous compound, uranium hexafluoride (UF_6). If the UF_6 vapour is allowed to deposit on the internal walls of the plant equipment in the presence of air, it will form smooth spherical particles roughly 1–2 micrometres in diameter. With time, these particles will lose their shape, coalesce and form a smooth film. This film, in turn, can shatter and form much bigger particles in the form of plates.[20] The shape of these particles might be useful in determining its history.

Microanalysis

Microanalysis is generally understood as the determination of the absolute or relative abundances (often expressed as a concentration) of very small amounts of chemical substances or isotopes. '*In situ* microanalysis' is defined as the 'direct analytical investigation of the microstructural domains of a solid by focused beams of particles and radiation'.[21] The IAEA also includes some destructive bulk analysis techniques in this category if they can be adapted to characterize the individual constituents of the bulk material.

Microanalysis is used if, for example, imaging indicates that a sample is heterogeneous. In the case of the powder seized in Munich, plutonium

[19] Schenkel, R. et al., 'From illicit trafficking to nuclear terrorism: the role of nuclear forensic science', International Atomic Energy Agency (IAEA) and Institute for Transuranium Elements (ITU), *Advances in Destructive and Non-destructive Analysis for Environmental Monitoring and Nuclear Forensics,* Proceedings of an International Conference, Karlsruhe, 21–23 Oct. 2002 (IAEA: Vienna, 2003), p. 13.

[20] Kaurov, G. A., Stebelkov, V. A., Kolesnikov, O. N. and Frolov, D.V., *Atlas of Uranium Microparticles from Industrial Dust at Nuclear Fuel Cycle Plants* (Ministry for Atomic Energy, Laboratory for Microparticle Analysis: Moscow, 2000), p. 11; and Stebelkov, V., Khoroshilov, V. and Stebelkov, Yu., 'Occurrence of particles with morphology characteristics which are typical for certain kinds of nuclear activity', IAEA-CN-184/82, *Symposium on International Safeguards: Preparing for Future Verification Challenges*, Vienna, 1–5 Nov. 2010 (IAEA: Vienna, 2010).

[21] McNaught, A. D. and Wilkinson, A., International Union of Pure and Applied Chemistry (IUPAC), *Compendium of Chemical Terminology*, 2nd edn (Blackwell Scientific Publications: Oxford, 1997).

oxide particles of different types distinguished by SEM were individually picked and microanalysed by mass spectrometry, which allowed their isotopic composition to be determined, their age to be calculated and the type of reactor that produced them to be identified.[22]

Restrictions on characterization

As with sample collection, sample characterization in nuclear forensic applications is often restricted by legal or regulatory documents. Some of the information on a sample's contents that is revealed by measurement can be seen as excessive: it may be too revealing or classified within a particular legal framework.

Restrictions applied to the radionuclide component of the CTBTO's International Monitoring System (IMS) are a good example of this. The IMS equipment collects both noble gases and radioactive particles by filtering them out of the air.[23] Once the gases and particles have been collected, the IMS is set up to look for radionuclides indicative of a nuclear explosion: it searches for four isotopes of xenon (i.e. ^{131m}Xe, ^{133}Xe, ^{133m}Xe, ^{135}Xe) in its noble gas monitoring stations, and screens the particles for 84 radionuclides (42 fission products and 42 activation products) defined as relevant.[24] The IMS is limited to the use of particular measurement techniques: high-resolution gamma spectrometry (HRGS) for analysis of radioactive particles on filters, and HRGS and beta–gamma coincidence spectrometry (BGCS) for measurement of noble gases.[25] Other measurement techniques, including perhaps more powerful ones such as mass spectrometry, are not permitted. This arrangement allows the CTBTO monitoring system to detect nuclear explosions without the risk of inferring sensitive information concerning nuclear weapon design.

III. Nuclear forensic interpretation

Signatures

Sample characteristics are determined by measurements in order to create a description specific enough to identify the material—that is, to differentiate the material in the sample from the totality of other materials. Such

[22] Schenkel et al. (note 19), p. 14.

[23] Dahlman, O., Mykkeltveit, S. and Haak, H., *Nuclear Test Ban: Converting Political Visions to Reality* (Springer: Dordrecht, 2009), pp. 47–50.

[24] De Geer, L.-E., *CTBT Relevant Radionuclides*, Comprehensive Nuclear-Test-Ban Treaty Organization (CTBTO) Provisional Technical Secretariat (PTS) Technical Report PTS/IDC-1999/02 (CTBTO: Vienna, Apr. 1999), pp. 24–25.

[25] Kalinowski, M. B. et al., 'The complexity of CTBT verification: taking noble gas monitoring as an example', *Complexity*, vol. 14, no. 1 (Sep./Oct. 2008), p. 93; and Auer, M. et al., 'Intercomparison experiments of systems for the measurement of xenon radionuclides in the atmosphere', *Applied Radiation and Isotopes*, vol. 60, no. 6 (June 2004), p. 866.

differentiation is the essence of nuclear forensic interpretation. It is done to the degree necessary to attain the objectives of the analysis. The characteristic or, most likely, the combination of characteristics that enables such differentiation or identification is called a signature.

A single signature might or might not be sufficient to answer the question that a nuclear forensic analyst has been presented with in a particular case. For example, simple measurement of the size and shape of a fuel pellet might be enough to determine its producer, if it is known that pellets of the measured shape are only manufactured at a certain facility. Additional signatures have to be searched for if the investigation demands knowledge of, for instance, manufacturing date. Each nuclear forensic signature would thus be informative of some part of the material's history, revealing specific information concerning the material. The material could contain a number of independent signatures carrying information about the same feature of interest.

There are two types of signature: comparative (also known as empirical) and predictive. Comparative signatures are discovered by systematic measurement of nuclear or radioactive materials. For example, oxygen in nature consists of three stable isotopes: 99.762 per cent is ^{16}O, 0.038 per cent is ^{17}O and 0.200 per cent is ^{18}O. However, slight isotopic fractionation leads to relative variations up to 5 per cent in the $^{18}O : {}^{16}O$ ratio in rainwater, and these variations depend on average annual temperature, average distance from the ocean and latitude—that is, they depend on location. During the nuclear fuel cycle, the oxygen from water—which is a common solvent in the fuel cycle—makes its way into the uranium dioxide (UO_2) produced at any plant. Determination of the $^{18}O : {}^{16}O$ ratio in UO_2 fuel pellets might therefore be helpful in determining the location where those pellets were produced.[26] Thus, the $^{18}O : {}^{16}O$ ratio is an example of a comparative (or empirical) signature.

Predictive signatures are produced from modelling chemical and physical processes involved in the nuclear fuel cycle, nuclear weapon manufacture and testing. For example, if the reactor type and its mode of operation, initial fuel composition and the time of fuel irradiation in the reactor are known, then the resulting amount and isotopic composition of plutonium produced in the fuel can be predicted by calculation.[27] If the measured isotopic composition of plutonium in a sample matches the calculated predictive signature, then this facilitates the determination of the origin of the plutonium.

[26] Mayer et al. (note 1), pp. 403–404.

[27] On plutonium isotopic composition calculations see chapter 5 in this volume. On the methods of calculating the plutonium-production rate see Albright, D., Berkhout, F. and Walker, W., SIPRI, *Plutonium and Highly Enriched Uranium 1996: World Inventories, Capabilities and Policies* (Oxford University Press: Oxford, 1997), pp. 461–63.

The process of interpretation

During nuclear forensic interpretation the analyst uses available material characteristics in an attempt to construct useful signatures. These signatures are then used to differentiate the material in a sample from the totality of irrelevant materials. This is often done by the process of elimination.[28] Since there is only a limited number of nuclear or radioactive materials and physical processes that can produce them, it is normally possible to enumerate a limited number of possible origins and histories of the material. Available signatures can then be used to eliminate some of those possibilities. If elimination leaves only one possibility, then the interpretation process has been successful. If elimination leaves more than one possibility, the analyst has to decide which additional signature would differentiate between the remaining alternatives and then has to return to the characterization step in order to measure the necessary material properties and to construct the additional signature needed. In some cases—for example, if the information necessary for further elimination is classified or has been lost—the interpretation process cannot be finalized, leaving more than one alternative.

Two types of interpretation are often distinguished: endogenic and exogenic.[29] Endogenic information is obtained by interpreting the raw data on a sample's characteristics using only general knowledge, such as the laws of physics. In contrast, exogenic information is impossible to derive from the sample alone.

Endogenic information is often either self-evident or can be obtained by straightforward calculation. For example, the 'age' of nuclear material (i.e. the time lapsed since the previous chemical purification) can be calculated directly from relative concentrations of radionuclides present in a sample.[30] No data other than the sample characteristics is necessary to obtain this result, and thus the information on the material's age is endogenic. In some frameworks, the use of endogenic information alone is sufficient. For example, a confirmed detection by CTBTO radionuclide monitoring stations and laboratories of particles containing a number of fission products strongly associated with nuclear explosions, such as molybdenum-99 and zirconium-95, would fulfil the purpose of those systems—establishing

[28] Redermeier, A., 'Fingerprinting of nuclear material for nuclear forensics', *ESARDA Bulletin*, no. 43 (Dec. 2009), p. 76.

[29] Mayer, K. and Wallenius, M., 'Nuclear forensic methods in safeguards', *ESARDA Bulletin*, no. 38 (June 2008), p. 45.

[30] Moody, K. J., Hutcheon, I. D. and Grant, P. M., *Nuclear Forensic Analysis* (CRC Press: Boca Raton, FL, 2005), pp. 178–217.

the nuclear nature of the explosion associated with those particles.[31] In the case of environmental sampling for the IAEA safeguards purposes, failure to detect isotopes inconsistent with the state's declaration is a meaningful result.

In many other important cases, such as trafficking investigations, endogenic information alone is insufficient. A more sophisticated type of interpretation produces exogenic information by matching sample characteristics to external reference data. Such reference data could be obtained from open literature, individuals' expertise, databases, archived samples, specially acquired samples or model calculations.[32] One example of the use of exogenic information for successful nuclear forensic interpretation was the verification of Iran's acknowledgement, made in August 2003, that traces of highly enriched uranium (HEU) found by the IAEA on Iranian territory originated from centrifuge components supplied by Pakistan. The IAEA was indeed able to match HEU particles found in Iran with particles obtained by swipe sampling of parts provided by Pakistan in May 2005.[33]

Libraries, databases and archives

As illustrated by the case of particles of Pakistani HEU found in Iran, it is sometimes possible to simply request a reference sample for nuclear forensic interpretation. States and international organizations have created and maintained a few nuclear forensic databases for the purpose of combating trafficking of nuclear materials. There is also a significant number of databases and repositories of reference data and materials created for unrelated purposes, such as quality control, safeguards or nuclear material accountancy. Some relevant data repositories in both categories are described below, along with their major features.

It is noteworthy that it took more than a year for the IAEA to obtain a reference sample of Pakistani HEU. In applications where timely and accurate nuclear forensic assessments are demanded—for example, the investigation of a nuclear terrorism event—such a delay would not be considered acceptable. It is thus no coincidence that the suggestion that nuclear forensic interpretation would be better served by an ability to refer to a database or sample archive has been actively pursued in the context of

[31] Matthews, K. M., *The CTBT Verification Significance of Particulate Radionuclides Detected by the International Monitoring System*, National Radiation Laboratory (NRL) Report no. 2005/1 (New Zealand Ministry of Health, NRL: Christchurch, 2005), pp. 5, 41.

[32] In the case of model calculations, models used to provide reference data have been built and validated using data independent of the sample in question, and therefore their use differs fundamentally from calculations used to obtain endogenic information.

[33] See chapter 9 in this volume.

combating nuclear smuggling and acts of nuclear terrorism.[34] This has crystalized into a proposal to establish a coordinated network of national libraries, described below.

Databases maintained by international organizations

The IAEA's Incident and Trafficking Database (ITDB), which was launched in 1995, is described as an information system to record and analyse incidents of 'illicit trafficking and other unauthorized activities and events involving nuclear and other radioactive material outside of regulatory control'.[35] As of December 2013, 125 states participated in it. Its information is confidential, and only some aggregated data and some data concerning incidents involving HEU and plutonium are released to the public.[36]

The ITDB's main sources of information are incident-notification forms (INFs) that are normally received by the IAEA from participating states via a national point of contact. The INFs consist of two parts. Part I provides basic data, specifying at a minimum 'the date and location of the incident; type, quantity, and physical form and chemical properties of materials involved; enrichment level and isotopic content of nuclear materials or activity levels of radioactive sources'.[37] Data categories included in Part I generally coincide with the list of material characteristics categories used by the IAEA in its nuclear forensic guidance documents.[38] Part II provides additional, usually more sensitive, details on the event, such as 'individuals or organizations involved, intended use, means of detection/discovery, containers, packaging, and labelling, possible origins of the materials', and other information that the reporting state might wish to include.[39]

The ITDB also reviews 'credible and relevant' open-source reports, information from which can either complement a submitted INF, or trigger a request for additional information to the state where the discovery occurred. The IAEA Division of Nuclear Security analyses information contained in the ITDB and provides participating states with quarterly and

[34] Luetzenkirchen, K. and Mayer, K., 'How a database of nuclear databases could help the effort to combat trafficking', *Nature*, 18 Jan. 2007, p. 256; and May, M., Davis, J. and Jeanloz R., 'Preparing for the worst', *Nature*, 26 Oct. 2006, pp. 907–908.

[35] International Atomic Energy Agency (IAEA), 'Incident and Trafficking Database (ITDB)', <http://www-ns.iaea.org/security/itdb.asp>. Until 2012 the ITDB was known as the Illicit Trafficking Database.

[36] International Atomic Energy Agency (IAEA), 'IAEA Illicit Trafficking Database (ITDB)', Fact sheet, 2014, <http://www-ns.iaea.org/downloads/security/itdb-fact-sheet.pdf>.

[37] Satterfield, J., 'International Atomic Energy Agency (IAEA) illicit trafficking database programme', Institute of Nuclear Materials Management, *46th Annual Meeting of the Institute of Nuclear Materials Management (INMM 46)*, Phoenix, AZ, 10–14 July 2005 (INMM: Deerfield, IL, 2006), pp. 3–4.

[38] See section II above; and IAEA (note 1), pp. 29–30.

[39] Satterfield (note 37), p. 3.

annual reports assessing 'threats, patterns and trends' of illicit trafficking.[40] Since 2008 the IAEA has also maintained the Malicious Acts Database (MAD), which is designed to complement the ITDB. The MAD includes open-source data on 'threats, attempts, plots and fully or partially executed activities' aimed at intentionally defeating nuclear security procedures or systems at nuclear facilities.[41]

The IAEA also has a number of data repositories created for safeguards purposes. First, on the basis of national nuclear materials accountancy systems, each state provides the IAEA with declarations that describe its nuclear material inventories, flows and balances (i.e. differences between incoming and outgoing material flows).[42] Such a declaration should cover every batch of material and include some data on its physical form and chemical composition, as well as its weight and, in the cases of plutonium and uranium, isotopic composition.[43] Second, the IAEA collects every year, also for accounting purposes, thousands non-destructive assay results collected by inspectors in the field and hundreds of destructive assay results provided by its Safeguards Analytical Laboratories (SAL) and the Network of Analytical Laboratories (NWAL)—a network of 20 laboratories in various IAEA member states—mainly in order to compare the element (e.g. uranium or plutonium) or isotope (e.g. uranium-235) masses with the corresponding declared values.[44] Third, the IAEA has an Environmental Sampling (ES) Database, which stores the data that results from measurements of environmental samples (most often dust particles collected by inspectors on cloth swipes) conducted by the laboratories participating in the NWAL. These laboratories use various mass-spectrometry techniques, electronic microscopy and gamma-spectrometry analysis in order to obtain the full range of particle characteristics, all of which are stored in the ES Database, alongside quality-control data, among other things. Information contained in these three IAEA databases is considered 'safeguards confidential' and cannot be shared by the IAEA. However, each state either has its own safeguards data or can collect it, and

[40] Satterfield (note 37), p. 5.

[41] Hoskins, R., Turkin, V. and Wesley, R., 'Nuclear security incident analysis: towards an integrated and comprehensive approach', IAEA-CN-166/061, Presentation at the International Symposium on Nuclear Security, Vienna, 30 Mar.–3 Apr. 2009, <http://www-pub.iaea.org/MTCD/Meetings/cn166_Presentations_n.asp>.

[42] Norman, C. et al., 'The importance of correctness: the role of nuclear material accountancy and nuclear material analysis in the state evaluation process', IAEA-CN-184/267, *Symposium on International Safeguards: Preparing for Future Verification Challenges*, Vienna, 1–5 Nov. 2010 (IAEA: Vienna, 2010).

[43] IAEA (note 7), p.48.

[44] Norman et al. (note 42), pp. 3–4.

can therefore decide to make it available to nuclear forensic investigations.[45]

The IAEA also has a number of publicly available databases and information resources grouped into a single portal, NUCLEUS.[46] Some of these could prove to be relevant for identification of nuclear or radioactive materials, such as the Nuclear Fuel Cycle Information System (NFCIS), Country Nuclear Power Profiles (CNPP), the Power Reactor Information System (PRIS), the Research Reactor Database (RRDB), the World Distribution of Uranium Deposits (UDEPO) database, the International Catalogue of Sealed Radioactive Sources and Devices (ICSRS), and the Net Enabled Waste Management Database (NEWMDB).

The Nuclear Energy Agency (NEA) of the Organisation for Economic Co-operation and Development (OECD) maintains the Spent Fuel Isotopic Composition Database, which contains detailed data on spent fuel from 14 power reactors.[47]

Databases maintained by states

A database containing detailed description of fresh nuclear fuels used in Europe and the former Soviet Union is operated by the Institute for Trans-uranium Elements (ITU), based in Karlsruhe, Germany (part of the European Commission's Joint Research Centre), and the Moscow-based High-Technology Scientific Research Institute for Inorganic Materials (VNIINM, also referred to as the Bochvar Institute, part of Rosatom, the Russian State Atomic Energy Corporation). The database, which became fully operational in December 1997, was created with the explicit purpose to assist the identification of nuclear materials intercepted in cases of trafficking.

The database has three parts: two 'restricted sectors' and one 'common sector'. The ITU and the VNIINM each operate a restricted sector, which contains restricted data that can be made available to the other party on a case-by-case basis in the form of query results. The common sector is available to authorized personnel at both facilities. The database contains information on fuel suppliers; reference data on commercial and research reactor fuels, including their physical, chemical and isotopic characteristics; 'Typical distributions of microstructure parameters', such as grain

[45] Kuhn et al. (note 6), p. 3; Donohue, D., 'Environmental sample analysis: advances and future trends', IAEA-CN-184/159, *Symposium on International Safeguards: Preparing for Future Verification Challenges*, Vienna, 1–5 Nov. 2010 (IAEA: Vienna, 2010), and Vilece, K., Hosoya, M. and Donohue, D., 'Evolution of safeguards analytical services', *Proceedings of the 31st ESARDA Annual Meeting*, Vilnius, 26–28 May 2009 (European Safeguards Research and Development Association: Ispra, 2009), p. 6.

[46] A catalogue of NUCLEUS databases is available at <http://nucleus.iaea.org/CIR/>.

[47] Organisation for Economic Co-operation and Development (OECD), Nuclear Energy Agency, 'SFCOMPO: Spent Fuel Isotopic Composition Database', <http://www.oecd-nea.org/sfcompo/>.

size in fuel pellets; 'Detailed specification limits and real ranges of isotope and impurity contents' characteristic for a certain supplier; and information on the analytical methods and equipment that produced the data.[48]

In trafficking cases involving less standard material than commercial or typical research nuclear fuel, such as experimental or discontinued fuel, a broad understanding of production processes and intended uses is required. In order to address this requirement, the ITU–VNIINM database has been complemented with an electronic archive of literature published in the Soviet Union and Russia between 1972 and 2001. The archive also contains a 'common sector', containing open publications that are difficult to access outside of Russia, and a 'separate sector', containing internal reports produced by organizations in the Soviet and Russian nuclear complex. The archive contains information on such 'exotic' fuel designs as an oxide fuel of non-standard geometry; metallic, carbide, nitride and carbonitride fuels; thermionic fuel elements and fuel for nuclear rocket engines; fuels and targets for minor actinide transmutations; and radionuclide sources and radioisotopic thermoelectric generators.[49]

The Nuclear Materials Information Program (NMIP) was formally established by the president of the United States on 28 August 2006.[50] The NMIP is envisaged as 'an integrated and continuously updated information management system' with three major components.[51] First, the NMIP was to consolidate information from all sources available to the US Government concerning nuclear materials both in the USA and worldwide. The system is designed to contain the full range of nuclear material characteristics relevant to nuclear forensics. Second, it was intended to create and maintain an archive of 'a national registry for identifying and tracking nuclear material samples that are held throughout the [USA]'.[52] Third, the NMIP has an

[48] Dolgov, Yu. et al., 'Installation of a database for identification of nuclear material of unknown origin at VNIINM Moscow', *Proceedings of the 21st ESARDA Annual Meeting*, Seville, 4–6 May 1999 (European Safeguards Research and Development Association: Ispra, 1999), p. 833.

[49] Dolgov, Y., Bibilashvili, Y. and Schubert, A., 'Development of an electronic archive on non-conventional fuels as an integral part of a nuclear forensics laboratory', IAEA and Institute for Transuranium Elements (note 19).

[50] The document establishing the NMIP, National Security and Homeland Security Presidential Directive NSPD-48/HSPD-17, is not public. It was described by Mowatt-Larsen, R., Director, Office of Intelligence and Counterintelligence, US Department of Energy, Statement before the US Senate Homeland Security and Governmental Affairs Committee, 2 Apr. 2008, <http://www.hsgac.senate.gov/hearings/nuclear-terrorism-assessing-the-threat-to-the-homeland>. See also US Government Accountability Office (GAO), *Nuclear Nonproliferation: Comprehensive U.S. Planning and Better Foreign Cooperation Needed to Secure Vulnerable Nuclear Materials Worldwide*, GAO-11-227 (GAO: Washington, DC, Dec. 2010).

[51] Mowatt-Larsen (note 50).

[52] Mowatt-Larsen (note 50).

international outreach component, designed to encourage other states to create similar systems covering their own material.[53]

On 22 December 2003 the Council of the European Union (EU) issued a directive intended to improve and harmonize control of 'high-activity sealed sources' (HASS) within the EU, which became known as the HASS directive.[54] It requires all producers of HASS, as defined by the directive, to keep records of all their sources, their transfers and location, and to regularly submit the records to the relevant national authority. The competent national authorities were directed to keep records of authorized holders of those sources and sources themselves, and 'keep the records up to date, taking transfers into account, among other factors'. The record of each source should contain the name of the radionuclide, characteristics of its physical and chemical form, its activity level at the date of manufacture or first being placed on the market, manufacturing date and the identity of the manufacturer, and other information. At the end of 2013 the HASS directive and the activities under it were consolidated into a new Basic Safety Standards Directive.[55]

A proposed network of national libraries

The most straightforward, although unrealistic, way to achieve quick availability of reference data would be the creation of a single worldwide nuclear forensic database containing every characteristic that could conceivably prove useful for an investigation. For a number of obvious reasons, including the proprietary or classified nature of much of the relevant data, this option has not been developed further. Instead, in 2008 the Nuclear Forensics International Technical Working Group (ITWG) in cooperation with the IAEA put forward the concept of a distributed network of national nuclear forensic libraries (NNFL) and an international directory listing these libraries and the formal procedures to put appropriate requests for reference information to the governments hosting them.[56]

[53] Brisson, M., 'Nuclear Materials Information Program', Institute of Nuclear Materials Management, *51st Annual Meeting of the Institute of Nuclear Materials Management 2010 (INMM 51)*, vol. 2 (INMM: Deerfield, IL, 2011).

[54] Council Directive 2003/122/Euratom of 22 Dec. 2003 on the control of high-activity sealed radioactive sources and orphan sources, *Official Journal of the European Union*, L346, 31 Dec. 2003.

[55] Council Directive 2013/59/Euratom of 5 Dec. 2013 laying down basic safety standards for protection against the dangers arising from exposure to ionising radiation, *Official Journal of the European Union*, L13, 17 Jan. 2014. On the HASS directive and its implementation see European Commission, Directorate-General for Energy, 'Radioactive sources', <http://ec.europa.eu/energy/nuclear/radiation_protection/radioactive_sources_en.htm>.

[56] Wacker, J. F. and Curry, M., *Proposed Framework for National Nuclear Forensics Libraries and International Directories* (Nuclear Forensics International Technical Working Group: 8 June 2011). See also Smith, D., de Oliveira, C. N. and Abedin-Zadeh, R., 'Recent activities at the International Atomic Energy Agency to advance nuclear forensics', and Wacker, J. F. and Curry, M. R., 'A concept for national nuclear forensic libraries', Institute of Nuclear Materials Management (note 53), vol. 2.

The concept envisages that the NNFL would contain data relevant to nuclear forensic investigations concerning 'nuclear or other radioactive material that either resides in or was manufactured by a particular country'.[57] The library may take the form of an electronic database or of an archive of actual samples, or be a combination of these two. The concept puts no restriction on the national libraries' structure but implies that compatible structures would greatly facilitate the response to queries.

Two main factors influence the necessary complexity of the national library: the size and age of national holdings of nuclear and radioactive materials and the rate of their production, transfers and disposal.[58] The concept envisages that a national library's size would be commensurate with the country's activities involving nuclear and radioactive materials. While it would ideally encompass all materials ever produced or stored by the state, the library can be as extensive as deemed practicable by the state creating it. However, the IAEA and the ITWG seem to agree that a certain amount of data is required: 'chemical form, physical form, isotopic composition (fissile, major and minor isotope abundance), elemental composition, production date, and physical location'.[59] The USA's NMIP is probably closest to the idea of the comprehensive national nuclear forensics library. It contributes directly to the development and promotion of the NNFL and is also closely linked to the US state-level nuclear material accounting databases.[60]

The concept suggests that each government would create a national point of contact and mechanisms for initiating and responding to international queries. Points of contact (officials in appropriate parts of government) would liaise with the international directory. The international directory of national libraries is envisaged as a listing of governments that have national libraries, a brief summary of its contents and information on points of contact. The directory would not contain detailed description of any material, nor would it include proprietary or sensitive information of any kind.[61]

It remains to be seen whether and to what extent this concept will be implemented. It received considerable political support at the 2010 Nuclear Security Summit in Washington, DC, and the 2010 IAEA General

57 Wacker and Curry, *Proposed Framework* (note 56), p. 3.

58 LaMont, S. et al., 'National nuclear forensics libraries: a suggested approach for country specific nuclear material databases', LA-UR-10-06586, Presentation at the International Workshop on Nuclear Forensics Following on Nuclear Security Summit, 5–6 Oct. 2010, Tokai, <http://www.jaea.go.jp/04/np/activity/2010-10-05/index_en.html>, p. 8.

59 Wacker and Curry, *Proposed Framework* (note 56), p. 5.

60 Ascanio, X., Beams, J. and Dunsworth, D., 'Inventory characterization for planning and executing effective nuclear material management, consolidation and disposition', Institute of Nuclear Materials Management (note 53), vol. 2.

61 Wacker and Curry, *Proposed Framework* (note 56), p. 9.

Conference and from the Global Initiative to Combat Nuclear Terrorism.[62] A number of states have invested considerable effort in practical implementation of this concept.[63] However, the possibility still exists that some states might prove to be slow or non-responsive in implementing the NNFL concept due to the significant cost of the effort, reluctance to diminish deniability in case of an incident, lack of psychological readiness for transparency, lack of clarity concerning the consequences of a successful attribution, low priority assigned to the problem of nuclear smuggling, or ostensible ability to successfully conduct interpretation by other means.

[62] White House, Office of the Press Secretary, 'Work plan of the Washington Nuclear Security Summit', 13 Apr. 2010, <http://www.whitehouse.gov/the-press-office/work-plan-washington-nuclear-security-summit>; International Atomic Energy Agency (IAEA), General Conference, 'Nuclear security', Resolution, 24 Sep. 2010, GC(54)/RES/8, 24 Sep. 2010; and Sonderman R., 'Global Initiative to Combat Nuclear Terrorism (GICNT) efforts in nuclear forensics', Presentation at the International Forum on Peaceful Use of Nuclear Energy and Nuclear Non-proliferation, 2–3 Feb. 2011, Tokyo, <http://www.jaea.go.jp/04/np/activity/2011-02-02/index_en.html>, p. 9.

[63] For reports on national implementation of the NNFL concept see *International Conference on Advances in Nuclear Forensics: Countering the Evolving Threat of Nuclear and Other Radioactive Material out of Regulatory Control*, Book of Extended Synopses, Vienna, 7–10 July 2014 (IAEA: Vienna, 2014).

Appendix 2A. Basic facts and definitions related to measurement

SOPHIE GRAPE

Measurement is a basic but essential tool in nuclear forensics. Two of the stages in the process described in chapter 2 depend on it: categorization and characterization. Indeed, at any point in the iterative nuclear forensic process, the analyst may have to return to take new measurements, perhaps using one of the spectroscopy methods described in chapters 3 and 4. The vocabulary used in measurement—such as the difference between the terms 'precise' and 'accurate'—will also influence the ways in which the reconstruction of a nuclear event can be interpreted.

This appendix reviews some of the basic facts and definitions related to measurement that are essential for a nuclear forensic analyst. It starts in section I by introducing the terms and definitions often used in the context of measurement. Section II then explains how to assess a measurand (i.e. the subject under investigation, such as a sample that is being measured) in terms of detection and determination, while section III describes an application to counting measurements.

This appendix does not fully cover all aspects of metrology—the scientific study of measurement. More detailed information can be found in the references cited below or in dedicated textbooks that deal with the topic of statistics and measurement techniques.[1]

I. Terms and definitions

A 1968 article by Lloyd A. Currie has come to be considered as a guiding document on measurement detection and determination.[2] At the time of its publication, the plethora of expressions used for determining limits of detection in radiochemistry caused confusion and frustration because most of the terms had approximately the same meaning. Today, international

[1] E.g. Bhattacharyya, G. K. and Johnson, R. A., *Statistical Concept and Methods* (Wiley: Chichester, 1977); Gilmore, G., *Practical Gamma-ray Spectrometry*, 2nd edn (Wiley: Chichester, 2008); and Bevington, P. R. and Robinson, D. K., *Data Reduction and Error Analysis for the Physical Sciences*, 3rd edn (McGraw Hill: New York, 2003).

[2] Currie, L. A., 'Limits for qualitative detection and quantitative determination: application to radiochemistry', *Analytical Chemistry*, vol. 40, no. 3 (Mar. 1968).

standards can be used as a guide.[3] The International Organization for Standardization (ISO) has provided documentation specifically on, for example, quantities and units (ISO 80000 1:2009), atomic and nuclear physics (ISO 80000-10:2009), vocabulary and symbols in statistics (ISO 3534-1), and on determination of characteristic limits for measurements of ionizing radiation (ISO 11929, which is heavily inspired by the work of Currie—see section II).[4] Some of these are general (e.g. on general international vocabulary), while some are more specialized (e.g. ISO 11929, which focuses on ionizing radiation measurements).[5]

For all experiments, the importance of an established measurement procedure cannot be underestimated. This procedure should be a description of all operations and steps needed for a certain measurement and a given method. It should describe the measurement equipment, the measurement principle, the sample nature and the analysis methodology.

Accuracy and precision[6]

The reason for measuring anything is to obtain a reliable description of it. A physical quantity may be directly measurable or may need to be measured indirectly. For example, the mass of a fuel pellet can be measured directly by taking its weight, while a sample's isotopic composition can be determined indirectly by measuring gamma radiation.

A good measurement has both good accuracy and good precision. Accuracy describes the ability to obtain a value that is close to the true value; precision describes the ability to reproduce approximately the same output given the same input (see figure 2A.1). The term sensitivity does not seem to be uniquely defined in the context of measurements and should not be used without an explanation of what it refers to. In general it seems to express the degree of response to a signal or some other input, but it may

[3] International Organization for Standardization (ISO), *Determination of the Characteristic Limits (Decision Threshold, Detection Limit and Limits of the Confidence Interval) for Measurements of Ionizing Radiation: Fundamentals and Application*, International Standard 11929:2010 (ISO: Geneva, 2010); and Currie, L. A., International Union of Pure and Applied Chemistry (IUPAC), 'Nomenclature in evaluation of analytical methods including detection and quantification capabilities' (IUPAC Recommendations 1995), *Pure and Applied Chemistry*, vol. 67, no 10 (1995).

[4] International Organization for Standardization (ISO) and International Electrotechnical Commission (IEC), *Quantities and Units*, part 1, *General*, ISO/IEC 80000-1:2009, and part 10, *Atomic and Nuclear Physics*, ISO/IEC 80000-10:2009 (ISO: Geneva, 2009); and International Organization for Standardization (ISO), *Statistics: Vocabulary and Symbols*, part 1, *General Statistical Terms and Terms Used in Probability*, ISO 3534-1:2006 (ISO: Geneva, 2006).

[5] International Organization for Standardization (ISO) and International Electrotechnical Commission (IEC), *International Vocabulary of Metrology: Basic and General Concepts and Associated Terms (VIM)*, ISO/IEC Guide 99:2007 (ISO: Geneva, 2007); and ISO 11929:2010 (note 3).

[6] Joint Committee for Guides in Metrology (JCGM), *International Vocabulary of Metrology: Basic and General Concepts and Associated Terms (VIM)*, JCGM 200:2008 (Bureau International des Poids et Mesures: Sèvres, 2008).

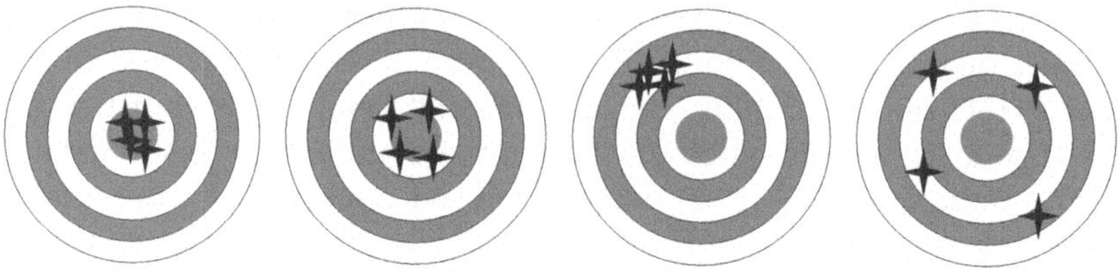

Figure 2A.1. Graphical representation of the terms accuracy and precision using a dartboard

also reflect the stability—that is, whether or not the signal is easily distorted.

Fundamental aspects of the measurand

The terms explained in this section are not only of importance to nuclear forensics but for presentation of experimental data in general. In an evaluation of data from measurements it is important to know the meaning of the stated information in order to interpret it correctly. The terms 'mean value', 'variance' and 'standard deviation' explained below are some of the most basic terms that are central to the understanding and interpretation of data.

Measuring the same variable, x, many times gives a number of observations, denoted x_i (where $i = 1, 2, 3, \ldots, n$). The mean value, $\overline{x}$, of these n individual measurements can then be calculated as

$$\overline{x} = \frac{\sum_{i=1}^{n} x_i}{n}.$$

The mean $\overline{x}$ is called an estimator of the limiting mean, μ. The limiting mean value is often also called the true mean value, because it is the value that the mean approaches as the number, n, of measurements increases towards infinity.

Even if a mean value is measured very accurately, it will not be possible to obtain its exact value. There will always be an uncertainty associated with the measurand. This uncertainty is expressed as a discrepancy between all the results that are obtained from measuring the same thing repeatedly. The origin of the uncertainty can be, for example, limitations in the theoretical description, the measurement equipment or simply fluctuations in the measurements. The uncertainty can be expressed in different ways; one of the most common is to state the value of the standard deviation. The standard deviation, σ, is equal to the square root of the variance, σ^2, where

$$\sigma^2 = \lim_{n\to\infty}\left[\frac{1}{n}\sum_{i=1}^{n}(x_i-\mu)^2\right].$$

The concept of a standard deviation should be interpreted such that for data taken from a normal distribution (see below), approximately 68 per cent of the data values will lie within $\pm\sigma$ of the limiting mean value μ.

One difficulty in calculating the standard deviation is that the above definition of σ^2 contains μ, the value of the limiting mean. Since this may not always be known, it does not seem possible to calculate σ. A solution is to estimate the limiting mean using the mean value from the sample, with the sample representing a subset of the full population. The estimated standard deviation of a population, denoted s, is then

$$s^2 = \frac{1}{n-1}\sum_{i=1}^{n}(x_i-\bar{x})^2 .$$

The reason for the factor $n-1$ in the denominator is that the number of independent measurements has decreased by 1 because of the calculation of the mean value $\bar{x}$.

Statistical distributions

A collection of random variables or events can be described using probability distributions. These distributions are functions that describe the possible range of the random variables or events, including the most probable values and the expected spread among them.

There are several different probability distributions but only three of them are mentioned here. Probably the best-known distribution is the symmetrical Gaussian, or normal, distribution. It has a typical bell shape, with many samples centred around the mean value and fewer samples at higher and lower values (see figure 2A.2). Because this distribution has many appealing and convenient properties, unknown distributions are often assumed to be normal and known distributions are sometimes approximated as normal. The normal distribution is particularly useful due to the central limit theorem, which states that a collection of sufficiently many random variables (data points) that each has a finite mean and variance will have a mean value that is approximately normally distributed.

It may be preferable to apply Student's t-distribution for small sample sizes of normally distributed data (see figure 2A.2). In such circumstances

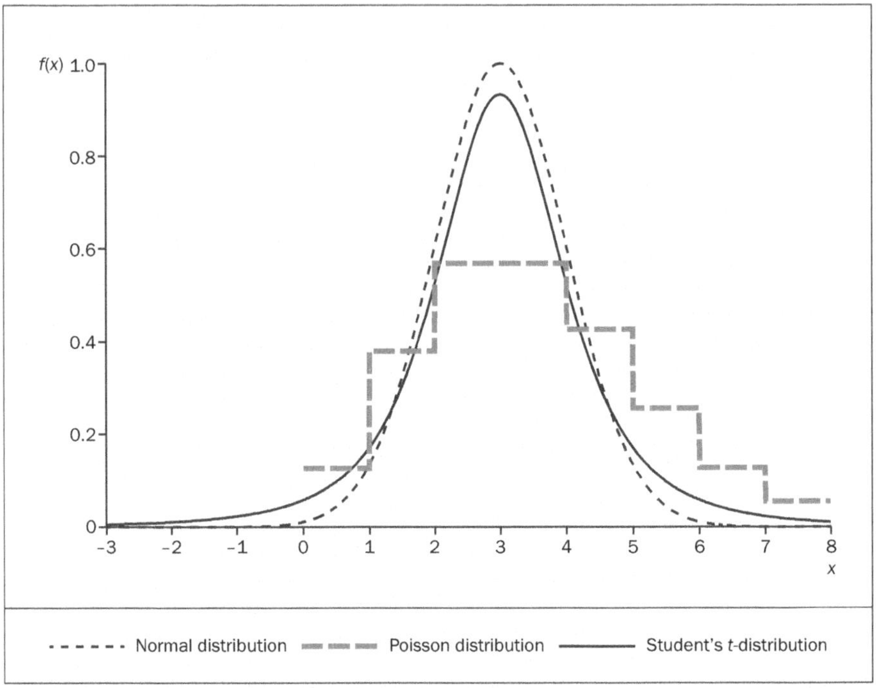

Figure 2A.2. The normal distribution, the Poisson distribution and Student's *t*-distribution with three independent parameters

Note: The expected value for all distributions is in this case 3 and all distributions have the same integral. The Poisson distribution is only valid for positive values of *x* and its asymmetric shape is clearly shown.

this distribution can give better estimations of the mean value and the standard error because it also takes into account the number of degrees of freedom. (This quantity is a measure of the number of independent data points present in the data set.) Compared to the normal distribution, the tails on the left and right sides of Student's *t*-distribution are larger (see figure 2A.2). As the sample size grows, the distribution approaches that of the normal.

Another distribution, common in physics measurements, is the asymmetric Poisson distribution, which is defined only for positive integer values. The asymmetry is seen as a larger tail to the upper side of the mean value, compared to the lower side (see figure 2A.2). The asymmetry is small for large mean values and large for small mean values, because there are fewer possible values for the random variable to take on below the mean value compared to above it. The Poisson distribution is often used to describe events that occur with constant probability within a certain time interval.

Measuring errors

In general, uncertainties or errors are often quoted together with the actual value in order to identify the precision with which the result has been determined. However, the two terms are not equivalent: 'error' implies that there is a difference between the observed or calculated value and the true value, while 'uncertainties' reflect discrepancies between results. Thus, the error can only be calculated if the true value is known; if it is not known, then the term uncertainty should be used.

Errors are often classified as either systematic or statistic.

Systematic errors are reproducible inaccuracies introduced by, for example, the measurement equipment, a bias in the system or a faulty calibration. The size of this error contribution is determined by the ability to understand, control, reduce and compensate for the inaccuracies that arise. Systematic errors are estimated through an analysis of the equipment and techniques used for the measurements.

Statistic errors do not arise due to lack of precision in different components but as a result of fluctuations in the number of collected counts during a finite time interval. Statistic errors manifest themselves through the difference in results that are obtained as the measurement is repeated. The standard deviation of statistical uncertainties can be estimated analytically for each observation—experimental determination is not always necessary. The number of counts in an observation (or any other experiment where data can be grouped in bins in a histogram or displayed in a frequency plot according to some criterion) is Poisson distributed. The standard deviation for this distribution is $\sigma = \sqrt{\mu}$, making the relative uncertainty, $\sigma/\mu = 1/\sqrt{\mu}$, smaller as the number of counts per time interval increases. Because the limiting mean value, μ, is rarely known, it is often approximated by the mean value, $\overline{x}$, obtained from measurements.

Although it is desirable for the errors of both types to be small, in practice it is often not possible to improve experimental results to any desired degree of accuracy. Common limitations include the time and effort needed to run or repeat a certain measurement, the degree of knowledge and understanding that is required to reduce systematic errors in a system and measuring components, or fluctuations in the result which are not of a statistical nature. The latter may be of unforeseen origin, such as rare background effects, unknown sample contamination, sudden malfunctions or simply careless mistakes.

Evaluations for safeguards applications

Three broad categories of error related to safeguards accountancy methods and safeguards applications can be identified: random errors, bias and short-term systematic errors.[7] These correspond, respectively, to statistic errors, systematic errors and errors that remain constant for a short time period while measurement conditions and settings vary in an unpredictable and long-term perspective.

Uncertainties are similarly expressed as either random uncertainty components or systematic uncertainty components, that is, as the standard deviations of the random errors and short-term systematic errors, respectively.

II. Assessment of a measurand

According to accepted standards, a qualitative assessment of a measurand should be expressed in terms of decision limit (or detection threshold), detection limit and limit of the confidence interval, by means of statistical tests and specified probabilities.[8] The measurement assessments proposed by Currie are common for all types of measurement and have therefore been exported to many fields. The assessments in the ISO standard are constructed by analogy to those of Currie, but are applied to the specialized field of ionizing radiation metrology with detailed procedures specified for calculations and measurements.[9] However, as the concepts of, for example, decision limits and detection limits are also applicable to other types of measurement, they also appear in other ISO documents.[10]

A fundamental difference between the assessments of Currie and ISO 11929 is that the former is based on so-called classical or frequentist statistics, while the latter is based on Bayesian statistics. The concepts of assessing the measurand are the same for the two approaches, but the ways in which the probability concept is viewed and uncertainties are treated are different. While classical statistics uses probability distributions to say something about unknown, true, measurand values, Bayesian analysis goes the opposite way and uses the measured value of the measurand to calculate the likelihood that it was obtained as a consequence of a certain probability distribution. This is a way of incorporating prior knowledge from

[7] This categorization follows the convention adopted by the Joint Committee for Guides in Metrology (JCGM), which consists of the ISO, the International Bureau of Weights and Measures (Bureau International des Poids et Mesures, BIPM), the International Electrotechnical Commission (IEC), the International Federation of Clinical Chemistry and Laboratory Medicine (IFCC), the International Union of Pure and Applied Chemistry (IUPAC), the International Union of Pure and Applied Physics (IUPAP), and the International Organization of Legal Metrology (Organisation Internationale de Métrologie Légale, OIML). ISO 3534-1:2006 (note 4); and ISO/IEC Guide 99:2007 (note 5).

[8] Currie (note 2); and ISO 11929:2010 (note 3).

[9] ISO 11929:2010 (note 3).

[10] One such document describes how the decision limit and the detection limit from experimental data can be estimated. International Organization for Standardization (ISO), *Capability of Detection*, part 2, *Methodology in the Linear Calibration Case*, ISO 11843-2:2000 (ISO: Geneva, 2000).

theory or earlier measurements into the equations.[11] The equations presented here on decision limit, detection limit and determination limit are compact, and in their form apply to both types of statistical interpretation.

Currie classification scheme and data interpretation

In order to say anything about any measurements, it is important to be aware of how the measurements were performed and what their outcomes were. Currie presented a classification scheme intended to provide the analyst with uniquely defined measurement limits such that the data and the results can be correctly interpreted.[12] According to Currie, three limiting levels must be defined: (*a*) a decision limit or critical level, L_C, above which an observed net signal (i.e. a signal with background subtracted) may be reliably recognized as detected; (*b*) a detection limit, L_D, above which a true net signal a priori is expected to lead to detection; and (*c*) a determination limit, L_Q, above which the measurement precision is sufficient for a quantitative determination with a given uncertainty.

The first limit, L_C, is considered an a posteriori limit because it is relevant only after a signal has been detected. The second limit, L_D, is considered an a priori limit because it characterizes the detection procedure before any detection has taken place. Assuming that the measurement procedure is followed, the smallest value of the measurand that ensures a specified probability of being detected is denoted the detection limit.

Hypothesis testing

An a posteriori problem is concerned with interpretation of a detected signal; for this purpose a decision tool called hypothesis testing is available. The purpose of applying hypothesis testing is to investigate the consistency of a theory and data and help the analyst to correctly interpret the data and results.

The general idea of hypothesis testing is the wish to investigate whether a signal or a set of data has a certain characteristic or not. The so-called null hypothesis, or starting point, is that the characteristic feature is absent in the data and the alternative hypothesis is that it is present.[13]

[11] Both brief introductions to Bayesian statistics and detailed textbooks on the subject exist elsewhere. E.g. Weise, K. et al., 'Bayesian decision threshold, detection limit and confidence limits in ionising-radiation measurement', *Radiation Protection Dosimetry*, vol. 121, no. 1 (Dec. 2006); Bolstad, W. M., *Introduction to Bayesian Statistics*, 2nd edn (Wiley: Chichester, 2007); and Hoff, P. D., *A First Course in Bayesian Statistical Methods* (Springer: Dordrecht, 2009).

[12] Currie (note 2).

[13] An example of how the Currie hypothesis testing procedure can be applied in practice to spectral data from gamma spectroscopy is presented in De Geer, L.-E., *A Decent Currie at the PTS: Detection Limit Concepts in the PTS Radionuclide Software*, Technical Paper CTBT/PTS/TP/2005-1 (Comprehensive Nuclear-Test-Ban Treaty Organization: Vienna, Aug. 2005); and De Geer, L.-E.,

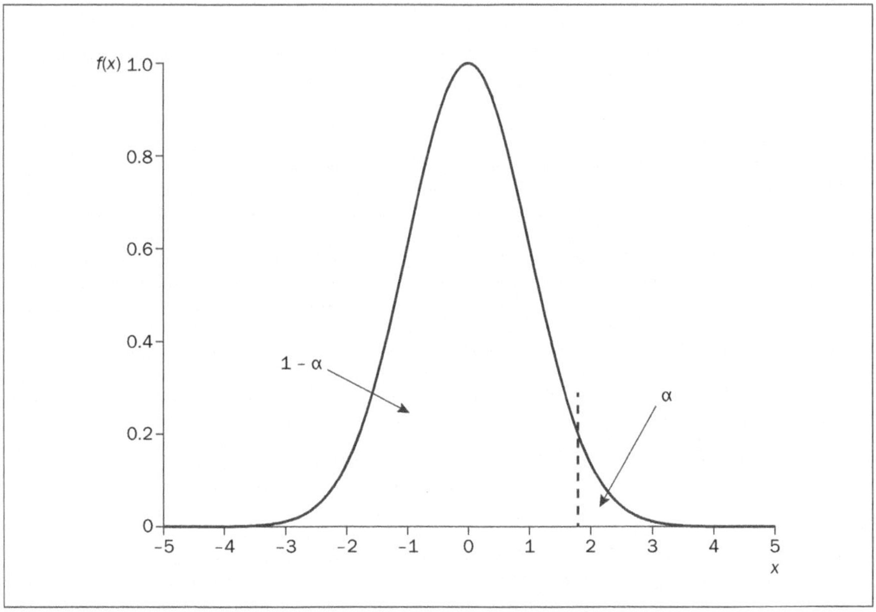

Figure 2A.3. The standard normal distribution showing the maximum acceptable error of the first kind

Note: The vertical line indicates the value of the parameter $k_{1-\alpha}$. The integral of the distribution below this limit is equal to $1 - \alpha$.

In hypothesis testing, two types of error can be made. An error of the first kind (denoted α) represents drawing false conclusions that a feature or physical effect is present when in reality it is not; an error of the second kind (denoted β) represents drawing the wrong conclusion that the physical effect is absent when in reality it is present. When the acceptable risk values of the two types of error have been decided, the decision and detection limits can be calculated. A frequently cited choice of risks, especially in ISO documents, is $\alpha = \beta = 0.05$, even though the values are sometimes too routinely decided, without adequate thought. However, in general gamma spectroscopy, applying those risks could generate several hundred false peaks in what in reality is an empty spectrum.[14] A so-called rejection probability function, γ, can also be constructed. It describes the possibility that the null hypothesis is rejected for a chosen value of the random parameter. Depending on how the null hypothesis is constructed, γ is equal to either α or $1 - \beta$.

'Currie detection limits in gamma-ray spectroscopy', *Applied Radiation and Isotopes*, vol. 61, nos 2–3 (Sep./Oct. 2004).

[14] Bolstad (note 11); and Hoff (note 11).

A guideline value of the measurand, y_r, should be quoted along with the choices of the errors of both kinds.[15] This provides an additional assessment of whether or not a certain measurement procedure satisfies requirements for scientific, legal or other reasons.

Determining the critical limit and the detection limit

The critical level L_C, above which an observed net signal may be considered detected in an a posteriori problem, can be determined using two terms. The first term requires knowledge of the maximum acceptable level of error of the first kind, α, and the second term is the standard deviation, denoted σ_0, of the measured net signal as the true net signal, μ_S, is zero. The critical level is then

$$L_C = k_{1-\alpha}\,\sigma_0 .$$

Assuming that the normal distribution is standardized (e.g. that it is centred around the value 0 with a standard deviation of 1), the parameter $k_{1-\alpha}$ marks the limit above which the integral (i.e. the area below the curve) is equal to α (see figure 2A.3).

If the measured signal exceeds the critical level L_C, it is labelled 'detected'. If the signal is below the threshold, it is not possible to conclude that it is actually there with the accepted risk, but it is also not possible to conclude that it is absent. After the establishment of L_C, the a priori detection limit L_D may be established, with

$$L_D = L_C + k_{1-\beta}\,\sigma_D .$$

The standard deviation σ_D characterizes the probability distribution of the net signal as $\mu_S = L_D$ and $k_{1-\beta}$ is the value below which the integral of the normal distribution is β. The detection limit represents the smallest value for which the risk of drawing the wrong conclusion that the physical effect is absent does not exceed the specified probability. The detection limit L_D is a characteristic of the measurement process itself; longer measurement times and more counts will reduce it.

An analyst should be alert to the risks of trusting stated limits without understanding how they were attained. For example, the use of low background detectors may under- or overestimate the critical limit and the detection limit.[16] This may be the result of the use of a certain statistical approach or of performing inapplicable approximations.

[15] ISO 11929:2010 (note 3).

[16] Hurtgen, C., Jerome, S. and Woods, M., 'Revisiting Currie: how low can you go?', *Applied Radiation and Isotopes*, vol. 53, nos 1–2 (July 2000); and Bernasconi, G., Greaves, E. D. and Sajo-Bohus, L., 'New approach in assessing the lower detection limit in low level radiation counting', *Nuclear Instruments and Methods in Physics Research, Section A: Accelerators, Spectrometers, Detectors and Associated Equipment*, vol. 307, nos 2–3 (Oct. 1991).

It should also be noted that for the calculations of the decision limit and the detection limit, the standard uncertainty as a function of the true signal (of the measurand) is needed. This can either be explicitly specified or it may follow as an interpretation of data from several measurements. For the latter alternative, there must be previous measurements of the same kind, carried out on different samples with differing activities but in other aspects under similar conditions.[17]

Confidence intervals for detection

It seems almost deceivingly simple to compare an experimental value with a decision limit in order to draw a qualitative decision. But with what level of confidence is the result consistent with the expected value?

If the true signal value S and its standard deviation σ are unknown, then the estimated net signal $\overline{S}$ and its standard deviation s can be used to create an interval of values for μ_S corresponding to a certain level of confidence, $1-\gamma$. In general terms, a confidence interval contains the 'true' value of the measurand with the probability $1-\gamma$. A high probability results in a large interval and a small probability results in a narrow interval. The limits of the interval are often chosen symmetrically such that the probabilities that the measurand is smaller than the lower limit or larger than the upper limit are both equal to $\gamma/2$. The definition of end-point values of the interval depends on the probability distribution applied to the specific case. For example, if the Student's t-distribution is applied, the symmetrical end-point values of the confidence interval are calculated as

$$\overline{S} \pm t_{1-\gamma/2} \frac{s}{\sqrt{n}}.$$

The interval end-point values for this example depend on the chosen level of confidence, the estimated signal value, the standard deviation and the so-called standard variable $t_{1-\gamma/2}$, which can be found in a table. It should be noted that for other statistical distributions, the notations of the standard variable as well as its numerical values differ from those of Student's t-distribution.

The symmetrical confidence level can be used if the measured signal S (or its estimation $\overline{S}$) is considered to be detected—that is, if it is larger than L_C. If the measured signal is smaller than L_C, only an upper limit corresponding to the one-sided confidence interval can be specified for the experimental result. The upper limit is in that case is $L_C + S$ (or $L_C + \overline{S}$). The maximal upper limit is $2L_C = L_D$.

[17] ISO 11929:2010 (note 3).

Quantitative analysis and the determination limit

The third limit suggested by Currie, the determination limit, marks the beginning of the region where a qualitative analysis of the detected signal can be meaningfully made because the measurement precision is high enough. The determination limit is defined as

$$L_Q = k_Q \, \sigma_Q,$$

with L_Q being the true value of the net signal value, σ_Q its standard deviation and k_Q a multiplier with the property that its inverse equals the specified relative standard deviation produced by estimates of the signal. A common default value for k_Q is 10.

III. Application to counting measurements

In applications where counting statistics measurements are performed, for example spectroscopic measurements, the signal is often detected on top of a background level. The background consists of events without relevance in a certain region of interest in the spectrum and it needs to be subtracted before further analysis can be performed.

There are many ways to construct a background function. The simplest is to choose a constant background level, if there is no reason to suspect another dependence. Other choices are a linear background, which is often the case for gamma radiation detection, a slightly curved background (described by a second- to fourth-order polynomial) or a strongly curved parabola, which may be the case for alpha radiation detection.[18]

As a measure of how well the selected background (or signal) function fits the spectrum, a test quantity χ^2 can be calculated. It can be expressed in terms of the theoretical or expected values ($\mathrm{Exp}(x_i)$) and the observed or measured values ($\mathrm{Obs}(x_i)$):

$$\chi^2 = \sum_{i=1}^{M} \frac{(\mathrm{Obs}(x_i) - \mathrm{Exp}(x_i))^2}{\mathrm{Obs}(x_i)},$$

where M is the number of channels in the background region. The smaller the test variable is, the better the fit. It is desirable that

$$\left|\chi^2 - M + m\right| \le k_{1-\delta/2}\sqrt{2(M-m)}, \qquad (*)$$

where m denotes the number of random variables used as input to the model. If it is not, then the fit does not agree with the data and it should be

[18] Details on background determination, intervals and other specifications can be found in e.g. ISO 11929:2010 (note 3).

investigated whether the background interval can be adjusted until it is. A significance level of $\delta = 0.05$ is commonly used.

How good is the fit to data?

It is often desirable for an analyst to obtain a so-called fit function—a mathematical description of the data distribution—that accurately reproduces the measured data. In order to obtain a good agreement between the fit function and the measured data, physicists or analysts use the 'spectrum unfolding' procedure. This is simply the fitting of estimates of parameters to the data until a good agreement is found. The estimated values may depend not only on the value of the measurand, but also on values for certain constants, among other things.

A recommended spectrum unfolding method is the least-squares fit, which takes the uncertainty of each data point into consideration. The method is based on minimizing deviation between observed and expected values for measured quantities that stem from a Gaussian distribution. The variable being minimized is the 'goodness-of-fit' parameter, χ^2, where

$$\chi^2 = \sum_{i=1}^{M} \left(\frac{y_i - y(x_i)}{\sigma_i} \right)^2$$

(which is constructed in analogy to the above expression for χ^2), with y_i being the measured value, $y(x_i)$ the expected value and σ_i the standard deviation for the observation. The fit and the data can be said to be consistent on a certain level of confidence if the condition (*) is fulfilled.

3. Inorganic mass spectrometry as a tool of destructive nuclear forensic analysis

KLAUS MAYER, MARIA WALLENIUS, ZSOLT VARGA, MAGNUS HEDBERG AND NICOLE ERDMANN

Mass spectrometry is a well-established, highly versatile technique that offers high sensitivity, high selectivity and high precision in combination with the potential for high accuracy. Credible nuclear forensic conclusions, however, need to be based on validated procedures and on measurement techniques that are well understood. Nuclear forensic investigations typically start with non-destructive determination (i.e. high-resolution gamma spectrometry) of the radionuclides present in the sample and a visual inspection, followed by optical microscopy of the material.[1] Subsequently, samples are taken for electron microscopy and for chemical analysis. Mass spectrometry is certainly the most prominent and versatile analytical methodology than can be applied.

A number of variants of mass spectrometry can be used in nuclear forensics, each able to provide valuable information to nuclear scientists analysing nuclear material that enables the drawing of conclusions in support of non-proliferation and law enforcement investigations (see table 3.1). Thermal ionization mass spectrometry (TIMS), inductively coupled plasma mass spectrometry (ICP-MS) and secondary ion mass spectrometry (SIMS) allow determination of key parameters such as isotopic composition of major and minor constituents and the concentration of chemical impurities in the nuclear material. More sophisticated techniques, such as accelerator mass spectrometry (AMS) and resonance ionization mass spectrometry (RIMS), are currently being investigated for their applicability to nuclear forensics challenges.

This chapter provides some details on different mass spectrometric techniques, outlines their general principles and limitations, and illustrates their application in nuclear forensic investigations. Sections I–III describe the most prominent mass spectrometric techniques: TIMS, ICP-MS and SIMS, respectively. Section IV describes two techniques, AMS and RIMS, that are applied only in special cases.

[1] On gamma spectrometry see chapter 4 in this volume.

Table 3.1. Nuclear forensic measurement methods

The list of measurement methods is non-exhaustive and gives priority to mass spectrometric techniques.

Parameter	Measurement method	Interpretation by comparison against
Uranium isotope ratios	TIMS, ICP-MS, SIMS, AMS, RIMS	Database
Plutonium isotope ratios	TIMS, ICP-MS, SIMS, RIMS, gamma spectrometry	Model calculations, database
Metallic impurities (concentrations, patterns)	ICP-MS, ICP-OES, GD-MS	Process knowledge, database, known samples
Stable isotope ratios	ICP-MS, GC-MS	Database, known samples
Macroscopic appearance	Optical microscopy	Process knowledge database, known samples
Microscopic appearance	SEM, TEM	Process knowledge database, known samples
Radioisotopes	Gamma spectrometry, liquid scintillation counting, ICP-MS	Model calculations
Non-metallic impurities	Gas chromatography	Process knowledge database, known samples

AMS = accelerator mass spectrometry; GC-MS = gas chromatography coupled to mass spectrometry; GD-MS = glow-discharge mass spectrometry; ICP-MS = inductively coupled plasma mass spectrometry; ICP-OES: inductively coupled plasma optical emission spectrometry; RIMS = resonance ionization mass spectrometry; SEM = scanning electron microscopy; SIMS = secondary ion mass spectrometry; TEM = transmission electron microscopy; TIMS = thermal ionization mass spectrometry.

I. Thermal ionization mass spectrometry

TIMS is widely applied for high-precision measurement of isotope ratios for elements with reasonably low ionization potential. Nuclear materials such as uranium and plutonium satisfy this requirement and the methodologies for uranium and plutonium isotope ratio measurement by TIMS were established in the early days of nuclear technology. With improvements in instrumentation and with progress in data-processing techniques, the quality of results has largely improved and today the potential of TIMS as a powerful analytical technique is being exploited. TIMS offers high selectivity, high sensitivity, specificity and has proven to be highly precise and to have the potential for being very accurate. For uranium and plutonium measurements in particular, accurate information on isotopic composition is useful, as it helps in distinguishing between materials of different batches and different origins, intended for different applications.

Isotope dilution TIMS (ID-TIMS) is a variant of TIMS that enables the exact quantification of an element of interest in a sample. With ID-TIMS (or more generally with isotope dilution mass spectrometry) the quantity of an element present in a material is determined from the change produced in the isotopic composition of the element when a known amount of 'spike' (i.e. an enriched isotope of the same chemical element) is added.[2] ID-TIMS is suited for the highest possible accuracy and, since it is essentially a single-element technique, it is particularly applicable to the quantification of uranium and plutonium. However, the following discussion focuses on the application of TIMS to determining isotopic composition since that often proves to be the more useful 'fingerprint' in nuclear forensics.

Principle and general description

TIMS is a well-established technique.[3] It is particularly suitable for chemical elements with a low first-ionization potential. Prior to the actual measurement, a sample-preparation step is required: this consists of the chemical separation of the element of interest from other elements (e.g. matrix materials or impurities). The purified sample is then deposited on a metal ribbon, called a filament.

Typically, amounts ranging from a few micrograms down to picograms of material are deposited on a filament. The filament consists of a refractory, high work-function metal such as rhenium, tantalum, platinum or tungsten. The filament is then heated by passing an electrical current through it. This leads to vaporization of the sample, to atomization and finally to the ionization of the atoms at a hot surface (which is actually the filament). This process of removing an electron from the outer shell by thermal energy led to the name 'thermal ionization'. For uranium samples, the U^+ ions are formed and accelerated by applying a high voltage and subsequent mass separation (e.g. between $^{234}U^+$, $^{235}U^+$, $^{236}U^+$ and $^{238}U^+$) by means of a mass analyser—that is, a magnetic field, an electrostatic field or a quadrupole. In uranium and plutonium measurements, thermal ionization ion sources are typically combined with magnetic sector fields for mass separation and with Faraday cups or secondary electron multipliers for ion detection.

[2] For detailed descriptions of the technique see e.g. Fassett, J. D. and Paulsen, P. J., 'Isotope dilution mass spectrometry for accurate elemental analysis', *Analytical Chemistry*, vol. 61, no. 10 (May 1989); De Bièvre, P., 'Isotope dilution mass spectrometry: what can it contribute to accuracy in trace analysis?', *Fresenius' Journal of Analytical Chemistry*, vol. 337, no. 7 (Aug. 1990); and Gopalan, K., 'Isotope dilution mass spectrometry', eds S. K. Aggarwal and H. C. Jain, *Introduction to Mass Spectrometry* (Indian Society for Mass Spectrometry: Mumbai, 1997).

[3] TIMS is described in a number of textbooks, e.g. de Laeter, J. R., *Applications of Inorganic Mass Spectrometry* (Wiley: New York, 2001); Platzner, I. T. *Modern Isotope Ratio Mass Spectrometry* (Wiley: New York, 1997); and Aggarwal, S. K. and Jain, H. C., 'Introduction to mass spectrometry', eds Aggarwal and Jain (note 2).

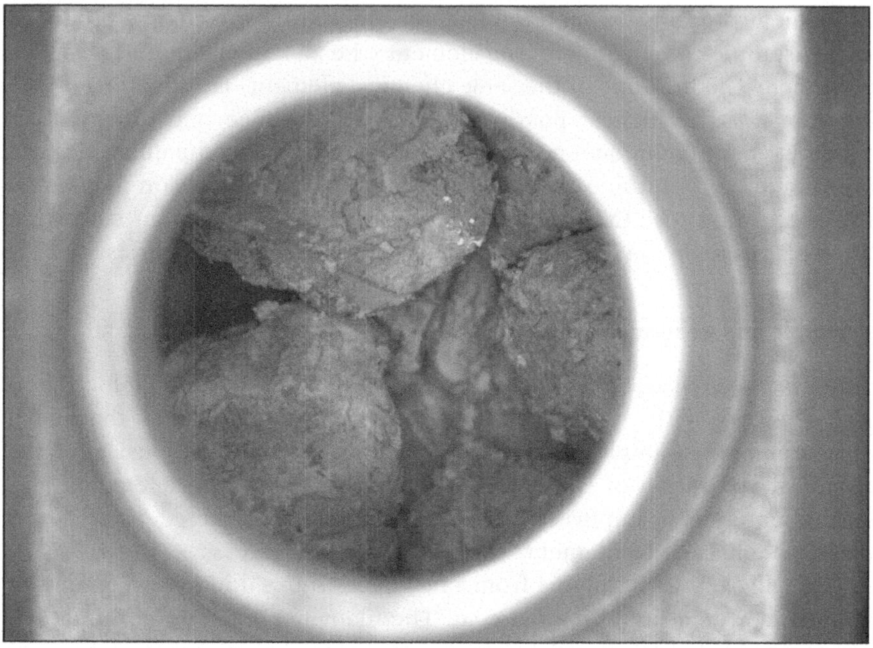

Figure 3.1. Sample of uranium ore concentrate seized at Rotterdam in 2003, after arrival at the nuclear forensics laboratory

Source: European Commission, Joint Research Centre (JRC), Institute for Transuranium Elements (ITU).

Application of TIMS in nuclear forensics

TIMS is a key measurement technique in nuclear forensic investigations. Subsamples for TIMS are dissolved and then subjected to dilution and to chemical separation. A small sample of the purified fraction is then loaded onto a rhenium filament.

The uranium isotopic composition provides information on the intended use of the material. Depleted uranium (i.e. where the ratio $^{235}U : U$ is less than 0.7 per cent) is often encountered in metallic form in shielding material of strong radioactive sources. Natural uranium (i.e. where $^{235}U : U$ is 0.7 per cent) is mined in large quantities and serves as a feed material in the nuclear fuel cycle. Some reactor types operate with fuel of natural isotopic composition. Most power reactors around the world are so-called light water reactors and operate with low-enriched uranium (LEU)—more precisely, with ^{235}U enrichments up to 4.5 per cent. Research reactors typically require higher enrichments, up to 90 per cent. The minor abundant isotopes also provide useful forensic information. The $^{234}U : ^{238}U$ isotope-abundance ratio in natural uranium shows small but measureable variations that correlate with the geographic origin of the uranium.[4] ^{236}U is present in natural uranium only in quantities irrelevant for this example, but it is produced in much larger quantities in a reactor by neutron capture of ^{235}U. Thus, the presence of this isotope in a uranium sample points at the irradiation history of uranium.

The application of TIMS measurements can be illustrated using the example of uranium ore concentrate (UOC) that was seized in 2003 at the harbour of Rotterdam, the Netherlands. Several kilograms of material were detected among a shipload of scrap metal arriving from Jordan. Some external indicators suggested that Iraq could be the origin of the material. Nuclear forensic analysis was requested by the competent authority in order to acquire more information on the history of the material and prove (or disprove) this hypothesis. A sample (see figure 3.1) was taken to the Institute for Transuranium Elements (ITU) for further analysis.

The measured $^{235}U : ^{238}U$ and $^{234}U : ^{238}U$ isotope-abundance ratios were $(7.253 \pm 0.013) \times 10^{-3}$ and $(5.522 \pm 0.072) \times 10^{-5}$, respectively, identical to those of natural uranium. The anthropogenic isotope ^{236}U was initially not detectable by TIMS using a Finnigan MAT 261 instrument (since the $^{236}U : ^{238}U$ isotope-abundance ratio was below 3×10^{-7}). A more recent measurement using a Triton mass spectrometer (by Thermo Scientific of Bremen, Germany) equipped with decelerating device and ion-counting capability revealed a $^{236}U : ^{238}U$ isotope-abundance ratio of $(1.545 \pm 0.002) \times 10^{-7}$. This result changed the picture significantly: the traces of ^{236}U indicate that the material was contaminated with previously irradiated (and recycled) uranium. This clearly indicates that the material was not produced simply by mining but other nuclear fuel cycle stages, such as irradiation in a reactor, were involved. The $^{234}U : ^{235}U$ isotope-abundance ratio, which shows small variations in nature, was 0.00761 ± 0.00011. This value agrees well with the published $^{234}U : ^{235}U$ isotope-abundance ratio in undeclared UO_4 produced at al-Qaim, Iraq, using phosphorite quarried in Akashat, northern Iraq (0.00765 ± 0.00002). It is different from the ratio found in natural uranium material confiscated in Iraq deriving from a declared source (i.e. imported from Italy, with a $^{234}U : ^{235}U$ ratio of 0.00742–0.00757).[5] The observed $^{234}U : ^{238}U$ isotope-abundance ratio suggests that the ore from which the uranium was mined was a low-

[4] Richter, S. et al., 'Isotopic "fingerprints" for natural uranium ore samples', *International Journal of Mass Spectrometry*, vol. 193, no. 1 (1999). On nuclear forensic signatures see chapters 5 and 6 in this volume.

[5] International Atomic Energy Agency (IAEA), Fourth semi-annual report on the implementation by the IAEA of the plan for the destruction, removal or rendering harmless of items listed in paragraph 12 of UN Security Council Resolution 687 (1991), IAEA report to the United Nations Security Council, S/25983, 21 June 1993.

temperature redox deposit.[6] Such a deposit is consistent with the phosphorite deposit in northern Iraq.

II. Inductively coupled plasma mass spectrometry

ICP-MS is at present the most frequently used mass spectrometric technique for both concentration and isotope ratio measurements, even if the analyte is present down to a sample level of femtograms per gram. This powerful analytical technique is also increasingly used for the measurement of long-lived radionuclides, providing a complementary tool to the traditional radioanalytical techniques. As ICP-MS is not only capable of measuring elemental concentration but also provides isotopic information on the material, it is a highly versatile technique in nuclear forensics. This is also supported by ICP-MS's high sensitivity (which involves low sample consumption), good precision and accuracy as well as the high diversity of sample types and introduction methods that the method allows.[7]

Principle and general description

ICP-MS uses inductively coupled plasma as an ion source. The sample solution can be introduced into the plasma (most frequently argon) by various sample-introduction systems (e.g. pneumatic nebulizer, ultrasonic nebulizer, electrothermal vaporization or laser ablation), where the sample decomposes into its atomic constituents in the plasma at a temperature of approximately 6000–8000 kelvin and is ionized with a high degree of ionization (i.e. ionization efficiency is higher than 90 per cent for most chemical elements) with a low fraction of multiply charged ions. The positively charged ions are extracted from the plasma operating at atmospheric pressure into the high vacuum region of the mass spectrometer via an interface. Several types of analyser can be used for the separation of the ions, such as quadrupole analysers, time-of-flight (TOF) analysers or combinations of electrostatic and magnetic sector field analysers (so-called double-focusing instruments).

Although ICP-MS instruments with quadrupole analysers (ICP-QMS) are cheaper, more robust and easier to operate, the better detection limits (approximately 1–3 orders of magnitude) of double-focusing sector-field analysers (ICP-SFMS) means that they are the main instruments used for

[6] Buchholz, B. A. et al., 'Investigating uranium isotopic distributions in environmental samples using AMS and MC-ICPMS', *Nuclear Instruments and Methods in Physics Research, Section B: Beam Interactions with Materials and Atoms*, vol. 259, no. 1 (June 2007).

[7] Becker, J. S. and Dietze, H.-J., 'Inorganic trace analysis by mass spectrometry', *Spectrochimica Acta, Part B: Atomic Spectroscopy*, vol. 53, no. 11 (Oct. 1998); and Montaser, A. and Golightly, D. W. (eds), *Inductively Coupled Plasmas in Analytical Atomic Spectrometry*, 2nd edn (VCH: New York, 1992).

ultratrace-level measurements (typically below nanogram per gram). Moreover, such mass analysers can achieve higher mass resolution (denoted as *R*, or sometimes as $m / \Delta m$). Most commercially available ICP-SFMS instruments can operate with a mass resolution up to 10 000. Use of the higher mass resolution is useful for the separation of the analyte peak from spectral interference, for instance, separating the $^{56}Fe^{+}$ analyte peak from the interfering $^{40}Ar\,^{16}O^{+}$ peak at mass $m / z = 56$. However, it also results in lower ion transmission, and thus lower sensitivity. After the separation of the analyte ions according to their mass-to-charge ratios (m / z), the ions are detected and counted. Use of multiple detectors (so-called multi-collector ICP-SFMS instruments, MC-ICP-SFMS) improves the precision of the measurement (usually expressed in relative standard deviation, RSD) as it detects the isotopes of interest simultaneously and eliminates the of the sample-introduction and ion source.[8]

The major problem in trace-level analysis by ICP-MS is the appearance of isobaric interferences in the investigated mass region. These isobaric interferences derive from polyatomic ions, which are generated in the plasma by the combination of the matrix elements in the sample (e.g. bismuth, lead, platinum, mercury or uranium), the elements of the solvent (e.g. hydrogen, oxygen or nitrogen in the case of nitric acid, HNO_3) and the elements of the plasma (e.g. argon, carbon or oxygen). These elements form polyatomic ions that appear in the mass spectrum causing elevated background and, consequently, false results. In nuclear forensics the hydride (e.g. $^{238}U\,^{1}H^{+}$ or $^{238}U\,^{1}H_2^{+}$) and oxide (e.g. $^{206}Pb\,^{16}O_2^{+}$) interferences are of high concern for trace-level plutonium and minor ^{236}U analysis. In order to eliminate most hydride and oxide interferences, special sample-introduction devices have been developed for the removal of solvents and for the generation of dry aerosol. The principal of these instruments is the same: the wet aerosol produced by the conventional pneumatic or ultrasonic nebulizer is driven through a heated chamber, where the solvent evaporates. In the subsequent step, the solvent vapour is removed either by condensation or by membrane desolvation. These instruments not only decrease the oxide and hydride background but also have better transfer efficiency of the analyte, and thus better sensitivity than conventional nebulizers.[9] However, the complete removal of these interferences, especially if they are present in large quantities, should better be accomplished by chemical sample preparation.

In cases when dissolution of the sample should be avoided (e.g. analysis of highly radioactive materials or of confiscated samples that are evidence

[8] eds Montaser and Golightly (note 7).

[9] Zoriy, M. V. et al., 'Reduction of UH^{+} formation for $^{236}U/^{238}U$ isotope ratio measurements at ultratrace level in double focusing sector field ICP-MS using D_2O as solvent', *Journal of Analytical Atomic Spectrometry*, vol. 19, no. 3 (2004).

in a forensic investigations), sample introduction by laser ablation (LA-ICP-MS) is a possible option.[10] In this case, a small portion of the investigated material is evaporated by a high-energy laser beam from a confined surface. The laser beam diameter is in the micrometre range. The ablated material is transferred into the plasma of the ICP-MS instrument as an aerosol using a carrier gas. Laser ablation offers an easy option for direct isotope ratio measurements of solid materials, as the ablation and ionization properties of isotopes of the same element are identical.[11] For concentration measurements, however, the different ablation and ionization efficiencies have to be taken into account, and these depend on the analyte, the matrix and the laser ablation characteristics.[12]

Laser ablation coupled to ICP-SFMS fulfils several requirements desirable for nuclear forensics: it has excellent detection capabilities down to the nanogram per gram range; it is a quasi-non-destructive technique (less than a microgram amount of sample is consumed for the analysis); and it is not necessary to carry out lengthy and hazardous chemical manipulations with the nuclear materials, which require special laboratory conditions and expertise. This also implies minimization of generated nuclear (radioactive) waste. The disadvantages of LA-ICP-MS methods are the cumbersome calibration for concentration measurement and usually the limitation of the maximum volume of the ablation chamber, which holds the sample. The fact that LA-ICP-MS is a surface analytical technique can be exploited to obtain isotopic and elemental information from the sample at micrometre scale. However, it can result in significantly different results from bulk analysis due to the sampling.

The most important advantage of ICP-MS is the versatility: it can be used for concentration measurement as well as for isotope ratio determination of various elements. It is a very sensitive measurement tool, down to 10^{-12}–10^{-15} g g^{-1} concentration level for almost all elements. The technique can be used to identify the type of the material and can also provide isotopic information. Compared to TIMS it requires usually less tedious sample prepar-

[10] Günther, D., Horn, I. and Hattendorf, B., 'Recent trends and developments in laser ablation-ICP-mass spectrometry', *Fresenius' Journal of Analytical Chemistry*, vol. 368, no. 1 (Aug. 2000); and Günther, D., Jackson, S. E. and Longerich, H. P., 'Laser ablation and arc/spark solid sample introduction into inductively coupled plasma mass spectrometers', *Spectrochimica Acta, Part B: Atomic Spectroscopy*, vol. 54, nos 3–4 (Apr. 1999).

[11] Guillong, M. et al., 'A laser ablation system for the analysis of radioactive samples using inductively coupled plasma mass spectrometry', *Journal of Analytical Atomic Spectrometry*, vol. 22, no. 4 (2007); and Stefánka, Z., Katona, R. and Varga, Z., 'Laser ablation assisted ICP-MS as a tool for rapid categorization of seized uranium oxide materials based on isotopic composition determination', *Journal of Analytical Atomic Spectrometry*, vol. 23, no. 7 (2008).

[12] Leloup, C. et al., 'Quantitative analysis for impurities in uranium by laser ablation inductively coupled plasma mass spectrometry: improvements in the experimental setup', *Journal of Analytical Atomic Spectrometry*, vol. 12, no. 9 (1997); and Varga, Z. and Surányi, G., 'Detection of previous neutron irradiation and reprocessing of uranium materials for nuclear forensic purposes', *Applied Radiation and Isotopes*, vol. 67, no. 4 (Apr. 2009).

ation and can be used also for elements with high ionization potential. Since it is a destructive analytical technique, sample preparation has to be carefully planned, especially if a limited amount of material is available. In this case, laser ablation ICP-MS is one of the possible solutions.

Determination of elemental composition and impurities

The elemental composition of the investigated nuclear forensic evidence is a major characteristic of the material, providing hints on the possible origin and intended use of the material. ICP-MS is capable of detecting most elements (the few exceptions include hydrogen, oxygen, nitrogen and argon). This is especially important in nuclear forensics for establishing the history of illicit material. The composition of the sample can be given after the measurement of the major constituents (e.g. identification of the uranium or plutonium compound, types of UOC, determination of alloys or various metals) that already suggest the possible source and use of the material.[13]

Generally, the minor constituents and trace elements give more detailed bases for the origin assessment. Trace elemental impurities are derived either as a residual of the feed material due to incomplete purification (so-called source-material inherited signatures) or from metallurgical production (e.g. by the added chemicals or contamination; so-called process inherited signatures). These subtle variations can be used to compare samples with similar composition to verify the possibly same origin or process.[14]

In recent years impurity analysis has been increasingly used to compare various uranium ores or ore concentrates to identify the source of an unknown material.[15] Such measurements, supported by multivariate statistical techniques, can give evidence either for verification of a declared origin or for origin assessment. For instance, elevated levels of

[13] Wallenius, M., Mayer, K. and Ray, I., 'Nuclear forensic investigations: two case studies', *Forensic Science International*, vol. 156, no. 1 (Jan. 2006); and Mayer, K. et al., 'Application of isotopic fingerprinting in nuclear forensic investigations: a case study', IAEA-CN-98/11, International Atomic Energy Agency (IAEA) and Institute for Transuranium Elements (ITU), *Advances in Destructive and Non-destructive Analysis for Environmental Monitoring and Nuclear Forensics*, Proceedings of an International Conference, Karlsruhe, 21–23 Oct. 2002 (IAEA: Vienna, 2003).

[14] Wallenius et al. (note 13).

[15] Varga, Z., Wallenius, M. and Mayer, K., 'Origin assessment of uranium ore concentrates based on their rare-earth elemental impurity pattern', *Radiochimica Acta*, vol. 98, no. 12 (Dec. 2010); Keegan, E. et al., 'The provenance of Australian uranium ore concentrates by elemental and isotopic analysis', *Applied Geochemistry*, vol. 23, no. 4 (Apr. 2008); Švedkauskaite-LeGore, J. et al., 'Investigation of the isotopic composition of lead and of trace elements concentrations in natural uranium materials as a signature in nuclear forensics', *Radiochimica Acta*, vol. 95, no. 10 (Oct. 2007); and Švedkauskaite-LeGore, J. et al., 'Investigation of the sample characteristics needed for the determination of the origin of uranium-bearing materials', *Journal of Radioanalytical and Nuclear Chemistry*, vol. 278, no. 1 (Oct. 2008).

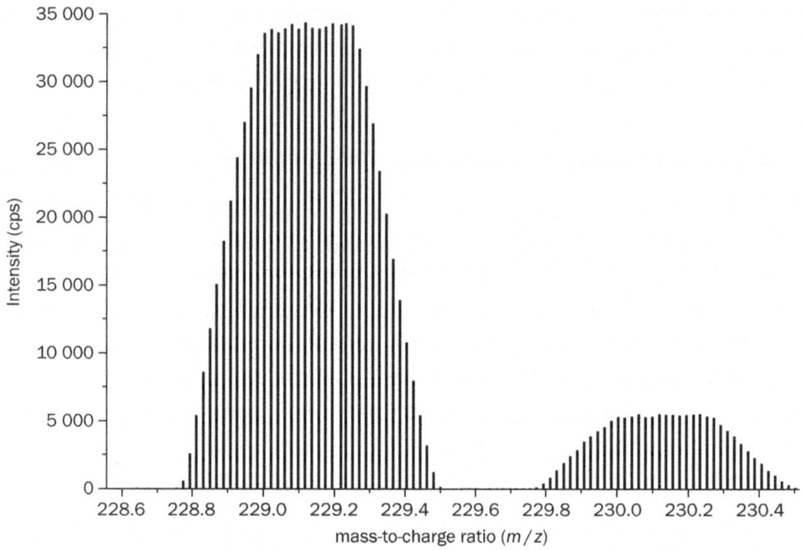

Figure 3.2. A typical mass spectrum of the thorium fraction, obtained from confiscated material for age measurement

cps = counts per second; *m* = mass; *z* = charge.

Note: The ^{230}Th analyte (daughter product, approximately 5 pg g^{-1} in the measured sample) was quantified using a ^{229}Th tracer.

Source: European Commission, Joint Research Centre (JRC), Institute for Transuranium Elements (ITU).

phosphorous have been found to be an indicator of UOC produced as a by-product of the fertilizer industry, while a higher level of thorium and its decay product ^{208}Pb are indicative of yellowcake milled from thorium-rich ores (e.g. quartz-pebble conglomerate).[16]

In the case of the UOC seized at Rotterdam (described in section I above), an elevated level of phosphorus indicated the possible use of phosphate rock as feed material. This was also indicated by the rare-earth element pattern, which was measured by ICP-SFMS: a flat, shale-like pattern without cerium or europium anomaly, similar to patterns previously found in UOC produced from certain sandstone-type deposits and in reworked sedimentary phosphorites.[17]

[16] Varga, Z. et al., 'Origin assessment of uranium ore concentrates based on their rare-earth elemental impurity pattern', *Radiochimica Acta*, vol. 98, no. 12 (Dec. 2010); and Varga, Z. et al., 'Application of lead and strontium isotope ratio measurements for the origin assessment of uranium ore concentrates', *Analytical Chemistry*, vol. 81, no. 20 (Oct. 2009).

[17] On nuclear forensic signatures see chapters 5 and 6 in this volume.

Several approaches have been used to measure a sample using ICP-MS as the types of the material vary extensively. Due to the high uranium or plutonium content in the sample, the most frequently used approach is matrix-matched calibration, which quantitatively takes into account the interactions between matrix components and analytes and its effect on the measurement result. In case of interference or if the analyte is present in very low concentration, prior chemical separation and pre-concentration can be used to enhance the measurement capabilities. This is typically the case for rare-earth elements, trace-level activation or fission products (e.g. indicating the use of a reprocessed materials), or for the decay products of the host nuclear material.[18]

The measurement of decay products allows the determination of the 'age' of the material, which actually describes the time elapsed since the last chemical separation.[19] The method is based on the assumption that the base nuclear material (i.e. the parent nuclide) is completely separated from its daughter radionuclides during production. The quantity of these daughter nuclides in the material subsequently starts to increase due to the radioactive decay of uranium or plutonium. By measuring the parent-to-daughter ratio (e.g. $^{234}U : ^{230}Th$ or $^{239}Pu : ^{235}U$), and using the equations of the radioactive decay, the time elapsed since the last chemical separation can be calculated. Compared to the other applicable techniques, such as alpha spectrometry or TIMS, ICP-MS has the advantage of a better detection limit (the parent-to-daughter ratios of uranium materials are typically in the picogram to nanogram per gram range). The mass spectrum of the measured thorium isotopes of a real sample provides a good example (see figure 3.2). The concentration of ^{230}Th in the sample was measured using ICP-MS to be approximately 5 $pg\,g^{-1}$, and it was quantified using a ^{229}Th isotope tracer (which is usually not present in typical uranium fuel materials). Based on the $^{234}U : ^{230}Th$ ratio, the age of the material was calculated to be 47.6 ± 0.6 years (with a reference date of 22 December 2009), which means that the production of the material was calculated to have taken place in May 1962 (±7 months).

As the decay products are at trace level, chemical separation is necessary in most cases. However, due to the excellent detection limits of ICP-SFMS instruments, direct measurement methods have also been proposed either

[18] On rare-earth elements see Varga et al. (note 15); and Varga, Z. et al., 'Determination of rare-earth elements in uranium-bearing materials by inductively coupled plasma mass spectrometry', *Talanta*, vol. 80, no. 5 (Mar. 2009). On trace-level activation or fission products see Varga and Surányi (note 12). On the decay products of the host nuclear material see Varga and Surányi (note 12); and Varga, Z., Wallenius, M. and Mayer, K., 'Age determination of uranium samples by inductively coupled plasma mass spectrometry using direct measurement and spectral deconvolution', *Journal of Analytical Atomic Spectrometry*, vol. 25, no. 12 (Dec. 2010); and Wallenius, M. and Mayer, K., 'Age determination of plutonium material in nuclear forensics by thermal ionisation mass spectrometry', *Fresenius' Journal of Analytical Chemistry*, vol. 366, no. 3 (Feb. 2000).

[19] See also chapter 5 in this volume.

Table 3.2. Typical detection limits and precisions of inductively coupled plasma mass spectrometry instruments for fissile nuclides

Instrument	Typical detection limit (fg g^{-1})	Precision (RSD%)
ICP-MS with a quadrupole analyser	10–600	*0.1–0.5*
ICP-MS with a quadrupole analyser and collision cell	3–10	*0.07–0.1*
ICP-MS with a time-of-flight analyser	100–1000	*0.1–1*
ICP-MS with a double-focusing sector field analyser	0.02–1	*0.02–0.3*
Multi-collector ICP-MS with a double-focusing sector field analyser	0.6–0.2	*0.002–0.05*

ICP-MS = inductively coupled plasma mass spectrometry; fg g^{-1} = femtograms per gram; RSD = relative standard deviation.

Sources: Becker, J. S., 'Mass spectrometry of long-lived radionuclides', *Spectrochimica Acta, Part B: Atomic Spectroscopy*, vol. 58, no. 10 (Oct. 2003); Becker, J. S., 'Inductively coupled plasma mass spectrometry (ICP-MS) and laser ablation ICP-MS for isotope analysis of long-lived radionuclides', *International Journal of Mass Spectrometry*, vol. 242, nos 2–3 (Apr. 2005); Agarande, M. et al., 'Sector field inductively coupled plasma mass spectrometry, another tool for plutonium isotopes and plutonium isotope ratios determination in environmental matrices', *Journal of Environmental Radioactivity*, vol. 72, nos 1–2 (2004); and Vanhaecke, F. et al., 'Applicability of high-resolution ICP-mass spectrometry for isotope ratio measurements', *Analytical Chemistry*, vol. 69, no. 2 (Jan. 1997).

by LA-ICP-MS (without sample preparation) or using post-measurement spectral deconvolution of the mass spectrum (after measurement using only the dissolved material, without chemical separation).[20]

The capabilities of ICP-MS instruments for isotope-ratio measurements are primarily determined by the sample introduction, the analyser and the detection system, since the ion source is the same. Different ICP-MS instruments for uranium and plutonium analysis have varying typical detection limits and precision (see table 3.2).

Isotope ratio measurements by multi-collector ICP-MS

Multi-collector ICP-MS (MC-ICP-MS) aims to provide high-precision isotope ratio measurements. To achieve this, MC-ICP-MS instruments combine the most suitable combination of ion source (a plasma), mass analyser (a double-focusing set-up of electrostatic analyser and magnetic sector field) and a detection system enabling the simultaneous detection of ion beams for different mass-to-charge ratios. The double-focusing mass analyser is well established in ICP-SFMS and the multi-collector array has a proven record in isotope ratio measurements by TIMS. Usually, the

[20] Varga, Wallenius and Mayer (note 18).

detector array consists of two different types of detector—Faraday cups and ion counters (using either discrete dynode secondary electron multipliers, a Daly detector or micro-channeltrons)—in order to cover a large range of ion current intensities.

Although the benefits of MC-ICP-MS are best exploited in isotope ratio measurements, these instruments can in principle also be used to measure elemental concentrations. As pointed out above, the plasma source allows measuring the isotopic composition of almost all chemical elements using MC-ICP-MS, while TIMS is applicable only to elements with a low first-ionization potential. MC-ICP-MS is thus the preferred measurement technique in geology and archaeology, for example, where elements with high ionization potentials that are present only at low concentrations often have to be analysed. Moreover, the often time-consuming chemical separation can be avoided when using MC-ICP-MS combined with laser ablation.

Isotopic analysis of uranium, lead and strontium is of particular interest for nuclear forensics. In the case of uranium, the $^{234}U : ^{238}U$ isotope-abundance ratio varies in nature mainly due to the combined effects of alpha-recoil from mineral grains during ^{238}U decay and subsequent damage to the crystal lattice. As a result, loosely bound ^{234}U is more susceptible to preferential extraction (leaching) from minerals than lattice-bound ^{238}U. In geology the $^{234}U : ^{238}U$ isotope-abundance ratio is measured, for example, to monitor weathering rates and river- and ground-water flow patterns. The $^{234}U : ^{238}U$ isotope-abundance ratio is also a necessary component in determining the age of uranium in samples.[21] The isotopic composition of lead is also commonly determined for use in age dating. In addition, it can provide clues of geochemical provenance and serve as evidence for transport, redeposition and preferential leaching processes as well as in geochemical studies.[22] A common application, but from a different discipline, is lead isotopic analysis in forensic studies when bullets are analysed in investigations of shooting incidents.[23]

In the case of strontium, the $^{87}Sr : ^{86}Sr$ isotope-abundance ratio, which is widely used in geology for age determination, can also be used for origin

[21] Andersen, M. B. et al., 'Toward epsilon levels of measurement precision on $^{234}U/^{238}U$ by using MC-ICPMS', *International Journal of Mass Spectrometry*, vol. 237, nos 2–3 (Oct. 2004).

[22] Ehrlich, S. et al., 'Lead and uranium isotopic behavior in diagenetic and epigenetic manganese nodules, Timna Basin, Israel, determined by MC-ICP-MS', *Applied Geochemistry*, vol. 19, no. 12 (Dec. 2004); Weiss, D. J., 'Accurate and precise Pb isotope ratio measurements in environmental samples by MC-ICP-MS', *International Journal of Mass Spectrometry*, vol. 232, no. 3 (Apr. 2004); and Cocherie, A. and Robert, M., 'Direct measurement of lead isotope ratios in low concentration environmental samples by MC-ICP-MS and multi-ion counting', *Chemical Geology*, vol. 243, nos 1–2 (Aug. 2007).

[23] Zeichner, A. et al., 'Application of lead isotope analysis in shooting incident investigations', *Forensic Science International*, vol. 158, no. 1 (Apr. 2006); and Buttigieg, G. A. et al., 'Lead isotope ratio determination for the forensic analysis of military small arms projectiles', *Analytical Chemistry*, vol. 75, no. 19 (Oct. 2003).

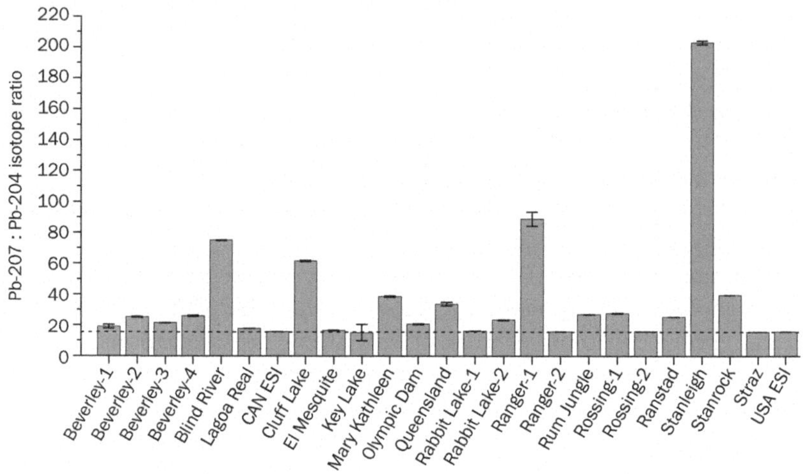

Figure 3.3. The lead-207 : lead-204 isotope-abundance ratio measured in selected uranium ore concentrate samples

Note: The horizontal line marks the present-day natural $^{207}Pb : ^{204}Pb$ isotope-abundance ratio of 15.8 ± 1.1.

Source: Reprinted with permission from Varga, Z. et al., 'Application of lead and strontium isotope ratio measurements for the origin assessment of uranium ore concentrates', *Analytical Chemistry*, vol. 81, no. 20 (Sep. 2009).

assessment, for example in archaeology to examine human migration or in determining food authenticity to distinguish products from various geographic origins.[24] The variation of the $^{87}Sr : ^{86}Sr$ isotope-abundance ratio is due to the decay of long-lived ^{87}Rb (with a half-life of 4.8×10^{10} years) to ^{87}Sr and the highly variable Rb : Sr elemental ratio in nature.

Similar to the above applications, precise isotopic composition analysis of uranium, lead and strontium is needed in nuclear forensics to obtain information on the origin of uranium. These elements were chosen as their isotopic compositions are presumed to reflect that of the ore and to give information on the type and age of the ore body, and hence on the source material.

A study by Bruce Buchholz and others provides an example of the variations in the minor isotope ratios of uranium, that is $^{234}U : ^{238}U$ and

[24] Buzon, M. R., Simonetti, A. and Creaser, R. A., 'Migration in the Nile Valley during the New Kingdom period: a preliminary strontium isotope study', *Journal of Archaeological Science*, vol. 34, no. 9 (Sep. 2007); and García-Ruiz, S. et al., 'Evaluation of strontium isotope abundance ratios in combination with multi-elemental analysis as a possible tool to study the geographical origin of ciders', *Analytica Chimica Acta*, vol. 590, no. 1 (May 2007).

$^{236}U : ^{238}U$, as measured by MC-ICP-MS in natural uranium samples from various locations.[25] Three out of the four stable lead isotopes, ^{206}Pb, ^{207}Pb and ^{208}Pb, are the final decay products of the decay series of ^{238}U, ^{235}U and ^{232}Th, respectively. Depending on the history of the uranium deposit (e.g. fractionation of the U and Pb due to the weathering), its age and the initial U : Th elemental ratio in the ore, differences in the radiogenic lead composition have been observed.[26] Besides detecting differences between the uranium deposits (see figure 3.3), isotopic composition of lead can be used in nuclear forensics (similar to geology) to estimate the age of the deposit, and therefore to limit the types of deposit that should be considered when looking for the origin of the natural uranium. However, as lead is a common element, careful blank control (i.e. measurement of blank samples alongside the real ones to check for artificially introduced contamination or to trace its source) has to be implemented during the analysis in order to ensure unbiased results.

The $^{87}Sr : ^{86}Sr$ isotope-abundance ratio can also be used in nuclear forensics, because it is a characteristic of the geological background of uranium deposits. While a drawback in lead and uranium isotope analysis is the variation in isotope ratios within a single mine site, analysis of several natural uranium samples has shown much less variation in the $^{87}Sr : ^{86}Sr$ ratio (see figure 3.4).

Example of the use of ICP-MS in nuclear forensics

A good example of the use of above-described methods in nuclear forensics is provided by the analysis of the yellowcake seized at Rotterdam (see above). The uranium isotopic composition was measured with a Nu Plasma MC-ICP-MS (manufactured by Nu Instruments, Wrexham, UK) using ion counting for the minor isotopes ^{234}U and ^{236}U and Faraday cups for the major isotopes ^{235}U and ^{238}U.[27] The mass 239 (i.e. $^{238}U + {}^{1}H$) was also measured by ion counting in order to correct the hydrogen contribution at mass position 236. The following isotope-abundance ratios were found: $^{234}U : ^{238}U = 5.502 \pm 0.066 \times 10^{-5}$, $^{235}U : ^{238}U = 7.253 \pm 0.013 \times 10^{-3}$ and $^{236}U : ^{238}U = 1.4 \pm 0.2 \times 10^{-7}$. This isotopic composition shows that the material is of natural origin, however a small abundance of ^{236}U suggests a minute contamination with previously irradiated and reprocessed uranium.

[25] Buchholz et al. (note 6).

[26] Svedkauskaite-LeGore at al., 'Investigation of the isotopic composition of lead and of trace elements concentrations in natural uranium materials as a signature in nuclear forensics' (note 15); and Varga, Z. et al., 'Application of lead and strontium isotope ratio measurements for the origin assessment of uranium ore concentrates', *Analytical Chemistry*, vol. 81, no. 20 (Sep. 2009).

[27] Varga et al., 'Analysis of uranium ore concentrates for origin assessment', *Proceedings in Radiochemistry*, vol. 1, no. 1 (Sep. 2011).

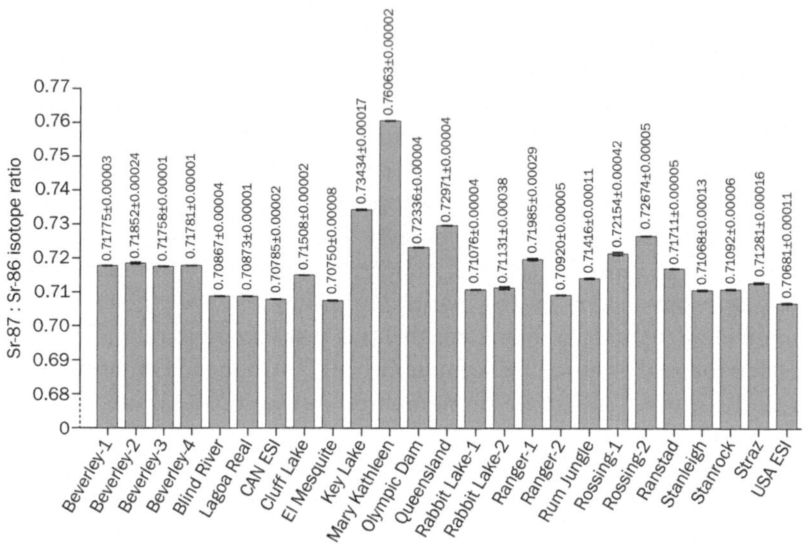

Figure 3.4. The strontium-87 : strontium-86 isotope-abundance ratio of selected uranium ore concentrate samples

Source: Reprinted with permission from Varga, Z. et al., 'Application of lead and strontium isotope ratio measurements for the origin assessment of uranium ore concentrates', *Analytical Chemistry*, vol. 81, no. 20 (Sep. 2009).

The isotope composition of lead was measured to be 1.357 ± 0.005 per cent ^{204}Pb, 25.91 ± 0.21 per cent ^{206}Pb, 21.03 ± 0.11 per cent ^{207}Pb and 51.70 ± 0.13 per cent ^{208}Pb), making it indistinguishable from the natural composition of lead. This indicates that the uranium most probably originates from a young uranium deposit containing low grade ore, which leads to a minimum ingrowth of radiogenic lead. The ^{87}Sr : ^{86}Sr ratio was measured to be 0.708.[28] This value is relatively low (see figure 3.4), which indicates that there is rubidium incompatibility (i.e. low Rb content in the mineral), as observed, for example, for phosphorite-based materials. All these signatures support the initial hypothesis that the material originated in the Middle East, most likely from Iraq.

[28] Varga et al., 'Application of lead and strontium isotope ratio measurements for the origin assessment of uranium ore concentrates' (note 16).

III. Secondary ion mass spectrometry

SIMS is a powerful analytical technique for investigating the surfaces of solid materials. The technique provides information on the spatial as well as depth-dependent concentration of constituents of interest. It has been of inestimable value to cosmochemistry (i.e. for particles of cosmic dust) and for characterization of interfaces and surface layers. SIMS offers the possibility to measure elemental composition and isotope ratios and it allows ion imaging (i.e. visualizing the spatial distribution of selected isotopes on the surface). For almost two decades, SIMS has also been used in nuclear safeguards for analysing uranium microparticles.

Principle and general description

In SIMS, the surface of the sample under investigation is bombarded with a focused beam of primary ions. High-energy ions ranging from 0.5 to 20 kiloelectronvolts (keV), such as O^-, O_2^- or Cs^+, may be used. Under this bombardment, atoms are sputtered off the surface. A fraction of the sputtered species is emitted as electrically charged particles (secondary ions). These secondary ions are passed to a mass analyser and then to a detections system. Magnetic sector field, TOF or quadrupole mass analysers may be used. Double-focusing instruments combining an electrostatic analyser with a magnetic sector field offer high resolution (300 up to 30 000) and allow high-precision analysis of thin films or particles.

SIMS has gained significant attention over the past 15 years for application in the area of particle analysis. Today, protocols for measuring individual particles in the micrometre range are established and tools for identifying particles of interest (e.g. uranium-containing particles) have been developed. State-of-the-art SIMS instruments allow the simultaneous detection of all uranium isotopes, while operating at high mass resolution.

Classical safeguards application of uranium particle analysis by SIMS

SIMS has been a mainstay technique for more than a decade in safeguards applications to detect undeclared nuclear activities.[29] The main application has been in the search for particles that were released through the handling of nuclear material and then collected by IAEA inspectors on swipes as part

[29] Ranebo, Y. et al., 'Improved isotopic SIMS measurements of uranium particles for nuclear safeguard purposes', *Journal of Analytical Atomic Spectrometry*, vol. 24, no. 3 (Mar. 2009).

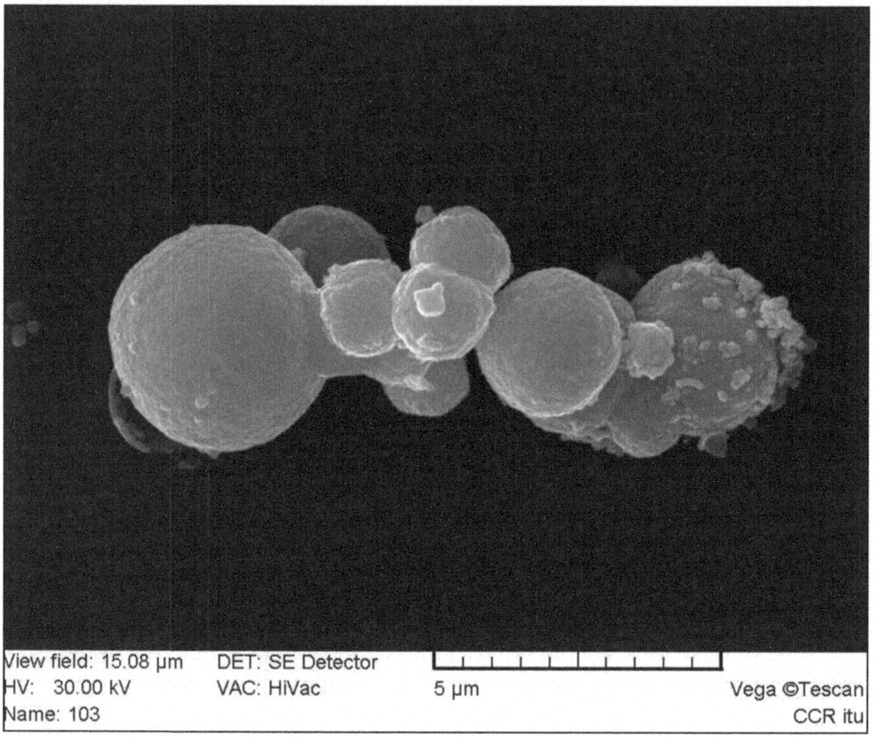

Figure 3.5. Scanning electron microscope picture of a conglomerate of uranium particles

Source: H. Thiele, European Commission, Joint Research Centre (JRC), Institute for Transuranium Elements (ITU).

of environmental sampling (see figure 3.5). The basics of these particle analyses for safeguards purposes are the following.

1. Fine particulate material or aerosols are often released in the handling of nuclear material. Normal nuclear safety monitoring (using radiometric techniques) cannot detect this material as the radioactivity is below background levels.
2. The particles are representative of the original material and their composition provides specific information about the source.
3. The released particles are highly mobile and can be found in many locations within a nuclear facility.
4. It is difficult to clean up and remove the released particles.
5. Samples taken (e.g. by a safeguards inspector) at a facility that has been operating over a long period can provide an insight into the entire history of the operation.

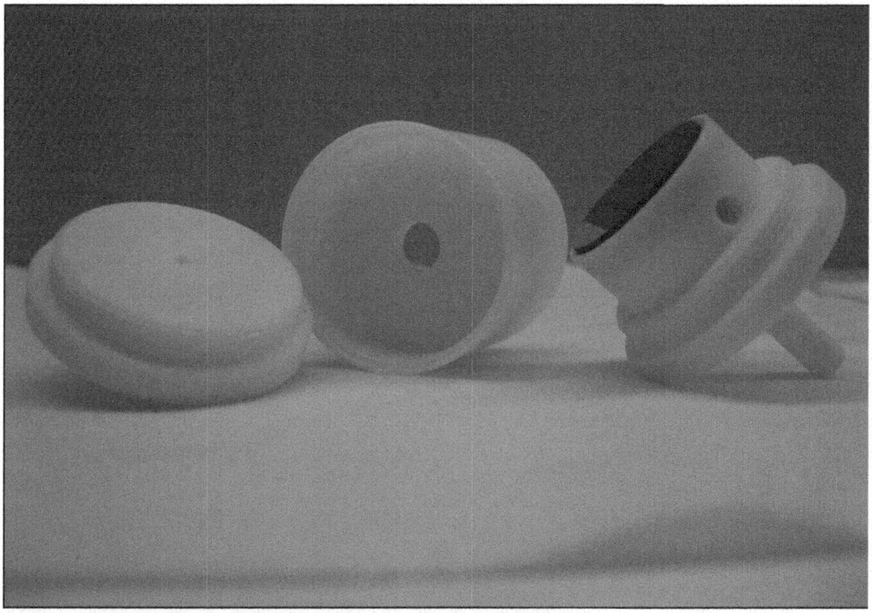

Figure 3.6. Vacuum impactor

Note: The particles are sucked through a small hole and sprayed onto a graphite planchet. The disassembled impactor has the graphite planchet in the holder to the right.

The analytical SIMS work can be separated into three main tasks that must be performed with the best balance of speed, sensitivity, and the highest levels of accuracy and quality control. The first task is to remove particles from a cotton swipe and to place them on a planchet with a homogeneous dispersion. The particles are removed with the use of a vacuum impactor technique, which pulls particles from the sample swipe with a vacuum suction and sprays them directly onto a sample planchet.[30] This has to be done in a clean laboratory following a strict procedure to avoid contamination of the traces of uranium in the sample (see figure 3.6).

The second task is to search through the particles deposited on the planchet to find the particles of interest. One of the fundamental strengths of SIMS is its ability to quickly perform particle searches of deposits containing millions of particles in order to find the ones of interest in a short time with a high detection capability. This is done by fast screening measurements using the unique ion-imaging capabilities of the instrument. An important improvement that has recently been made is the development of new automated particle measurement (APM) software that can make the screening measurements in a fully automated sequence. These measure-

[30] Esaka, F. et al., 'Particle isolation for analysis of uranium minor isotopes in individual particles by secondary ion mass spectrometry', *Talanta*, vol. 71, no. 3 (Feb. 2007).

Figure 3.7. A large geometry SIMS instrument

Note: This instrument, a Cameca IMS 1280 at the NORDSIM facility in Stockholm, Sweden, is physically five times larger than the conventional SIMS used today. This detector system can be seen in the front of the picture.

Source: M. Whitehouse, NORDSIM.

ments provide both the exact location of the uranium particles and a first estimate of their enrichment.[31]

The final task is to perform accurate and precise measurements to determine the isotopic composition of the particles selected. This requires the highest possible ion yield in the uranium measurements. SIMS has a higher ion yield for uranium compared to commercial TIMS instruments, and so it can perform analyses of smaller particles with a lower uncertainty. A new development is the use of large geometry SIMS (LG-SIMS) instruments for safeguards applications (see figure 3.7). The basic property of the LG-SIMS that provides the improved performance in uranium particle analysis is its high sensitivity at high mass resolution, as well as the multi-collector elec-

[31] Hedberg, P. M. L. et al., 'Improved particle location and isotopic screening measurements of sub-micron sized particles by secondary ion mass spectrometry', *Journal of Analytical Atomic Spectrometry*, vol. 26, no. 2 (Feb. 2011).

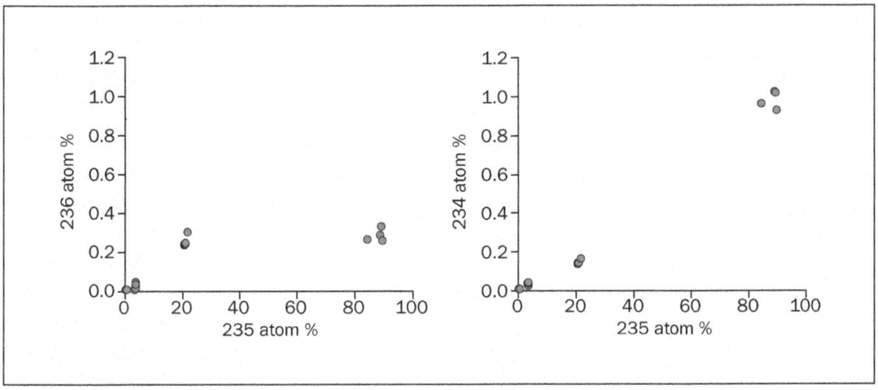

Figure 3.8. Plot of individual particle analysis showing uranium-236 versus uranium-235 and uranium-234 versus uranium-235

Notes: Four groups of enrichment can be seen in the plot of ^{234}U versus ^{235}U. In the plot of ^{236}U versus ^{235}U, the 90 per cent enriched material has a distinct $^{236}U:^{235}U$ isotope-abundance ratio compared to the other materials.

Source: European Commission, Joint Research Centre (JRC), Institute for Transuranium Elements (ITU).

tron multiplier detector system. Common molecular interferences that can hamper the measurement in normal SIMS analysis are removed effortlessly, thus improving measurements of minor isotopes.

Examples of forensic application of SIMS analysis

There have been a number of forensic cases were there is a need for uranium particle analysis to complement the more traditional bulk measurements. In one recent case, in 2012 a European steel-producing company found that a number of pieces of imported scrap metal showed an enhanced level of radiation. The material turned out to be contaminated by uranium. Measurements from one piece of the contaminant material were made by TIMS to establish the isotopic composition of the material. TIMS measurement suggested a ^{235}U enrichment of 9.0333 ± 0.0052 per cent (by weight) for this material. Other analyses were made including impurity measurements, morphology studies by scanning electron microscopy (SEM) and age determination. As such an enrichment is not very common, the sample was ground into a powder and SIMS analysis was applied to individual particles of the material to check the isotopic homogeneity (see figure 3.8). The SIMS analysis revealed that the material was in fact a mixture of different materials with different history and possibly different origin. Several groups of material were identified with different enrichments (0.5 per cent, 3.6 per cent, 21 per cent and 90 per cent ^{235}U enrich-

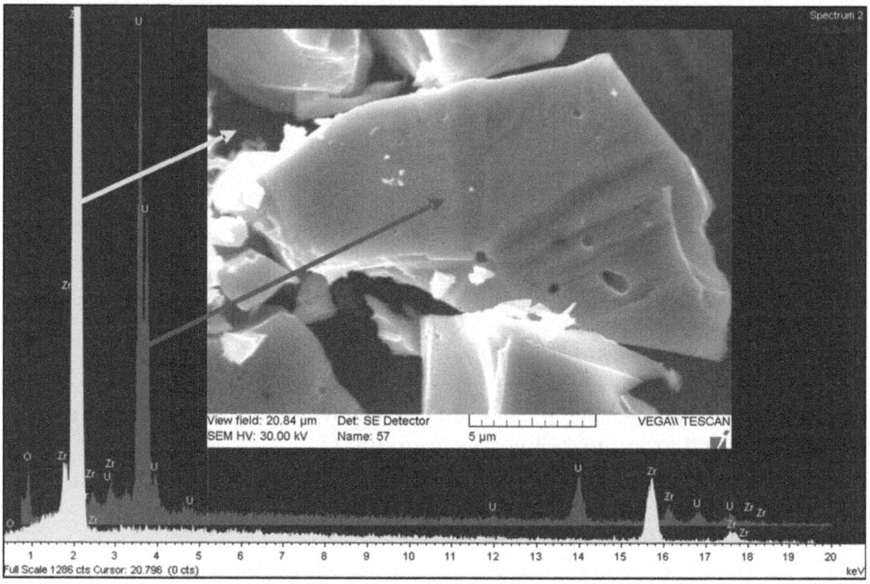

Figure 3.9. The typical structure of a particle from a nuclear fuel

Note: Two EDX spectra show the uranium in the fuel and the zirconium from a cladding material.

Source: H. Thiele, European Commission, Joint Research Centre (JRC), Institute for Transuranium Elements (ITU).

ment). Without the SIMS analysis it would not have been possible to detect that the material was in fact a mixture of several materials.

Another example is when SEM analysis finds a specific morphology on a particle such as a fuel material (see figure 3.9). SIMS can be used to relocate the same particle after the SEM analysis and determine the isotopic composition of a specific material, thus morphology, elemental composition and isotopic composition can be derived from an individual particle.

IV. Other mass spectrometric techniques

Specific problems may require more sophisticated analytical techniques. Extremely low isotope abundances or ultratrace quantities of a chemical element lead to measurement challenges due to the huge excess of matrix material. Two techniques that are suitable for addressing these challenges are resonance ionization mass spectrometry and accelerator mass spectrometry.

The first category of challenge—an excess of other isotopes of the element of interest—is particularly relevant to uranium isotopes, especially ^{236}U

and ^{233}U, where a measurement technique is needed that can discriminate between a variety of molecular and atomic species as well as tails of neighbouring masses that interfere at nearly the same mass position as the isotopes of interest. AMS offers sufficiently high abundance sensitivity for measuring extremely small isotope ratios. The second category—an excess of another chemical element—is relevant for minute traces of plutonium present in a huge excess of uranium. Here, RIMS has proved to be highly selective and to offer sufficient sensitivity.

Resonance ionization mass spectrometry

The development of lasers as strong and quasi-monochromatic light sources with good tunability has brought significant progress in ultratrace analysis. The optical excitation of atoms by resonant absorption of laser light in a step-wise process followed by the photo-ionization of the excited atom can nowadays be performed in a very efficient way. Combined with selective and nearly background-free detection of the resulting photo-ions in mass spectrometers, this technique is the basis of RIMS.[32] The RIMS method has been applied in recent years for the isotopically selective ultra-trace analysis of long-lived radionuclides with detection limits down to 10^6 atoms (equivalent to approximately 0.4 fg) and isotopic selectivities of up to 10^{13}.[33]

RIMS offers several outstanding properties. First, it has almost complete suppression of atomic or molecular isobaric interferences due to the element-selective optical excitation and ionization processes. Second, it has high overall efficiency resulting in low detection limits in the femtogram range (or *c.* 10^6 atoms, 0.4 fg) due to the high cross section of optical excitations. Third, it has good isotopic resolution, based on the selective ion detection in a mass spectrometer. Fourth, it has the feasibility of ultra-high isotopic selectivity by taking advantage of the isotope shift in the atomic transitions using narrow-band lasers, in addition to the abundance sensitivity of the mass spectrometer.

[32] Hurst, G. S. and Payne, M. G., *Principles and Applications of Resonance Ionization Spectroscopy*, (Hilger Publications: Bristol, PA, 1988); and Letokhov, V. S., *Laser Photoionization Spectroscopy* (Academic Press: Orlando, FL, 1987).

[33] Erdmann, N. et al. 'Resonance ionisation mass spectrometry for trace analysis of long-lived radionuclides', ed. P. P. Povinec, *Analysis of Environmental Radionuclides*, Radioactivity in the Environment no. 11 (Elsevier: Amsterdam, 2008); Trautmann, N., Passler, G. and Wendt, K. D. A., 'Ultratrace analysis and isotope ratio measurements of long-lived radioisotopes by resonance ionization mass spectrometry (RIMS)', *Analytical and Bioanalytical Chemistry*, vol. 378, no. 2 (Jan. 2004); and Wendt, K. and Trautmann, N., 'Recent developments in isotope ratio measurements by resonance ionization mas spectrometry', *International Journal of Mass Spectrometry*, vol. 242, nos 2–3 (Apr. 2005). Isotopic selectivity is the capability to selectively ionize a given isotope without ionizing another isotope of the same element.

Principle and general description

RIMS generally requires neutral atoms in the gas phase in order to realise the selective optical excitation and ionization steps of the element under investigation. Thus, after a chemical separation procedure, the resulting analyte must be efficiently vaporized and atomized within the RIMS apparatus. A number of different atomization techniques can be used, depending basically on the particular task to be accomplished and the laser system in use.[34]

The key feature of RIMS is the resonant excitation of atoms in several steps by properly tuned laser light followed by the ionization step. Two- to three-step excitation or ionization processes with ultraviolet, visible or infrared light are used. Typically starting from the ground state or a low-lying state, the atoms are step-wise excited to a high-lying state by absorption of one or two resonant photons. The highly excited atom is finally ionized by absorption of an additional photon, which either non-resonantly raises the electron energy beyond the ionization limit to the continuum or, alternatively, resonantly populates an auto-ionizing state, that is, a bound state lying energetically above the first-ionization potential. This auto-ionizing state immediately decays through emission of an electron and formation of a residual positive ion. As a third alternative, high-lying Rydberg states can be populated resonantly and subsequently ionized, for example, by application of an electric field, by far infrared photons or by collisions with buffer gas atoms. Due to rather high cross sections for the photon–atom interactions, near 100 per cent efficiency of the optical excitation and ionization can be reached with state-of-the-art laser systems.

Numerous types of tunable laser have been used for resonant excitation in RIMS applications. Typical RIMS applications, primarily addressing isobaric suppression with only moderate demands on isotope selectivity, involve pulsed laser systems, like high-repetition-rate copper-vapour-pumped dye lasers or Nd:YAG-pumped titanium:sapphire lasers (combined with frequency-conversion processes in non-linear optical media), which cover a wide spectral range. The pulsed laser systems can easily be combined with a TOF mass spectrometer. For a good overall efficiency, the temporal and spatial overlap of the continuously evaporated atoms from the sample with the laser pulse structure must be considered; that is, high-repetition-rate pulsed lasers, which deliver about 5000–25 000 pulses per second, are required.[35]

[34] Erdmann et al. (note 33).

[35] Grüning, C. et al., 'Resonance ionization mass spectrometry for ultratrace analysis of plutonium with a new solid state laser system', *International Journal of Mass Spectrometry*, vol. 235, no. 2 (July 2004).

High optical selectivity requires narrow transitions and a small laser bandwidth. For such applications, CW-lasers stabilized to a bandwidth of less than 1 megahertz replace the powerful pulsed lasers. Optical isotopic selectivities as high as 10^8 have been realized, in combination with a detection limit of less than 10^6 atoms (0.4 fg).[36]

The use of RIMS for nuclear forensic analysis

The analysis of the smallest traces of plutonium in uranium samples illustrates the outstanding property of RIMS: its extremely high elemental selectivity with almost complete suppression of isobars from other elements. The isotopic fingerprint for plutonium of different origins (e.g. weapon-grade, reactor-grade, fallout) is quite distinct, so a medium resolution of the mass spectrometer is normally sufficient for source attribution. However, in samples with a high excess of uranium, the standard mass spectrometric techniques suffer from isobaric interferences, especially for atom ratios $^{238}Pu : ^{238}U$ (and $^{241}Pu : ^{241}Am$), even after a chemical separation of uranium and plutonium (and americium). With RIMS, these isobaric interferences can be suppressed by many orders of magnitude, enabling the isotopic analysis of ultratraces of plutonium in uranium samples.

This advantage of RIMS is illustrated by the example of a metal plate of natural uranium from the German nuclear programme of the early 1940s. Nuclear forensic investigation of the plate was carried out in order to prove its authenticity and to provide further information on the history of the material (i.e. isotope ratio measurements of uranium and strontium, impurity measurements, and determination of the last chemical separation of the uranium).[37] Comparison with ore samples taken from Jáchymov in the Czech Republic (also known as Joachimsthal, the presumed source of the uranium) and with literature values showed that the sample under investigation was manufactured using such ore as starting material. To check for any evidence of an exposure of the uranium to a significant neutron dose, the sample was analysed by RIMS for its ^{239}Pu content, which is produced by neutron capture of ^{238}U. Several small (gram-size) subsamples were cut off and sent to the Institute for Nuclear Chemistry of Johannes Gutenberg University Mainz for analysis by RIMS.[38]

As a first step, the sample was dissolved and known amounts of ^{244}Pu were added as a tracer to determine the overall efficiency. Plutonium was

[36] Müller, P. et al., '^{41}Ca ultratrace determination with isotopic selectivity >1012 by diode-laser-based RIMS', *Fresenius' Journal of Analytical Chemistry*, vol. 370, no. 5 (July 2001).

[37] Mayer, K. et al., 'Analysis of uranium metal samples from Germany's World War II nuclear program: a case between science history and nuclear forensics', Institute of Nuclear Materials Management (INMM), *50th Annual Meeting of the Institute of Nuclear Materials Management 2009 (INMM 50)*, vol. 1 (INMM: Deerfield, IL, 2009).

[38] Grüning et al. (note 35).

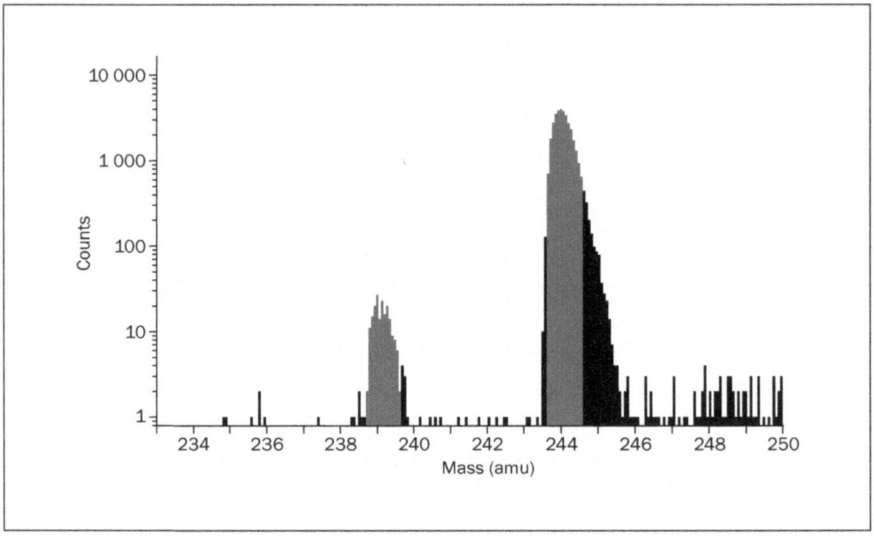

Figure 3.10. Resonance ionization mass spectrometry spectrum of plutonium from a metal plate of natural uranium from the German nuclear programme of the 1940s

amu = atomic mass unit.

Counts in the channels marked in grey are used for data evaluation.

Source: Institute for Nuclear Chemistry, Johannes Gutenberg University Mainz.

then chemically separated from uranium using extraction chromatographic columns. The remaining plutonium fraction still contained a huge excess of uranium. For the RIMS analysis, a filament was prepared by electrodeposition of plutonium as hydroxide on a tantalum foil followed by sputtering the deposit with a thin titanium layer. With tantalum as backing material and titanium as reducing agent it is possible to produce an atomic beam of plutonium with high efficiency by resistive heating.[39] The atoms were ionized by a three-step resonant excitation using tunable titanium sapphire lasers (the photons for the first excitation step were obtained by frequency doubling), pumped by a pulsed, high-repetition-rate (7 kilohertz) frequency-doubled Nd:YAG laser. The ions were mass-selectively detected with a TOF spectrometer and a multi-channel plate detector.

The two peaks of ^{239}Pu and ^{244}Pu (added as tracer to determine the overall yield) can clearly be seen in the resulting plutonium mass spectrum (see figure 3.10), but no background due to ^{238}U appears at mass 238, demonstrating the superior elemental selectivity of RIMS. The counts at mass 239

[39] Eichler, B. et al., 'An atomic beam source for actinide elements: concept and realization', *Radiochimica Acta*, vol. 79, no. 4 (Nov. 1997).

correspond to 3×10^7 atoms of ^{239}Pu (15 fg), leading to an average plutonium content in the samples of 1.4×10^{-14} grams of plutonium per gram of uranium. For comparison, the concentration of plutonium in a sample of Joachimsthal ore is approximately 5 times higher. This can be explained by the fact that the uranium material used for production of the metal plate was purified from decay products of uranium, including thorium. At that time, no information about plutonium and its chemical behaviour was available in Germany and it can be assumed that during the purification process part of the plutonium was removed together with thorium, provided it was in the tetravalent form. As a conclusion, it can be said that the samples analysed with RIMS show no evidence that plutonium was formed by neutron capture to a larger extent from the uranium in the sample.

Accelerator mass spectrometry

The fundamental difference between AMS and conventional mass spectrometric techniques is one of energy. Ions are accelerated to attain energies in the megaelectronvolt (MeV) range. AMS comprises an ion source, an ion accelerator, multiple selection stages for energy, momentum, velocity and atomic charge, and finally identification of nuclear mass and charge with an ion detector. The use of charge, energy and mass signatures are the basis for this type of ultrasensitive mass spectrometry.

The principle of AMS can be illustrated using the example of measuring the abundance of the isotope ^{236}U in a uranium sample. The uranium is first sputtered with a Cs^+ beam, and in a first analysis $^{236}U\,O^-$ (with a mass of *c.* 252 atomic mass units) is selected; this ion is accompanied by molecular isobars, the most abundant being $^{235}U\,^{16}O\,H^-$. While conventional mass spectrometry (i.e. ICP-MS and TIMS) cannot suppress this background sufficiently, AMS destroys the molecules by acceleration to high energies (up to several MeV) and stripping to a high positive charge state in a dilute gas cell. After another stage of acceleration, $^{236}U^{5+}$ is selected in a second mass analysis. The ions are detected in a TOF spectrometer and a final ionization chamber.

Other ions can reach the detectors only on irregular trajectories, for example due to charge exchange on residual gas or scattering on surface in the spectrometer. This allows detection limits for the $^{236}U : ^{238}U$ isotope abundance ratio of below 10^{-11} to be reached, which implies that for uranium samples the detection limit corresponds to approximately 0.1 pg. Recent investigations have shown that small differences in the $^{236}U : ^{238}U$ isotope ratio (typical values for natural uranium range from a few parts in 10^{-10} to a few parts in 10^{-11}) in natural uranium can be used as a nuclear forensic signature.[40]

[40] Srncik, M. et al., 'Investigation of the $^{236}U/^{238}U$ isotope abundance ratio in uranium ores and yellow cake samples', *Radiochimica Acta*, vol. 99, no. 6 (June 2011).

4. Gamma spectrometry as a tool of non-destructive nuclear forensic analysis

SOPHIE GRAPE

The use of gamma spectrometry in nuclear forensic applications is often motivated by the desire to either cheaply, quickly and non-destructively categorize radioactive materials in terms of their isotopic composition or to characterize them in terms of their elemental composition. It may also be possible to use gamma spectrometry for more sophisticated tasks, such as determining the age of samples and investigating whether or not the material has been produced from spent nuclear fuel.[1]

The physical principles that underlie gamma spectrometry are briefly explained in section I of this chapter. Section II describes the major components of gamma spectrometry equipment that are most relevant to nuclear forensics. The requirements and limitations that can be put on gamma spectrometers by nuclear forensic applications are discussed in section III, as well as equipment features that can be adjusted in order to meet these requirements. Finally, section IV lists some more exotic applications of gamma spectrometry to demonstrate the versatility of this technique.

I. The principle of gamma spectrometry

The physics behind gamma spectrometry

All nuclei are made of nucleons: protons and neutrons. The nucleons are kept together by means of the strong force, from which the binding energy can be derived. Light and stable nuclei have approximately as many protons as neutrons, while heavier nuclei tend to accumulate neutrons in order to compensate for the increasing Coulomb repulsion of the protons. This affects the short-range binding energy, which, when equally divided between all nucleons, decreases for nuclei heavier than iron.

Radioactive decays occur in unstable nuclei and as a result alpha, beta or gamma radiation is emitted. Heavy nuclei with a surplus of neutrons often decay by ejecting an ionized helium nucleus, that is, an alpha particle. Other radioactive nuclei decay by emitting beta particles, which are either

[1] Nguyen, C. T. and Zsigrai, J., 'Basic characterization of highly enriched uranium by gamma spectrometry', *Nuclear Instruments and Methods in Physics Research, Section B: Beam Interactions with Materials and Atoms*, vol. 246, no. 2 (May 2006).

electrons or positrons. These decays are called weak decays, because they are governed by the weak force in nature. In the case of a beta minus decay, a neutron decays to a proton, an electron and an antineutrino. Decays where gamma radiation is emitted often follow from an initial alpha or beta decay, where the daughter nucleus is left in an excited state. The emitted photons have energies that are characteristic for each nuclide.

Electromagnetic interactions with matter

Many different materials and detector types can be used for photon detection, but the interaction principle with matter is the same: electromagnetic radiation enters the detector, interacts with the atoms in the detector material and thereby transfers energy to loosely bound electrons which either directly or indirectly give rise to an electric current that can be measured.

Photons have no electric charge or mass and can travel far without losing energy. When they do interact with matter, it is via the photoelectric effect, Compton scattering or pair production. Low-energy photons interact primarily via the photoelectric effect, whereby the photon is completely absorbed by an atom and one of the atomic electrons is given the full photon energy minus the binding energy of the electron. For photons with slightly higher energies, Compton scattering is the dominant interaction process. Here, the photon scatters off an atomic electron, resulting in a photon with lower energy than the initial one as well as a moving electron, carrying the energy lost by the initial photon. When the energy of the initial photon is higher than two times the rest mass of an electron, pair production may take place. This is an interaction whereby the photon disappears and an electron–positron pair is created. For conservation of momentum, this may only happen in the presence of a nucleus that can absorb the recoil.

Measurement principles

A gamma spectrometer consists of a detector used for measuring photon energies, possibly a preamplifier, an amplifier, an analogue-to-digital converter (ADC) and a multichannel analyser (MCA). The preamplifier converts a charge pulse to a voltage pulse, while the amplifier provides a voltage gain. The ADC transforms the signal to a digital pulse that is used as input to the MCA. The MCA sorts the pulses according to their pulse height, which is proportional to the photon interaction energy. A resulting pulse-height spectrum, or energy spectrum, is obtained where the different gamma energies from the source are displayed as peaks of different intensities.

The gamma spectrum

All nuclei have specific energy levels and therefore also specific gamma energies. Assuming a large enough gamma spectrometer with sufficient ability to detect, collect and separate incoming gamma energies from each other, it is therefore possible to identify the elements in a radioactive sample by measuring its gamma spectrum and comparing the peaks with those from known samples. This technique is commonly used to determine the uranium enrichment and plutonium isotopic composition of nuclear material samples.

The interactions that take place in the detector determine what can be seen in the energy spectrum. If the full energy of the incoming photon is deposited in the detector via the photoelectric effect, the photo peak (or the full energy peak) is visible. If the photon Compton scatters on nuclei in the material and some of the energy escapes the detector, a smaller amount of energy is left in the detector and this will contribute to the Compton continuum part of the spectrum.

Energy regimes of interest

In general, the detection efficiency of the gamma spectrometer varies with incoming photon energy. It is therefore important to know what energy regime will be measured while the detector system is being mounted. A nuclear forensic analyst is often looking for uranium and plutonium isotopes. Sometimes these can be measured directly; in other cases they must be estimated by looking at the daughter particles. The amount of ^{235}U is directly proportional to the enrichment and is seen via the characteristic gamma energies of 143.8, 163.4, 185.7, 194.9 and 205.3 kiloelectronvolts (keV). The abundance of this isotope is used to categorize the samples as either low-enriched uranium (LEU), highly enriched uranium (HEU), natural uranium or depleted uranium. Although ^{238}U does not emit gamma rays, its metastable daughter ^{234m}Pa is an emitter of 250.3, 766.4 and 1001 keV photons.

Uranium isotope concentrations are commonly expressed in the sample as a ratio to ^{238}U. For example, this is the case when the age of samples needs to be determined using the ^{234}U gamma emission at 120.9 keV. The abundance of this isotope may also give information on where the ore was mined since abundances vary slightly depending on geographical origin. A ratio with ^{238}U is also calculated for the low-abundant isotope ^{236}U (49.37 keV), while certain isotopes such as ^{233}U are instead often related to ^{229}Th and ^{232}U for age determination and sample composition.[2]

[2] Ramebäck, H. et al., 'Basic characterization of ^{233}U: determination of age and ^{232}U content using sector field ICP-MS, gamma spectrometry and alpha spectrometry', *Nuclear Instruments and Methods in Physics, Research Section B: Beam Interactions with Materials and Atoms*, vol. 266, no. 5 (2008), pp. 807–12.

Most plutonium isotopes (from ^{238}Pu up to ^{242}Pu) also emit gamma rays. The isotope ^{239}Pu, which has non-peaceful applications, sends out gamma energies of 129.4 and 413.7 keV, while ^{240}Pu emits gamma at energies of 45.24 and 160.3 keV. The isotope ^{241}Pu is seen through the gamma line at 59.54 keV of its daughter nuclide ^{241}Am. Other isotopes such as ^{237}Np, ^{60}Co, ^{95}Zr, ^{106}Ru, ^{134}Cs and ^{137}Cs can also be seen in spectra of spent nuclear fuel or nuclear explosion debris. They have long half-lives and can represent an important radiological hazard.

Methods of measurement

The evaluation of a spectrum often either makes use of the infinite thickness method or the intrinsic calibration method.[3]

The infinite thickness method is applied for uranium enrichment determination and requires a reference sample. The count rate of the most prominent gamma transition of ^{235}U, at 185.7 keV, is measured for both samples and the count rates are compared.

For the intrinsic calibration method, no reference sample is needed. Isotopic ratios are instead determined directly from the measured spectrum using gamma and X-rays for many isotopes and by calculating the effect of the detector efficiency, self-absorption and attenuation in the equipment. If the enrichment of fuel rods is known, this method can be used to calculate ^{235}U depletion and ^{239}Pu build-up.[4]

II. Gamma spectrometry equipment

Sample analyses where immediate results are not required are often sent to laboratories where a sophisticated analysis can be performed. Results can in some cases be expected after 24 hours and in other cases after several months. The measurement time depends on, among other things, the nuclide concentration of the sample and can be chosen to ensure that the relative uncertainty is less than some predetermined level.[5] The International Atomic Energy Agency (IAEA) Safeguards Analytical Laboratory in Seibersdorf, Austria, performs gamma activity measurements. Many laboratories around the world (in e.g. Australia, France, Germany, Japan, Russia, the United Kingdom and the United States) are connected to the IAEA Network of Analytical Laboratories (NWAL) and provide options for

[3] Mayer, K. and Wallenius, M., 'Nuclear forensic methods in safeguards', *ESARDA Bulletin*, no. 38 (June 2008).

[4] Sáfar, J., Almási, I. and Lakosi, L., 'Estimating plutonium buildup from the ^{137}Cs/^{238}U ratio for slightly irradiated low enriched uranium', *Nuclear Instruments and Methods in Physics Research, Section B: Beam Interactions with Materials and Atoms*, vol. 94, no. 4 (1994), pp. 537–39.

[5] Jodłowski, P. and Kalita, S. J., 'Gamma-ray spectrometry laboratory for high-precision measurements of radionuclide concentrations in environmental samples', *Nukleonika*, vol. 55, no. 2 (2010), pp. 143–48.

sample analysis using, for example, gamma spectrometry or 'round robin' exercises where identical samples are sent to participating laboratories for comparison.[6]

In case a laboratory analysis is not convenient, fixed, portable or handheld instruments can be used to investigate samples at the scene. An example of a fixed solution is a radiation portal monitor (RPM), which is a gamma radiation detector that is typically placed at a border crossing point with the purpose of indicating the presence of nuclear or radioactive material. Radionuclides that are of interest at such a location include ^{233}U, ^{235}U and ^{238}U in different forms of uranium and in recycled material as well as ^{237}Np and ^{239}Pu in forms ranging from reactor-grade to weapon-grade.[7]

Portable or handheld devices must, in contrast to laboratory instruments, be easy to hold and move. To ensure this, they typically contain sodium iodide (NaI) or other room temperature alternatives, such as cadmium zinc telluride (CdZnTe or CZT). For scanning of equipment or people with these types of detector, short measurement times (of less than a second) are sometimes both needed and desired.[8]

Detector types

For nuclear forensic applications basically only two types of detector are used: scintillators and semi-conductors. For optimal results on the isotopic content of a sample using non-destructive gamma spectrometry, the semiconductor high-purity germanium (HPGe) is preferable because of the excellent energy resolution (i.e. the ability to detect and separate incoming energies from each other—see section III below). The disadvantage is the need for detector cooling, which makes the system somewhat less mobile. Scintillators do not require cooling and can easily be embedded in handheld devices used in the field.

Two large companies provide a large spectrum of commercially available radiation detectors for nuclear forensic purposes: Canberra and ORTEC. Both suppliers provide laboratory systems, cryogenically and electrically cooled germanium detector systems as well as handheld equipment. They also provide detector systems for many specialized areas or applications, such as environmental or air monitoring, homeland security and nuclear safeguards.

[6] Nguyen and Zsigrai (note 1); and IAEA, 'Tools for nuclear inspection', Fact sheet, 2011, <http://www.iaea.org/Publications/Factsheets/English/inspectors.pdf>.

[7] For an example of this in practice see chapter 9 in this volume.

[8] International Atomic Energy Agency (IAEA), *Combating Illicit Trafficking in Nuclear and Other Radioactive Material*, IAEA Nuclear Security Series no. 6, (IAEA: Vienna, 2007).

Scintillators

Scintillators are a commonly used type of photon detector. They have good detection efficiency due to the high-density materials used. In a scintillator, incoming radiation collides with and excites the atoms in the detector material. The excited atoms rapidly return to their ground state while emitting visible, or near-visible, light that is converted into an electronic pulse that is measured.

Among the many different types of scintillator material, the most common categories are organic and inorganic scintillators. Organic scintillators can be natural or manufactured.

With an organic scintillator, the incoming light either excites the electrons of the organic molecules or the molecules start vibrating.[9] As the excited molecules return to their ground states, they may emit light, although some transitions instead generate heat. Typical organic scintillators are anthracene and stilbene. Both materials are fragile and difficult to obtain in larger sizes, and in addition the scintillation efficiency is dependent on the orientation of the ionizing particle with respect to the crystal axis.[10] As an alternative, organic scintillators are also available in the form of solid plastic that can be formed to practically any shape. They are considerably less fragile than natural scintillators and rather inexpensive, however they may have considerable self-absorption and light attenuation.[11]

In inorganic scintillators, the scintillation mechanism and the available energy states are determined by the crystal lattice of the material. The electrons in the material are confined to discrete energy bands. Two such bands are the valence band and the conduction band. The valence band is in general the fullest band in an insulator such as NaI and the conduction band holds electrons with sufficient energy to move through the crystal. Incoming radiation may excite electrons from the lower lying valence band up to the conduction band. Immediately upon returning to the valence band, photons are emitted with wavelengths corresponding to the visible range.[12] This is called fluorescence. There are also other, competing, processes that can take place within the crystal. For example, thermal excitations can cause the slow light component, phosphorescence, to be emitted. This light is often a source of so-called afterglow in scintillators. Typical inorganic scintillators are sodium iodide, caesium iodide, caesium fluoride and barium fluoride. Sodium iodide is the most widely used and serves as a reference for many other detector materials. It has a high light

[9] Krane, K. S., *Introductory Nuclear Physics* (Wiley: Chichester, 1988).
[10] Knoll, G. F., *Radiation Detection and Measurement*, 3rd edn (Wiley: New York, 2000).
[11] Knoll (note 10).
[12] Knoll (note 10).

yield, is relatively cheap and easy to obtain in large sizes, and has been chosen for many of the handheld devices of the IAEA. However, NaI is hygroscopic and needs to be carefully encapsulated to avoid contact with air and humidity. Its energy resolution is not comparable to that of a semiconductor, although resolutions around 6.5 per cent at 662 keV are common.[13]

The choice of detector material is dependent on the demands of the application; the detection efficiency, time and energy resolution, decay time, light output, and so on differ greatly between different materials. The photon detection efficiency of a material increases with higher mass numbers and therefore inorganic scintillators tend to have higher detection efficiency than organic scintillators. This allows for shorter measurement times and a better temporal resolution. Inorganic scintillators are also relatively cheap and mobile. In cases where very large scintillators are required or where radiation damage in solid structures is undesired, liquid solutions may be an option. However, the energy resolution of scintillators is inferior to that of the semiconductor HPGe detectors described below. This depends mainly on transmission losses in the photo multiplier tube (PMT), which translates the light into an electric pulse, and the low quantum efficiency of the photocathode. These losses lead to a relatively small number of detected events per emitted photon.

Research on scintillators for nuclear forensic applications

Research on new scintillator materials for nuclear forensic application is ongoing and during recent years materials such as cerium-doped lanthanum chrolide ($LaCl_3$:Ce), cerium-doped lanthanum bromide ($LaBr_3$:Ce), cadmium tungstate ($CdWO_4$) and scheelite ($CaWO_4$) have become popular.

For border-monitoring equipment, both $LaCl_3$:Ce and $LaBr_3$:Ce have been tested and their energy resolution above 120 keV has been shown to be superior to NaI. An energy resolution of 2.9 per cent at 662 keV has been observed for both $LaCl_3$:Ce and $LaBr_3$:Ce.[14] Other studies have found resolutions of 3.5 per cent and 2.9 per cent for $LaCl_3$:Ce, and 2.6 per cent and 3.2 per cent for $LaBr_3$:Ce at 662 keV.[15] Studies by the US National Nuclear

[13] Kinloch, D. R. et al., 'New developments in cadmium tungstate', *IEEE Transactions in Nuclear Science*, vol. 41, no. 4 (1994), p. 752.

[14] International Atomic Energy Agency (IAEA), *Improvement of Technical Measures to Detect and Respond to Illicit Trafficking of Nuclear and Radioactive Materials*, IAEA-TECDOC-1596-CD (IAEA: Vienna, July 2008).

[15] Iltis, A. et al., 'Lanthanum halide scintillators: properties and applications', *Nuclear Instruments and Methods in Physics Research, Section A: Accelerators, Spectrometers, Detectors and Associated Equipment*, vol. 563, no. 2 (2006), pp. 359–63; Moses, W. W., 'Current trends in scintillator detectors and materials', *Nuclear Instruments and Methods in Physics Research, Section A: Accelerators, Spectrometers, Detectors and Associated Equipment*, vol. 487, nos 1–2 (2002), pp. 123–28; and Shah, K. S. et al., '$LaCl_3$:Ce scintillator for γ-ray detection', *Nuclear Instruments and Methods in Physics*

Security Administration (NNSA) have focused on these two materials together with cerium bromide ($CeBr_3$) and thallium-doped NaI (NaI:Tl) for detection of gamma rays and neutrons emitted by special fissionable materials and other radioactive materials that may be connected to nuclear fuel cycle processes.[16] Small samples of $CaWO_4$ have revealed a high light yield and the material seems promising for border-control purposes.[17]

Semiconductors

Semiconductors are important in gamma spectrometry as they offer higher energy resolution than many other gamma spectrometers. Semiconductors are materials with an electric conductivity between that of a conductor and an insulator. If the energy of the electrons in the material is increased by, for example, impinging gamma rays, electrons may start to move and the material becomes conductive. For some semiconductors, the thermal energy available at room temperature may even be sufficient for this to happen. Different semiconductor materials need different amounts of energy to become conductive; the exact amount can be affected by adding small traces of various elements to the material. This process is called doping.

Germanium detectors are perhaps the most widely known and used semiconductor for high-resolution gamma spectrometry. Many geometrical configurations exist such as well detectors, planar detectors, semi-planar detectors and coaxial detectors. Well detectors are suitable for small samples and they have the largest geometrical detection efficiency.[18] The coaxial type has the broadest energy range and detects gamma rays below 5 keV and up to several megaelectronvolts. These detectors can be made in many sizes with large surface areas that make the probability of interaction high. Semi-planar detectors have a better energy resolution than coaxial detectors, but their energy range (from below 5 keV up to a few hundred keV) is smaller.[19]

The good energy resolution of an HPGe detector depends mainly on two factors. The first is related to how radiation is absorbed in this type of

Research, Section A: Accelerators, Spectrometers, Detectors and Associated Equipment, vol. 505, nos 1–2 (2003), pp. 76–81.

16 Guss, P. et al., 'CeBr3 as a room-temperature, high-resolution gamma-ray detector', *Nuclear Instruments and Methods in Physics Research Section A: Accelerators, Spectrometers, Detectors and Associated Equipment*, vol. 608, no. 2 (2009), pp. 297–304.

17 International Atomic Energy Agency (note 14).

18 ORTEC, 'How to choose the right photon detector for your application', User Guide, June 2011, <http://www.ortec-online.com/Solutions/RadiationDetectors/Semiconductor-Photon-Detector-Categories.aspx>.

19 Princeton Gamma-Tech Instruments, June 2011, <http://www.pgt.com>; and Lin, G. and Zhouguo, H., 'Attribution of nuclear material by non-destructive radiation measurement methods', *Advances in Destructive and Non-Destructive Analysis for Environmental Monitoring and Nuclear Forensics* (IAEA: Vienna, 2003), pp. 129–32.

detector. Interacting photons directly give rise to electron–hole pairs, unlike the case with scintillators where interacting photons are converted to measurable electrons in several steps. In addition, the mean energy required to create an electron–hole pair is quite low in a semiconductor (hence the need for cooling) because the atoms are located closer together.[20] However, all types of germanium detector need cooling to lower the background from thermal excitations. Cooling is usually provided by liquid nitrogen, which makes the detector system less portable.

Research on semiconductor systems for nuclear forensic applications

Current high-resolution gamma spectrometry systems with energy resolutions around or below 1 per cent are based on low-temperature HPGe detectors, which sometimes limit their application in the field.[21] Recent development with cooling methods has yielded portable, mechanically cooled systems of varying weights and power consumptions. An example of a portable device is a 4.5 kilogram, 5×5 cm^2 high-purity coaxial germanium detector with an energy resolution of about 3.5 per cent at 662 keV.[22] However, there are other alternatives that provide energy resolutions even down to 0.2 per cent at 1332 keV or just below.[23]

Alternatives to the germanium detector systems are semiconductor materials such as CZT. This is a relatively new detector material in the nuclear forensic context, but it is increasing in popularity because it has higher detection efficiency than the scintillator NaI and yet it can be operated at room temperature. It can be conveniently installed in handheld devices and current progress in fabrication techniques improves both the detector resolution and the ability to use larger detector samples.[24] Its energy resolution at room temperature is approximately 2–3 per cent at 662 keV, which is even slightly better than $CeBr_3$, a very promising scintillator material.[25] An example of a device presently under development for nuclear forensic applications is a combined neutron (^3He tubes) and gamma detector (CZT) package, mounted in a lightweight body pack with a

20 Parker, R. P., 'Semiconductor nuclear radiation detectors', *Physics in Medicine and Biology*, vol. 15, no. 4 (1970).

21 Gilmore, G., *Practical Gamma-ray Spectrometry*, 2nd edn (Wiley: Chichester, 2008).

22 Becker, J. A. et al., 'Portable, low-power, mechanically cooled Ge spectrometer', *Nuclear Instruments and Methods in Physics Research, Section A: Accelerators, Spectrometers, Detectors and Associated Equipment*, vol. 505, nos 1–2 (2003), pp. 167–69.

23 Upp, D. L., Keyser, R. M. and Twomey, T. R., 'New cooling methods for HPGe detectors and associated electronics', *Journal of Radioanalytical and Nuclear Chemistry*, vol. 264, no. 1 (2005), pp. 121–26; and Rosenstock, W. et al., 'Recent improvements in on-site detection and identification of radioactive and nuclear material', International Atomic Energy Agency (IAEA), *International Symposium on Nuclear Security, 30 March–3 April 2009* (IAEA: Vienna, 2009).

24 International Atomic Energy Agency (IAEA), *Safeguards Techniques and Equipment: 2003 edition*, International Nuclear Verification Series no. 1 (IAEA: Vienna, 2003).

25 Shah K. S. et al., '$CeBr_3$ scintillators for gamma-ray spectroscopy', *IEEE Transactions on Nuclear Science*, vol. 52, no. 6 (2005).

wireless control unit.[26] The idea is to use the spectral capabilities of the gamma detector to identify the presence of radioisotopes and the neutron response as an indicator of nuclear materials. The design includes wireless transmission of sensor data that allows the inspector to perform other activities while the detector continuously screens for radiation.

III. Ability, demands and limitations of gamma detectors

It is difficult to set limits for the detection of gamma radiation with a gamma spectrometer in terms of minimum sample activity, weight, size or distance between source and detector because these issues tend to be related to each specific detector and so must be considered separately for each case. The minimum detectable activity is related to, for example, the chosen level of confidence for the signal detection of the application, the critical level and uncertainties in the measurement technique (as discussed in appendix 2A). In practice, the detection limit is different for every detector set-up and variations between equipment intended for laboratory analysis and handheld detectors for in-field measurements may be large. Absence of a detector signal may arise, for example, either as a consequence of investigating a non-radioactive sample or as a consequence of the activity being masked by, among other things, self-absorption and attenuation in samples of certain size or shape. In other cases it may not be the sample activity that is a problem, but rather the background radiation concealing the actual signal. The detector efficiency and design are also crucial for the ability to detect a signal with gamma spectrometry.

Disregarding these difficulties, demands and limitations on a gamma spectrometer generally concern the energy resolution that can be or needs to be provided, and the time available for measurement. The energy resolution of a gamma spectrometer depends on the type of detector being used (i.e. a scintillator or a semiconductor). However, some impact factors are important for all detector types. These include the energy of the incoming radiation, the amount and type of background radiation, the detector temperature, the detector proportionality, doping of the detector material, and the light-collection process. For scintillator detectors, also of importance are the intrinsic crystal energy resolution governed by the linearity of the response, the crystal decay time, radiation damage, and connections with the photo multiplier tube or other read-out systems. For semiconductors, the leakage current, gain instabilities as well as recombination

[26] Zendel, M., 'IAEA safeguards: challenges in detecting and verifying nuclear materials and activities', Paper presented at the 6th International Conference Tunable Laser Diode Spectroscopy, Reimes, 2007, <http://tdls.conncoll.edu/2007/zendelpaperreimsconferencefinal.pdf>.

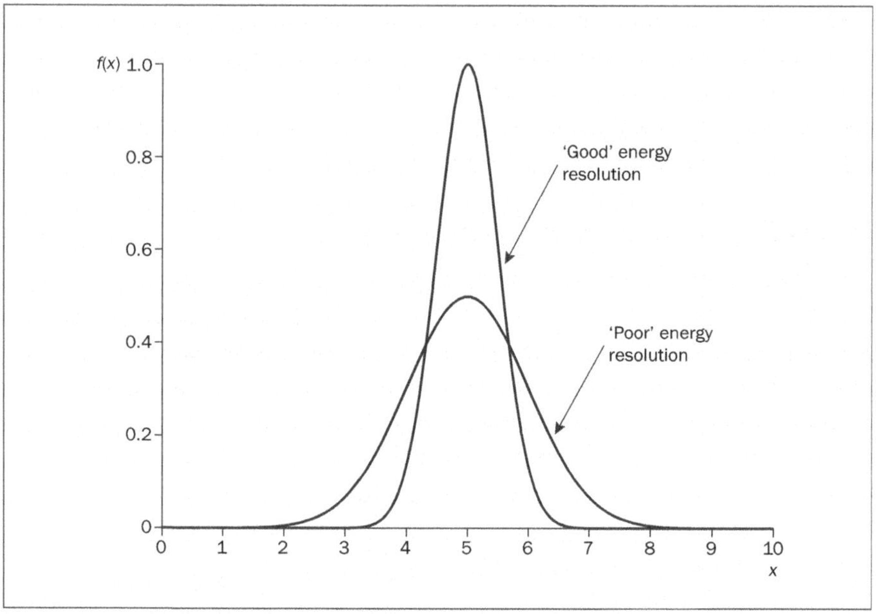

Figure 4.1. Example of two Gaussian distributions with the same centroid value but different resolutions

The 'good' resolution has a narrow distribution and the 'poor' resolution has a broad distribution.

and trapping of electron–hole pairs may become critical.[27] Some of these properties are determined by the detector material, while others may be improved by the user. The measurement time that is needed to obtain a result below a given uncertainty or that is required for a specific application is dependent on, for example, the detector efficiency, the count rate from the sample, the dead time in the system, the background radiation and so on. The following subsections discuss some of these impact factors further.

For gamma spectrometry within the field of safeguards or nuclear forensics, the purpose of the measurement is often to determine the isotopic content of a sample. This implies the need to distinguish photon peaks from one isotope from those of another in a spectrum.

The resolution of a detector is given by its response to an incoming energy. A material with a good (high) energy resolution is characterized by ability to measure both accurately and precisely (see appendix 2A). In everyday language this means that the detector should detect energies

[27] Mann, H. M., Bilger, H. R. and Sherman, I. S., 'Observations on the energy resolution of germanium detectors for 0.1-10 MeV gamma rays', *IEEE Transactions on Nuclear Science*, vol. 13, no. 3 (1966), pp. 252–64.

close to the 'true' value and that it should reproduce the output values for repeated measurements with the same input value. A good energy resolution is visualized in an energy spectrum by a narrow energy distribution centred on a pulse height value (the centroid), while a poor resolution results in a broader distribution centred on the same value (see figure 4.1). For poor energy resolutions, the uncertainty in determining the energy of the incoming radiation may be so large that the peak widths at different energies overlap to such a degree that it is not possible to distinguish the response to one energy from that of another. If this is the case, it is not possible to determine the location of the peak in the spectrum, nor its width: that is, there is no accuracy or precision in the measurement.

If two detectors have the same detection efficiency, the integrals below the curves are the same. The width of the detected peak reflects the fluctuations between deposited energy in the detector for the same incoming energy, on an event-by-event basis. For small fluctuations, the peak width is smaller and for the ideal case the peak will approach a sharp spike. In almost all measurements, the detected energy peaks are located on a background that also surrounds them. Before any analysis can take place, the background must be subtracted. A common measure of the peak width is the full width at half maximum (FWHM), which describes the peak width at a height that corresponds to half the maximum peak height. Because the widths of the peaks grow with increasing energy, it is common to give the relative energy resolution (i.e. energy resolution for a specific energy as a percentage of that energy) as a measure of the detector's performance. Semiconductors have, in general, the best energy resolution, in the order of less than 1 per cent. Many scintillators have resolutions in the order of 5–10 per cent.

Explanations for the peak width

The idealized thin spike in the detected energy spectrum becomes broadened in reality, due to several factors. The reasons for the fluctuations in deposited energy may include unstable operating conditions of the detector system (i.e. temperature, voltage etc.), sources of random noise, or in some cases varying levels of random noise and statistical noise from the measurement procedure.

For scintillators, three main factors add to the peak width: (*a*) contributions related to the statistics of the signal production in the detector, (*b*) properties related to the signal collection and linearity, and (*c*) electronic noise, drift or other detector system properties.[28] The first of these factors is generally responsible for the largest contribution because of the

[28] Gilmore (note 21).

limited and finite efficiency of the light-creation and -collection process, which often includes a PMT. The second factor is connected to the intrinsic energy resolution and it originates in the varying number of emitted photons as a response to incoming radiation. It is related to the internal structure of the scintillator such as the distribution of activator atoms, dopants and absorption and interaction of light in the detector material. This is also where the non-linearity of the response comes in.[29]

Scintillators are preferred in solutions where low cost, high detection efficiency, large detector geometries and room temperature operation are the dominating requirements. Because scintillators have worse energy resolution than germanium detectors, they tend to be better at verifying the presence of nuclear material rather than characterizing the isotopic composition of samples, especially with many radioactive nuclides present.

For a germanium detector, which is a semi-conductor detector, the resolution is in general governed by four types of uncertainty contribution. The first uncertainty is the uncertainty in the energy of the gamma ray leaving the nucleus. The second uncertainty comes from the charge production; that is, that incoming gamma rays create electron–hole pairs in the sensitive detector volume. The charge carriers must be collected in order to pass a pulse to the amplifier and variations in this number lead to a worse energy resolution. The size of this contribution may depend on the presence of trapping centres and radiation damage. The third uncertainty comes from the uncertainty in the collection of charge in the detector and the fourth comes from electronic noise.[30]

For all detector types, the number of charge carriers is subject to fluctuations even though the same energy is deposited in the detector. The generation of charge carriers is discrete and often described by the Poisson process, which means that for a total number of N charge carriers, the expected standard deviation describing statistical fluctuations is $\sqrt{N}$. If N is a large number, then the Poisson distribution can be approximated with the Gaussian distribution. For the latter distribution, the width parameter σ (see appendix 2A) is related to the FWHM via the relation $\mathrm{FWHM} = 2.35\sigma$. The relative energy resolution R, when considering only statistical fluctuations, is

$$R = \frac{\mathrm{FWHM}}{\mathrm{Centroid}} = \frac{2.35\sqrt{N}}{N} = \frac{2.35}{\sqrt{N}}.$$

[29] Gilmore (note 21); and Moszynski, M., 'Inorganic scintillation detectors in γ-ray spectrometry', *Nuclear Instruments and Methods in Physics Research, Section A: Accelerators, Spectrometers, Detectors and Associated Equipment*, vol. 505, nos 1–2, Proceedings of the 10th Symposium on Radiation Measurements and Applications (2003), pp. 101–10.

[30] Gilmore (note 21).

In fact, the charge generation is not fully a Poisson process, but that can be taken care of by the so-called Fano factor, F. This factor is less than 1 (*c.* 0.06 for germanium) and improves the relative energy resolution to $R = 2.35\sqrt{F}\sqrt{N}$. This means that a good energy resolution requires many charge carriers. This can be related to the impressive resolution of the semiconductor, where many charge carriers are produced per unit energy deposited in the detector, which contrasts with scintillator detectors, where losses of secondary electrons reduce the detection efficiency.

Measurement time

The choice of measurement time is often a result of a balancing act between keeping the time short to limit systematic errors and making it long enough that good statistics are obtained. Among the factors that affect the measurement times are detection efficiency, count rate and dead time.

Detection efficiency

Neutral particles such as photons or neutrons can travel long distances in materials before they are stopped or detected because they do not feel the Coulomb force of surrounding particles. This means that the detection efficiency for such particles tends to be lower than for charged particles. There are ways to improve the detection efficiency, but ultimately a low efficiency leads to long measurement or data collection times.

Absolute detection efficiency is commonly defined as the ratio of the number of recorded pulses divided by the number of emitted particles from the source. The value depends on the detector geometry, coverage and detector material characteristics. The intrinsic detector efficiency describes the ability of the detector to detect radiation, given that the radiation hits the front surface of the crystal as a parallel beam. The intrinsic efficiency depends on the detector material, the depth of the detector material in the direction of the incident radiation as well as transverse to it, and on the actual incident energy.

Total efficiency and peak efficiency can also be distinguished. The total efficiency is a measure of the probability that an incident photon creates a pulse in the detector. The amount of detected energy is irrelevant for this determination. In practice, this efficiency is difficult to obtain because a threshold must often be set to reduce the background, and hence the lowest energies rarely survive the recording. The peak efficiency is a measure of the percentage of events where all energy is deposited in the detector.

Count rate

One limitation for all detector types is that of the sample and background count rate. At count rates below a few kilohertz, effects on the results from count rate are in general low. At higher rates, problems such as degraded resolution, long counting times, erroneous peak to background ratios and inaccurate counting statistics may be experienced. The low resolution may occur because of pile-up, where the pulses occur so close in time that they climb on each other's tails.

The easiest solutions to these types of problem are increased distance between the sample and the detector as well as collimation and shielding or dilution of the sample. Limitations on the count rate are often given by the operating range of the preamplifier. Also the amplifier has a limit on the count rate to avoid a degraded resolution. The detector itself, the ADC and the MCA are in most cases not limiting factors.

Dead time

In gamma spectrometry, 'dead time' is the time after a detection of a gamma-ray during which the acquisition system is not able to record another event. It is set by a combination of the count rate and the minimum time needed by the electronics and the data-collection system to separate two pulses from each other. Because radioactive decays are random, there is a risk of missing new events if the system is busy processing old ones. This may become a severe problem for high count rates.

Two types of system response can be distinguished: paralysable and non-paralysable. A non-paralysable detector will not detect any new pulses while it is busy with a pulse, but when it is finished it will tend to the next incoming pulse. The paralysable detector will also not record new events while busy. However, it will register the pulses and extend the dead time by another period. Assuming a high count rate, this means that the detector will be constantly busy even though it is not recording any new events.

In practice, a mixture of the two types of system response is used. High dead times (in the order of 20 per cent or more) may cause the Poisson distribution-like behaviour of the true events in the detector to be distorted. This may influence the statistics of the recorded counts such that the expected variance from repeated measurements is reduced. The best solution to this problem is to try to reduce the count rate by, for example, increasing the shielding or the distance between the source and the detector or choosing a system with a shorter dead time.

Background radiation

In almost all types of gamma spectrometry measurement, background radiation constitutes a problem. It originates in cosmic radiation, radioactive impurities in the detector system, electronics or shielding, and external natural as well as anthropogenic radioactivity in the air or the surrounding environment.

Applications that require both well-separated peaks and well-determined gamma energies often make use of germanium detectors. This detector type is more efficient in the low-energy region, where it detects both the background and the signal well. For a cryogenically cooled germanium detector, several options for background reduction are available. Some options are to move components that contribute to the background outside the measuring system or to apply a filter for the background. The filtering is made by applying some kind of shielding to the experiment. Conventional shielding often uses lead because of its high density and high stopping power for gamma radiation, but for certain applications this may not be sufficient. Lead is almost always contaminated by ^{210}Pb, originating from either the lead ore or the refinement technology. This isotope has a half-life of 22.2 years and contributes to the background spectrum by emitting monoenergetic gamma rays of 46.5 keV and by adding a continuous bremsstrahlung contribution up to 1.16 MeV. The latter stems from the beta decay of ^{210}Bi, which is a daughter of ^{210}Pb. Fluorescent X-rays with energies around 74 keV and 85 keV may also appear as gamma rays strike the lead surface.[31] To minimize the background contribution of lead, old lead (e.g. ballast of sunken ships) can be used.[32]

In other cases, the shielding is done in different layers. The first layer can, with advantage, be made of high-purity electrolytic copper that may improve the spectrum especially in the low-energy region. Iron is also a good choice for the shielding. Pre-atomic era materials are preferable because more recently produced materials may be contaminated with cobalt.

Materials consisting of elements with high cross sections for (n,nγ) and (n,γ) reactions should be avoided as shielding material, as should hydrogen-rich materials that thermalize fast neutrons and absorb neutrons with sub-

[31] Verplancke, J., 'Low level gamma spectroscopy: low, lower, lowest', *Nuclear Instruments and Methods in Physics Research, Section A: Accelerators, Spectrometers, Detectors and Associated Equipment*, vol. 312, nos 1–2 (1992), pp. 174–82.

[32] Brodzinski, R. L. et al., 'Low-background germanium spectrometry: the bottom line three years later', *Journal of Radioanalytical and Nuclear Chemistry*, vol. 193, no. 1 (1995), pp. 61–70.

sequent emission of 2.22 MeV photons.[33] Shields thicker than necessary should also be avoided as they may add to the neutron multiplication.[34]

Other background contributions may come from gamma rays of ^{238}U, ^{235}U and ^{232}Th in standard aluminium that may be used in the equipment, or from ^{226}Ra, ^{228}Th and ^{40}K in the electrical components of preamplifiers. Additional contributions may come from ^{226}Ra, ^{137}Cs and ^{40}K in the molecular sieve or activated charcoal needed for the cryostat vacuum of a germanium detection system or even from the various (natural or synthesized) unstable isotopes of germanium created in the detector from cosmic radiation.[35] Radon gas may also interfere with the measurement as the isotopes ^{222}Rn (with half-life $T_{1/2}$ = 3.8 days), ^{219}Rn ($T_{1/2}$ = 4 seconds) and ^{220}Rn ($T_{1/2}$ = 56 seconds) diffuse into the measuring area and many of their daughter nuclides are quite gamma active. The influence of the gas is minimized by ventilating the measuring region near the detector.[36]

Background from cosmic radiation comes from either direct interactions or secondary induced reactions. The radiation consists mainly of protons, alpha particles and heavy nuclei. As they interact in the upper atmosphere, photons, electrons, muons and other particles are produced. Neutrons often arise in muon-induced reactions and are unwanted because they may in turn induce new reactions where gamma rays are produced. To reduce the contribution from cosmic radiation activation, the measuring equipment should not be transported by air as this increases the flux. Transportation as well as measurements should be conducted at ground level, or preferably below ground level. To further reduce the background, veto signals can be applied. These signals arise as cosmic radiation impinges on detectors placed around the detector system and lead to a stop in the data acquisition, at the price of an increase in system dead time.[37]

IV. Special applications of gamma spectrometry

Airborne and air-related measurements

Remote sensing of radioactive and nuclear materials is important both for health-related issues and for monitoring and surveillance of nuclear facilities.[38] Both passive and active systems are available for this purpose. Pas-

[33] For a definition of 'cross section' see the glossary in this volume.

[34] Verplancke (note 31).

[35] Verplancke (note 31).

[36] Núñez-Lagos, R. and Virto, A., 'Shielding and background reduction', *Applied Radiation and Isotopes*, Proceedings of the International Committee for Radionuclide Metrology Conference on Low-level Measurement Techniques, vol. 47, nos 9–10 (1996), pp. 1011–21.

[37] Núñez-Lagos and Virto (note 36).

[38] Mihalczo, J. T. et al., 'NMIS plus gamma spectroscopy for attributes of HEU, PU and HE detection', *Nuclear Instruments and Methods in Physics Research, Section B: Beam Interactions with*

sive systems are capable of determining the presence of plutonium, the fissile mass and the ^{220}Pu : ^{239}Pu ratio. Active systems with a deuterium–tritium source can also determine properties of HEU such as presence, fissile mass and enrichment.[39] Passive scanning can also be combined with imaging techniques for environmental monitoring.[40]

Another aspect of remote monitoring is surveillance via unmanned aerial vehicles for, for example, investigating radioactive plumes or material. NaI scintillators capable of detecting 10 Bq/m^3 have been mounted on small aeroplanes that are capable of downloading data to a ground station in real time. Studies have also been done with CZT detectors, registering iodine isotopes and low energy gamma emitters such as ruthenium and cerium.[41]

Gamma spectrometry systems such as SAUNA have also been developed with the verification of the 1996 Comprehensive Nuclear-Test-Ban Treaty (CTBT) in mind.[42] The design criterion of the instrument is that it should be able to detect 1 mBq/m^3 of ^{133}Xe or less during 24 hours of sampling.[43]

Underwater measurements

Gamma radiation can be shielded by many types of material but those with high density provide the most efficient shielding. Water is a cheap and relatively good shielding material, but despite this there are several underwater gamma spectrometers with applications ranging from mapping of sediments to tracing dumped nuclear waste. The IAEA Environment Laboratories in Monaco (IAEA-MEL) use, for example, both NaI:Tl scintillators and high-resolution HPGe detectors for investigating contaminations in marine environments close to nuclear facilities or at places where, for example, nuclear test have been performed.[44] The energy resolution of the NaI detector is below 7 per cent but, more importantly, it is operational down to depths of 2000 metres.[45]

Materials and Atoms, 5th Topical Meeting on Industrial Radiation and Radioisotope Measurement Applications, vol. 213 (2004), pp. 378–84.

[39] Caffrey, A. J. et al., 'Chemical warfare agent and high explosive identification by spectroscopy of neutron-induced gamma rays', *IEEE Transactions on Nuclear Science*, vol. 39, no. 5 (1992).

[40] Ziock, K. P. et al., 'Large area imaging detector for long-range, passive detection of fissile material', *IEEE Transactions on Nuclear Science*, vol. 51, no. 5 (2004).

[41] Kurvinen, K. et al., 'Design of a radiation surveillance unit for an unmanned aerial vehicle', *Journal of Environmental Radioactivity*, vol. 81, no. 1 (2005), pp. 1–10.

[42] Comprehensive Nuclear-Test-Ban Treaty (CTBT), opened for signature 24 Sep. 1996, not in force, <http://treaties.un.org/Pages/CTCTreaties.aspx?id=26>. See also chapters 8 and 9 in this volume.

[43] Schulze, J., Auer, M. and Werzi, R., 'Low level radioactivity measurement in support of the CTBTO', *Applied Radiation and Isotopes*, vol. 53, nos 1–2 (2000).

[44] Osvath, I. and Povinec, P. P., 'Seabed γ-ray spectrometry: applications at IAEA-MEL', *Journal of Environmental Radioactivity*, vol. 53, no. 3 (2001), pp. 335–49; and Krane (note 9).

[45] Osvath and Povinec (note 44).

Underground experiments

Probably the most spectacular limit of background reduction for gamma spectrometry is made by particle physics experiments looking for double-beta decay and proposed neutrinoless double-beta decays.[46] The first kind is a very rare process that has only been experimentally observed a few times, whereas the second is so far an experimentally unproven scenario. Background shields (as discussed above) have been applied with layered shielding of pure materials, anti-cosmic shielding, veto signals and underground laboratories. The background has been lowered to the limit where the germanium detector itself is the largest background contributor such that it needs to be enriched in the heavy isotopes, thereby depleting it of the isotopes ^{68}Ge and ^{76}Ge and leaving only ^{72}Ge, ^{73}Ge and ^{74}Ge.

The proposed Majorana and GERDA experiments are designed to determine whether the ultra-low background necessary to detect neutrinoless double-beta decays under certain conditions can be achieved. These and other experiments demonstrate that gamma spectrometry is capable of detecting rather rare events. They can, in principle, be applied to detection of radioactive isotopes in very low concentrations, perhaps even overlapping in some applications with mass spectrometric techniques in case if those are not available. For these experiments, the most serious background comes from cosmogenic production of ^{68}Ge and ^{60}Co in the germanium. The background has been estimated to be 0.01 counts/keV/kg/year for the Majorana experiment and 0.001 counts/keV/kg/year for the GERDA experiment.[47] For both experiments, an underground location, multilayer shielding of pure materials, and underground storage and manufacturing of material are foreseen.

[46] On double-beta decay see Brodzinski, R. L. et al., 'Ultra-low background germanium spectrometry: techniques and results', *Nuclear Physics B: Proceedings Supplements*, vol. 28, no. 1 (1992), pp. 415–19. On neutrinoless double beta decays see Aalseth, C. E. et al., 'The proposed Majorana ^{76}Ge double-beta decay experiment', *Nuclear Physics B: Proceedings Supplements*, Proceedings of the 8th International Workshop on Topics in Astroparticle and Underground Physics, vol. 138 (2005), pp. 217–20; and Schönert, S. et al., 'The GERmanium Detector Array (Gerda) for the search of neutrinoless ββ decays of 76Ge at LNGS', *Nuclear Physics B: Proceedings Supplements*, vol. 145 (2005), pp. 242–45.

[47] Aalseth et al. (note 46); and Maneschg, W. et al., 'Measurements of extremely low radioactivity levels in stainless steel for GERDA', *Nuclear Instruments and Methods in Physics Research, Section A: Accelerators, Spectrometers, Detectors and associated Equipment*, vol. 593, no. 3 (2008), pp. 448–53.

5. Sample characteristics and nuclear forensic signatures

KLAUS MAYER, MARIA WALLENIUS AND ZSOLT VARGA

The processes that lead to the generation, transformation or modification of nuclear material affect certain properties of the material. Consequently, nuclear material, like any other natural or industrial material, exhibits some characteristics that are related to the process that generated or modified it. If a relationship can be established between a measurable material parameter and the process that led to this parameter, it is termed a 'characteristic parameter'. Parameters include the chemical or isotopic composition of the nuclear material, chemical impurities, and the visual appearance and geometry of an item. Such parameters can provide valuable information for nuclear forensic analysis. Combining various characteristic parameters will increase confidence in the nuclear forensic conclusions drawn and enable the history of the material to be ascertained. Combinations of characteristic parameters are called 'nuclear forensic signatures'.

This chapter describes nuclear forensic signatures in the nuclear fuel cycle. (On signatures in post-explosion environments see chapters 6 and 8.) It follows a widely accepted classification, according to which a nuclear forensic signature can include four types of characteristic parameters: physical, chemical, elemental and isotopic.[1] Section I describes physical signatures in metallic uranium and plutonium, nuclear fuel pellets, and various powders and liquids common in the fuel cycle. Section II describes chemical signatures in the most important compounds of uranium and plutonium and discusses the most common non-nuclear chemicals pertinent to the fuel cycle. Section III reviews signatures obtainable from elements other than uranium, plutonium or thorium that are normally present in investigated material in widely varying concentrations. Section IV describes signatures that can be constructed by analysing the isotopic composition of both nuclear and non-radioactive materials.

I. Physical characteristics and signatures

The first step in nuclear forensic analysis usually involves visual inspection of the material, photographic documentation and, in the case of solid items,

[1] IAEA, *Nuclear Forensics Support: Reference Manual*, IAEA Nuclear Security Series no. 2, Technical Guidance (IAEA: Vienna, 2006), p. 29.

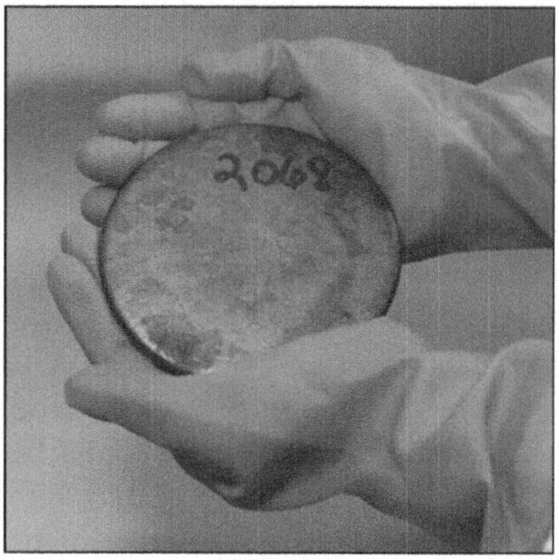

Uranium **Plutonium**

Figure 5.1. Uranium and plutonium metal in various shapes

Sources: United States Department of Energy; and Christensen, D., 'The future of plutonium technology', *Los Alamos Science*, no. 23 (1995), p. 170.

recording the dimensions and mass of the item. The observations made during this initial phase of examination allow some preliminary conclusions to be drawn and determine the further analytical steps. This section describes and illustrates the physical characteristics and signatures of nuclear material, looking in turn at metallic uranium and plutonium, fuel pellets, and powders and liquids related to the nuclear fuel cycle.

Metallic uranium and plutonium

Nuclear materials, such as uranium and plutonium, are used in metallic form only in specific applications—mainly related to military uses. The best-known application is in nuclear weapons. Another major military application of the metal form of uranium is in kinetic energy armour penetrators, which consists of depleted uranium alloyed with other elements, such as molybdenum or titanium. They take advantage of the high density of uranium metal: 19.1 grams/cm^3 versus 11.3 grams/cm^3 for lead. The high density and hardness of uranium metal also allows its use for tank armour.

In the civilian sector, depleted metallic uranium is used as a shielding material in containers for the storage and transport of highly radioactive materials. Although uranium metal itself is slightly radioactive, its high density makes it more effective than lead for shielding the radiation from strong radioactive sources. Similarly, it is used for shielding and collimation purposes in medical and industrial irradiators.[2] Depleted

[2] For the definition of 'collimation' see the glossary in this volume.

Germany **United States**

Figure 5.2. Metallic cubes made of natural uranium used in the first prototype reactors in Germany and the United States

Sources: Felix König; and US Department of Energy.

uranium metal has also been used for counterweights in aircraft. However, its use has been phased out in many newer aircraft.

The geometry of an item will provide the first clues about the intended use of the uranium or plutonium metal (see figure 5.1). In nuclear weapons, weapon-grade uranium or plutonium metal have well-defined shapes and sizes. In nuclear power reactors, the uranium fuel usually appears as uranium dioxide (UO_2) pellets (see below). However, the first prototype reactors in Germany and the United States used metallic uranium cubes or plates as fuel (see figure 5.2). The geometry of kinetic energy armour penetrators is generally similar to that of the of anti-tank ammunition, while uranium metal used in tank armour often takes the form of plates of depleted uranium. A Boeing 747-100, for example, may contain 400–1500 kg of uranium metal in the form of counterweights (see figure 5.3).

Fuel pellets

Macroscopic- and microscopic-scale physical parameters are useful nuclear forensic signatures that offer the possibility to distinguish between nuclear materials, such as nuclear fuel pellets, that have been produced by different processes.

In commercial nuclear power reactors, the fuel consists of uranium dioxide and in some cases of mixed uranium and plutonium dioxide (mixed oxide, MOX). Oxide fuels are preferred over metallic fuels for physical reasons (e.g. the higher melting point of the oxide compared to the metal, 2865 °C and 1132 °C, respectively), for chemical reasons (e.g. the tendency

Figure 5.3. Aircraft counterweight

Source: Theodore Gray, © 2014, <http://periodictable.com/>.

Figure 5.4. Examples of uranium oxide pellets used in nuclear power reactors

Sources: US Department of Energy (image *c.* 1958); and US Nuclear Regulatory Commission.

of uranium metal to oxidize), and for technological reasons (e.g. the swelling of uranium metal under neutron irradiation).

Uranium dioxide powder is compacted into cylindrical pellets and sintered at high temperatures (*c.* 1700 °C) to produce ceramic nuclear fuel with well-defined physical properties and chemical composition.[3] Subsequently, a grinding process is used to achieve the final geometry (in particular pellet diameter) within narrow tolerances (see figure 5.4). The fuel pellets are then stacked and filled into the cladding (metallic tubes that are typically made of a zirconium-based alloy), which are called fuel rods. The

[3] For the definition of 'sintering' see the glossary in this volume.

Table 5.1. The geometry of pellets used in different types of nuclear reactor

Reactor type	Diameter (mm)	Height (mm)	Central hole (mm)
VVER-440	7.6	9–12	1.6
VVER-1000	7.6	9–12	2.4
RBMK-1000	11.5	12–15	–
RBMK-1500	11.5	12–15	2.0
CANDU	12.1	13	–
PWR (western EU)	9.1	11	–
BWR (USA)	12	15	–

BWR = boiling water reactor; CANDU = Canadian deuterium–uranium reactor; EU = European Union; PWR= pressurized water reactor; RBMK = high-power channel reactor (*reaktor bolshoy moshchnosty kanalny*); VVER = water–water energetic reactors (*vodo-vodyanoi energetichesky reaktor*)

Sources: Pajo, L. et al., 'Identification of unknown nuclear fuel by impurities and physical parameters', *Journal of Radioanalytical Nuclear Chemistry*, vol. 250, no. 1 (2001); Information obtained from the seized material analysis at the Institute for Transuranium Elements (ITU), Karlsruhe; and Olander, D. et al., 'Uranium–zirconium hydride fuel properties', *Nuclear Engineering and Design*, vol. 239, no. 8 (2009).

finished fuel rods are grouped into fuel assemblies that are used to build up the core of a power reactor.

MOX fuel is a blend of plutonium and natural or depleted uranium, which behaves similarly to the enriched uranium feed and is therefore an alternative to low-enriched uranium (LEU) fuel used in light-water reactors (LWRs).

Different reactor types may use fuel of different ^{235}U enrichments, and pellet diameters may also vary with reactor type. Thus, the geometry of the pellet (e.g. diameter, height and possible central hole) is an important signature that often reveals sufficient information to enable identification of the reactor type for which a particular fuel pellet was intended (see table 5.1).

Since different manufacturers may produce pellets for a certain reactor type, pellet diameter may not serve as a useful discriminating factor. Additional physical parameters have to be examined in order to possibly distinguish the pellets produced by different manufacturers. Such distinguishing features can be found in the details of the top and bottom surface of the pellet. Often the parallel planes of the cylindrically shaped pellet are specifically tailored (for technological reasons). The sub-areas that are created are known as 'dishing', 'land' and 'chamfer', and the pellet diameter may be subdivided into the diameters of these sub-areas, giving three distinct and well-defined parameters. A difference of 0.2 millimetres in the chamfer width can be enough to differentiate the pellets of two fuel manufacturers.[4]

[4] Information obtained from seized material analysis at the Institute for Transuranium Elements (ITU), Karlsruhe.

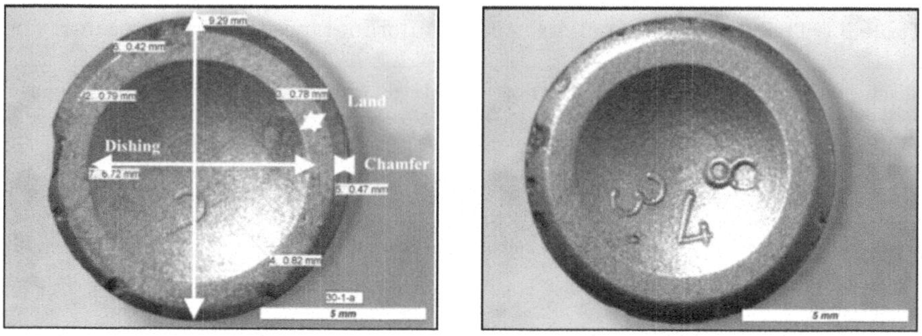

Measurements **Markings**

Figure 5.5. A uranium fuel pellet, showing dimensions and markings

Source: Information obtained from analysis of seized material at the European Commission, Joint Research Centre (JRC), Institute for Transuranium Elements (ITU), Karlsruhe.

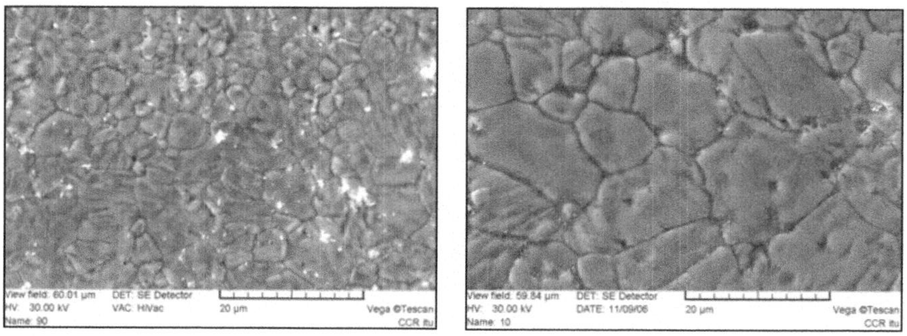

Figure 5.6. Scanning electron microscopy images of the grain morphology of two pellets from different manufacturers

Source: Information obtained from analysis of seized material at the European Commission, Joint Research Centre (JRC), Institute for Transuranium Elements (ITU), Karlsruhe.

The various grinding methods used to finish the pellets to their exact diameter cause differences in the pellets' surface in the micrometre range. Wet grinding is known to produce lower surface roughness (smoother surfaces) than the dry grinding method; therefore pellets originating from processes using different grinding methods can be distinguished from each other.[5]

In addition to geometric parameters, the markings found on nuclear pellets can be indicative of the intended use or the production process. In some cases, this can be quite clear, for example showing the nominal ^{235}U enrichment, forged on the top of a pellet and the press number used for pressing the uranium oxide powder into the 'green' (i.e. unsintered) pellets

[5] Pajo, L. et al., 'Identification of unknown nuclear fuel by impurities and physical parameters', *Journal of Radioanalytical and Nuclear Chemistry*, vol. 250, no. 1 (2001).

marked on the other side of the pellet (see figure 5.5). However, not all pellets bear such clear marking, although some batches of pellets carry other markings, such as geometric symbols, which may serve as distinguishing parameters.

Microscopic parameters also make it possible to distinguish between pellets produced by different manufacturers. The grain size distribution and the grain structure of uranium oxide fuel pellets indicate the different production processes used for manufacture of the material (see figure 5.6). Grain size can be affected by various factors, such as precipitation conditions, the starting material used prior to calcination (i.e. the type of uranium ore concentrate, UOC), the additives used and the calcination conditions.

Powders and liquids related to the nuclear fuel cycle

For nuclear forensic purposes, liquids are of marginal relevance, because only a few cases of illicit nuclear material have been reported involving liquid samples. However, knowledge of the liquids used for processing nuclear material (from mining to reprocessing) is extremely useful when analysing the residuals of solvents or extractants in nuclear material.

In contrast, powders have frequently been encountered in incidents involving illicit nuclear material. Certain physical characteristics of powders are typical of the nuclear fuel cycle, but it should be noted that powder morphology and its interpretation for nuclear forensic purposes is a complex issue. For comparative evaluation of powder samples that were produced using different processes, the microstructure of a material is an important characteristic. This can be illustrated by scanning electron microscopy (SEM) images of samples of UOC from two facilities—the Key Lake mill and the Blind River refinery, both in Canada. A lower magnification image of the Key Lake sample reveals particles of about 20 μm in diameter, while the Blind River sample has a much coarser, granular composition (see figure 5.7). At higher magnification the surface structure of the particles from Key Lake is visible (see figure 5.8). In contrast, the SEM image of the Blind River sample shows small crystals with a fairly smooth surface. The notable differences in the morphology of the two materials allows the use of this parameter as a distinguishing feature. Although both materials are UOC, they are chemically different.

The material from Key Lake was obtained by precipitation of uranium through the addition of ammonia, resulting in ammonia diuranate. The material was dried and then calcined at 750 °C to produce U_3O_8. In contrast, the UOC from Blind River was produced by precipitation of uranium through the addition of NaOH, resulting in $Na_2U_2O_7$, which was dried but not calcined, resulting in a granular product. The slow increase in pH

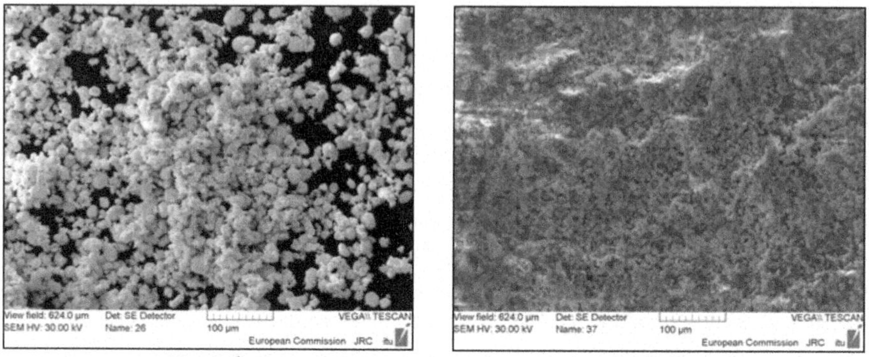

Key Lake **Blind River**

Figure 5.7. Lower magnification scanning electron microscopy images of two samples of uranium ore concentrate

Source: Unpublished data from the European Commission, Joint Research Centre (JRC), Institute for Transuranium Elements (ITU), Karlsruhe.

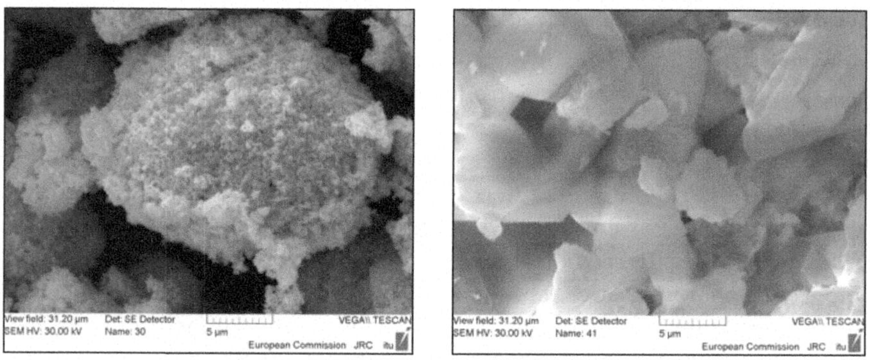

Key Lake **Blind River**

Figure 5.8. Higher magnification scanning electron microscopy images of two samples of uranium ore concentrate

Sources: Unpublished data from the European Commission, Joint Research Centre (JRC), Institute for Transuranium Elements (ITU), Karlsruhe.

during the steps preceding the actual precipitation allows the $Na_2U_2O_7$ crystals to grow, resulting in microcrystals (shown in figure 5.8).

Using the same methodology, powders such as UF_4, UO_2 and UO_3 can also be investigated and compared. As the examples illustrate, the evaluation of microstructures relies strongly on comparative evaluation based on phenomenological observations. As research progresses, the

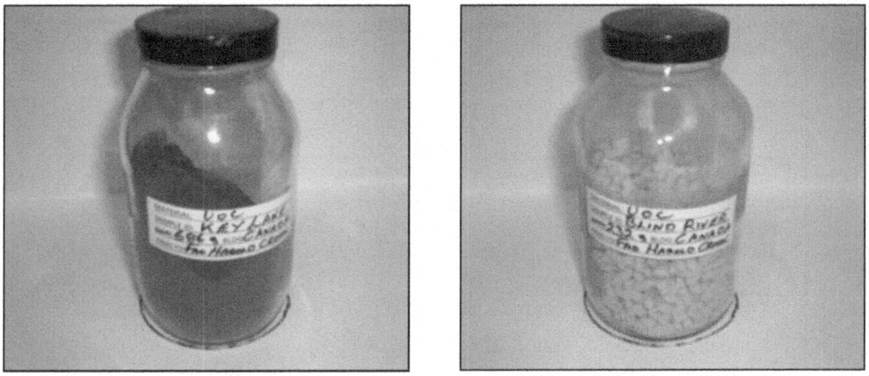

Key Lake **Blind River**

Figure 5.9. Photographs of two samples of uranium ore concentrate

Source: Unpublished data from the European Commission, Joint Research Centre (JRC), Institute for Transuranium Elements (ITU), Karlsruhe.

relationship between observable physical parameters—such as particle size distribution, surface structure, grain structure and inclusions—and the process that was used to produce these materials will become better understood.

Sometimes even the colour of the material can provide useful information and may serve as a distinguishing parameter. In the above example, the colour of the UOC from the two facilities differs significantly—the sample from Key Lake is dark grey, while that from Blind River is bright yellow (see figure 5.9). The dissimilarity in colour may not always be as pronounced as in this example, but it may permit initial conjecture about the chemical compound involved.

II. Chemical characteristics and signatures

Uranium and plutonium are encountered in different chemical forms at different stages of the nuclear fuel cycle. Most of the chemical compounds are fairly common and, at first glance, do not offer unique nuclear forensic signatures. The type of chemical compound encountered at a given stage of the nuclear fuel cycle, however, may provide useful hints on the processing technology used and thus permits excluding certain facilities as the origin of the material. The following three subsections present and examine different chemical compounds encountered in the nuclear fuel cycle, from uranium mining and milling to fuel fabrication and reprocessing of spent fuel.

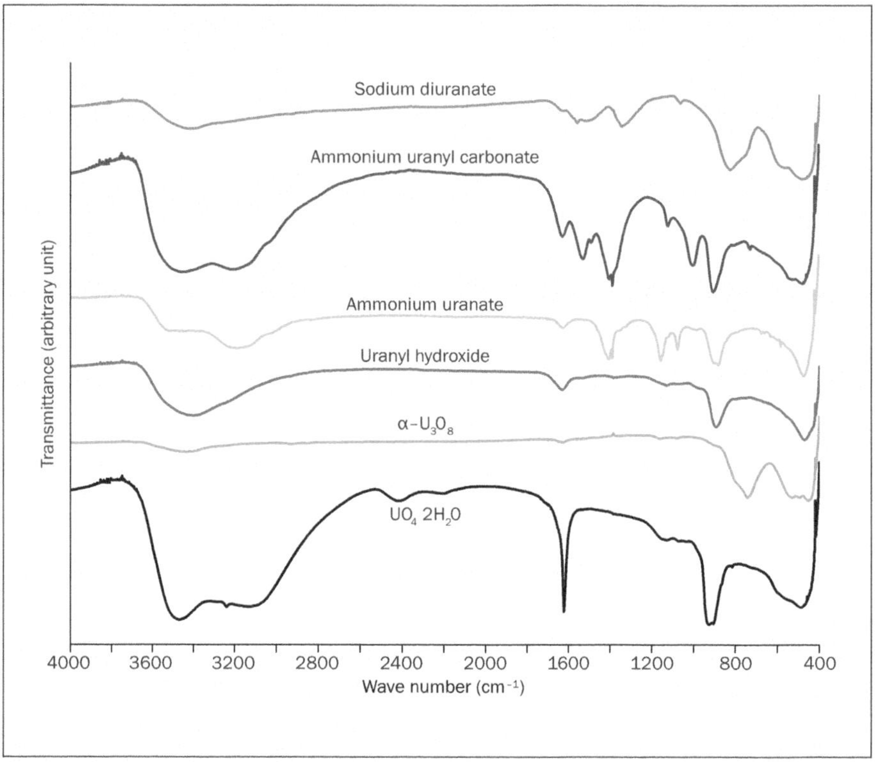

Figure 5.10. The Fourier-transform infrared spectroscopy spectra of chemical compounds found in uranium ore concentrate

Source: Varga, Z. et al., 'Characterization and classification of uranium ore concentrates (yellow cakes) using infrared spectrometry', *Radiochimica Acta*, vol. 99, no. 12 (2011).

Uranium compounds

The nuclear fuel cycle starts with the mining of uranium ore from which uranium is extracted and converted into UOC. As UOC is prepared by various extraction and refining methods, depending on the type of ore from which the uranium was mined, the respective chemical compositions may vary widely. Typically, UOC contains a high fraction of natural uranium (above 65 weight per cent uranium), which can be included in a wide range of chemical compounds, such as ammonium diuranate $(NH_4)_2U_2O_7$ (ADU), sodium diuranate $Na_2U_2O_7$, uranyl hydroxide $UO_2(OH)_2$, uranyl sulphate UO_2SO_4, uranyl peroxide UO_4 or ammonium uranyl carbonate $(UO_2CO_3{\cdot}2(NH_4)_2CO_3)$. The uranium can also be in oxide form, as UO_3 or U_3O_8, if the material has been calcined. Many of these compounds in their solid form are yellow to red in colour, and sometimes grey to black, depending on the degree of calcination. The colour alone provides a first indication of the type of compound present.

A simple yet appropriate experimental technique for characterizing these compounds is infrared (IR) spectroscopy. The first studies using it were carried out in the late 1950s and have been reviewed more recently.[6] The different UOC compounds can easily be distinguished using IR spectra (see figure 5.10).

Systematic studies of solid UOC material with a modern Fourier-transform infrared spectroscopy (FT-IR) instrument have revealed a feature of relevance to nuclear forensics: in addition to the absorption bands that arise from the bondings within the actual molecule, it is also possible to identify anionic impurities.[7] These detected impurities are indicative of the solution composition (so-called pregnant liquor), from which uranium is precipitated. These anionic impurities provide useful clues about the acids used to dissolve the uranium and for its back-extraction after pre-concentration. The most typical impurities identified are sulphate, nitrate and carbonate.

Sulphate, which is widely used for the leaching of the ore and in the subsequent purification steps (e.g. for uranium elution in ion exchange separation), often appears in the FT-IR spectrum. It has a well-detectable peak at 1120 cm^{-1}. Nitrate gives a sharp characteristic absorption band at 1384 cm^{-1}. Although nitrate is highly soluble and can easily be eliminated by subsequent washing steps, it remains measureable where nitrate solution is used for uranium stripping after the ion exchange purification. The FT-IR spectrum of a sample of UOC from the Denison mill at Elliot Lake, Ontario, Canada, illustrates this (see figure 5.11). In this case sulphuric acid is used for dissolution of the ore, followed by an ion exchange separation with nitrate elution. Ammonium diuranate is precipitated from the aqueous solution with ammonia, and both sulphate and nitrate can be detected in the sample.

Anionic impurities can also be determined by ion chromatography (IC). A comprehensive investigation of UOC samples originating from a variety of geographic origins has been conducted.[8] In the case of the Denison sample, the resulting information was similar to that offered by FT-IR (see figure 5.12). While IC provides better detection limits, it requires more effort in sample preparation.

[6] Jones, L. H., 'Determination of U-O bond distance in uranyl complexes from their infrared spectra', *Spectrochimica Acta*, vol. 15, no. 6 (1959); and Hausen, D. M., 'Characterizing and classifying uranium yellow cakes: a background', *JOM Journal of the Minerals, Metals and Materials Society*, vol. 50, no. 12 (1998).

[7] Varga, Z. et al., 'Characterization and classification of uranium ore concentrates (yellow cakes) using infrared spectrometry', *Radiochimica Acta*, vol. 99, no. 12 (2011).

[8] Badaut, V., Wallenius, M. and Mayer, K., 'Anion analysis in uranium ore concentrates by ion chromatography', *Journal of Radioanalytical and Nuclear Chemistry*, vol. 280, no. 1 (2009).

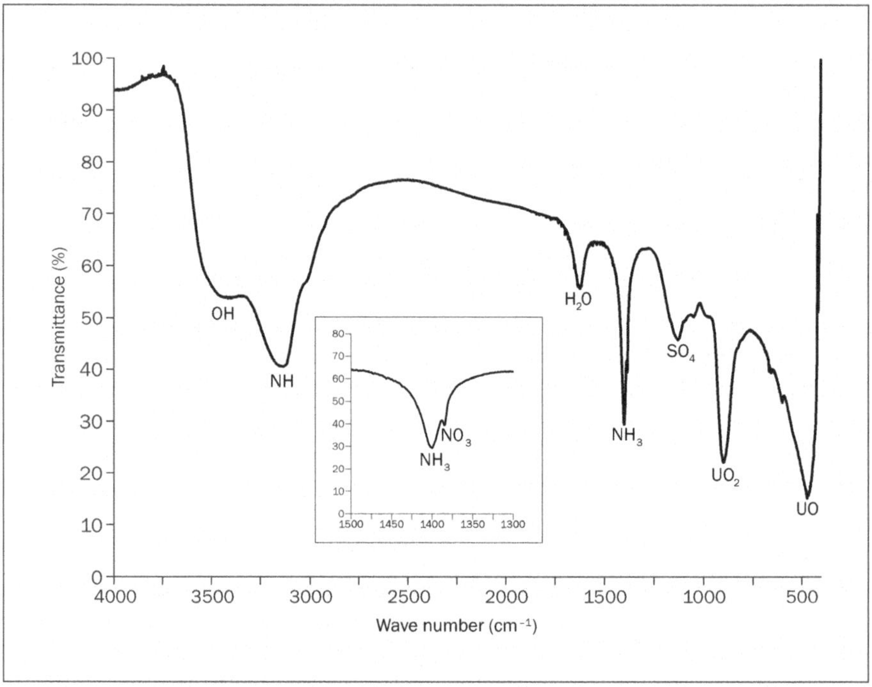

Figure 5.11. The infrared spectrum recorded for ammonium diuranate from a sample of uranium ore concentrate

Notes: The sample was obtained from the Denison mill at Elliot Lake, Ontario, Canada. As well as ammonium diuranate, the spectrum reveals absorption bands that can be attributed to sulphate and to nitrate.

Sources: Varga, Z. et al., 'Characterization and classification of uranium ore concentrates (yellow cakes) using infrared spectrometry', *Radiochimica Acta*, vol. 99, no. 12 (2011).

The microstructure of the material is another property that helps to differentiate UOC materials that were produced by different methods or under different conditions. The size and shape of the particles and their surface texture reflect the chemical and thermal conditions under which the solids were precipitated from the supernatant solution (as discussed in section I).

The production of uranium ore concentrates is followed by conversion of uranium oxide to uranium hexafluoride. Two basic processes, wet or dry, can be used for UF_6 production. In either case, UF_4 (a greenish powder) is obtained as an intermediate product, which is then oxidized by direct reaction with fluorine to UF_6. Depending on the production conditions, the particle size and microstructure of the UF_6 may vary, thus providing clues to the production process. UF_6 is a solid showing a high vapour pressure at room temperature. As it hydrolyses easily, it has to be kept dry, which is

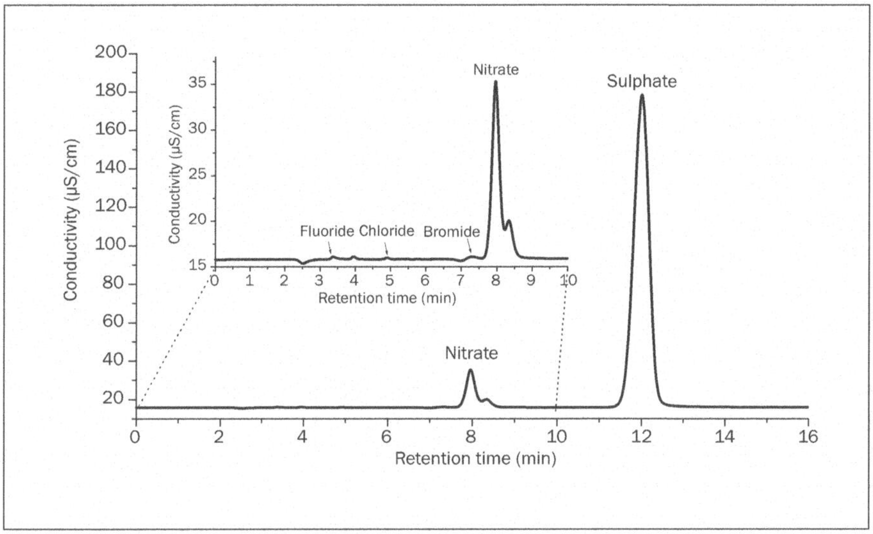

Figure 5.12. The measured ion chromatogram for ammonium diuranate from a sample of uranium ore concentrate

Notes: The sample was obtained from the Denison mill at Elliot Lake, Ontario, Canada. The presence of sulphate and nitrate used for material production can be detected. The inset figure shows the minor anionic components measured in a more concentrated sample.

Source: Unpublished result from the European Commission, Joint Research Centre (JRC), Institute for Transuranium Elements (ITU), Karlsruhe.

typically achieved by storing it in corrosion-resistant metal containers. At slightly elevated temperatures, UF_6 sublimes and can be chemically purified by (fractioned) distillation. The world's largest conversion facilities are located in Port Hope, Ontario, Canada; Springfields, Salwick, United Kingdom; Glazov (Udmurt Republic), Angarsk (Irkutsk Oblast) and Seversk (Tomsk Oblast) in Russia; Pierrelatte, France; Lanzhou, Gansu Province, China; and Metropolis, Illinois, USA.[9]

The elemental signatures (metallic impurities) present in natural uranium are gradually lost during the conversion process. When UF_6 is distilled, only those impurities that form volatile fluorides will follow the uranium. Hence, after distillation or after the isotopic enrichment process (which is based on gaseous UF_6) a completely new impurity pattern develops. The upper limits of the impurity content in UF_6 (to be used for nuclear fuels) are defined in appropriate standards, such as the American Society for Testing and Materials (ASTM) standard C996-10.

[9] World Nuclear Association, 'Uranium enrichment', Aug. 2014, <http://www.world-nuclear.org/info/inf28.html>.

Following isotopic enrichment of uranium (i.e. increasing the fraction of the fissile isotope ^{235}U in the uranium), the UF_6 is hydrolysed to UO_2F_2 and then precipitated and calcined to uranium oxide. The precipitation conditions will affect the microstructure of the resulting oxide as well as its sinterability. Studies are being performed in order to evaluate potential correlations between uranium oxide (powder) microstructure and precursor compounds or the preceding chemical treatment.

For the production of uranium fuels, the uranium oxide powder is reduced to UO_2, pressed to pellets and sintered. The uranium in such pellets may consist of natural uranium (for heavy-water moderated reactors) or of low enriched uranium (for light-water moderated reactors). At this stage of the nuclear fuel cycle, the chemical composition of the fuel does not provide useful hints for nuclear forensics. Most of the nuclear forensic information can be obtained by combining parameters such as pellet dimensions, enrichment level and chemical impurities.

Metallic uranium is encountered in applications such as counterweights in aircraft, as shielding material for highly radioactive sources in medical radiation therapy and industrial radiography equipment, and in containers used to transport radioactive materials. As described in section I, military uses include defensive armour plating and penetrators ammunition. While shielding materials typically consist of fairly pure uranium metal, depleted uranium penetrators often contain small amounts (0.5–3 per cent) of titanium or molybdenum. Uranium–aluminium alloys that contain up to 25 per cent uranium may indicate that they have been used in research reactors. Uranium–aluminium alloys can also be manufactured into 'targets', which are metal plates designed to be irradiated in a specialized reactor for the production of ^{99}Mo, an isotope with medical applications. These targets may include HEU or LEU. Pure uranium metal is also an essential ingredient for nuclear weapons: HEU is used for uranium weapons, while depleted uranium may be used as tamper for plutonium weapons or for thermonuclear warheads. However, in these latter cases there is little information available in the open literature that could be exploited for nuclear forensic purposes.

Plutonium compounds

Plutonium is obtained from reprocessing irradiated nuclear fuels, where it is produced from uranium by multiple neutron capture. In the civil nuclear fuel cycle, plutonium is typically encountered in the chemical form of plutonium nitrate solution or as solid PuO_2. Precipitation of plutonium oxalate from nitric acid solutions is the most commonly used intermediate for the production of PuO_2. Plutonium oxide is a greenish powder that is stable in ambient conditions. The microstructure of plutonium oxide varies with the chemical compound from which it was produced through calcination. However, using the same basic process (precipitation of plutonium oxlalate from nitric acid solution), the morphology and particle size of calcined PuO_2 have also been found to depend strongly on the method of mixing Pu and oxalate ions, the Pu valence, the reactant concentrations and the precipitation temperature.[10] The stoichiometry (i.e. the oxygen to metal ratio in the oxide) of plutonium oxide may vary and is often described as $Pu\,O_{2-x}$, particularly if the plutonium has been heat treated under reducing conditions (e.g. in a mixed argon–hydrogen (H_2) atmosphere).

Mixed oxides of uranium and plutonium form a particular group of compounds. The crystal structures (fluorite lattice) of UO_2 and PuO_2 are the same. Mixed oxides can be prepared by powder blending of UO_2 and PuO_2 or by co-precipitation of uranium and plutonium to form a mixed solution. The plutonium can be reactor- or weapon-grade material, and the uranium can be of natural isotopic composition, depleted or reprocessed uranium. MOX can be used in nuclear reactors as fuel. Powder blending and co-milling of uranium and plutonium oxides are widely used for fuel production. This process, however, results in fuel pellets that show inhomogeneities at the microscopic level, reflecting their production history. Light-water reactors typically accept MOX containing 5 weight per cent of plutonium, thus replacing LEU fuel. Certain types of fast breeder reactors use MOX with about 20 weight per cent of plutonium.

Pure plutonium metal exhibits six solid-state phase transformations before reaching its liquid state at a relatively low melting temperature of 640 °C. These phase transitions coincide with large volume expansions and contractions. These physical parameters limit the use of pure plutonium metal in technical applications. For military applications, a plutonium phase with high compressibility and easy machinability is required. The delta phase of plutonium, a high-temperature modification of plutonium metal, exhibits these desired properties. The addition of 0.8–1 weight per cent gallium stabilizes the δ phase at room temperature and at the same time drastically reduces the corrosion of plutonium. Consequently, the alloying components of plutonium metal provide useful clues to the intended use of the material.

Common non-nuclear chemicals in the fuel cycle

The non-nuclear chemicals associated with nuclear materials may serve to provide clues on the origin of nuclear materials and on the processes used

[10] Smith, P. K. et al., 'Effect of oxalate precipitation on PuO_2 microstructures', Paper presented at the Sixth International Materials Symposium, University of California, Berkeley, CA, 24–27 Aug. 1976, <http://www.osti.gov/scitech/biblio/7234786>.

Table 5.2. Examples of non-nuclear components in nuclear material

Concentration level	Typical examples	Origin of additives
Major constituents	0.8–1.0% Ga in Pu	δ phase stabilization of plutonium
	U–Al, U–Zr, UZrH	Higher fissile atom density in alloyed fuels
	Ceramic fuels (UN, UC)	Increasing heat conductivity and melting point
	Burnable poisons (Er, Gd)	Control of large amounts of excess fuel reactivity
Minor constituents	Several anions in uranium ore concentrates	Residue from the purification process
	C, N and P in UO_2 fuel	Impurity from production process
Trace components	Most transition metals (e.g. Fe, Ni, Co) in fuels	Impurity from production process
	Pu residue in reprocessed fuel	Impurity from the reprocessed feed
	U-236	Use of reprocessed fuel as feed material
	Rare-earth elements	Feed material (e.g. ore)

Sources: Coleman, G. H., *The Radiochemistry of Plutonium* (National Academy of Sciences, National Research Council: Springfield, VA, 1965); Choppin, G. R., Liljenzin, J.-O. and Rydberg, J., *Radiochemistry and Nuclear Chemistry*, 2nd edn (Butterworth-Heinemann: Oxford, 1995); Badaut, V., Wallenius, M. and Mayer, K., 'Anion analysis in uranium ore concentrates by ion chromatography', *Journal of Radioanalytical and Nuclear Chemistry*, vol. 280, no. 1 (2009); Edwards, C. R. and Oliver, A. J., 'Uranium processing: a review of current methods and technology', *JOM Journal of the Minerals, Metals and Materials Society*, vol. 52, no. 9 (2000); and Varga, Z. and Surányi, G., 'Detection of previous neutron irradiation and reprocessing of uranium materials for nuclear forensic purposes', *Applied Radiation and Isotopes*, vol. 67, no. 4 (Apr. 2009).

for their preparation. Chemical solvents and reagents are used throughout the nuclear fuel cycle for dissolution, extraction, purification, precipitation and solidification of the nuclear material. Moreover, certain chemicals are intentionally added in order to obtain defined material properties, such as grain growth or powder flowability.

The use of chemical solvents in the nuclear fuel cycle starts with the leaching of uranium ores and continues until the dissolving of spent nuclear fuels for reprocessing. In uranium mining, acids such as sulphuric acid or nitric acid are used and, depending on the type of ore, sodium carbonate solutions can also be used. Traces of these chemicals can be detected in the products of the mining and milling process and provide hints of the process used.[11] In the pre-concentration and purification steps, organic solvents and extractants are also used. Despite their low solubility in aqueous media, they can be detected in small amounts in the products of

[11] Badaut, Wallenius and Mayer (note 8).

the mining and milling processes. Molecules such as tri-n-butyl-phosphate (TBP) or ternary amines are typically used as extractants, while kerosene or other aliphatic compounds serve as solvents.

III. Elemental characteristics and signatures

In addition to the nuclear components (i.e. uranium, plutonium and thorium), numerous other elements are usually present in investigated illicit material, sometimes even at concentrations exceeding that of the fissile material. In nuclear forensic terminology these elements are often referred to as 'impurities', a result of the fact that most investigations focus on the fissile elements and their concentrations and isotopic composition. However, the other major or minor components of the sample under investigation carry indispensable information on the source and intended use of the material because these constituents determine the fundamental physical and chemical properties.

These additional components may have been added intentionally to the material to achieve certain material properties. For example, erbium or gadolinium are often mixed with uranium oxide fuel as so-called burnable neutron poisons in order to control excess fuel reactivity, or plutonium can be stabilized in the δ phase by alloying it with a small amount of another metal, such as 0.8–1.0 weight-per cent gallium. Chemical impurities in nuclear material can be present inadvertently as a result of residual trace elements from the initial feed materials or as residuals from chemicals added during the production process as well as from corrosion or abrasion of vessels and pipework. The measurement of these components for nuclear forensics is thus of utmost importance, as they can provide information on not only the intended use, but also the source or production facility. A sample of nuclear material, therefore, discloses a complex set of information via its composition. These non-nuclear components or additives can be present at major (in the per cent range), minor (at the mg g^{-1} level) or trace (below 10^{-6} g g^{-1}) levels (see table 5.2).

Based on the origin of these components, two major categories can be distinguished: process and source-related signatures.

If the component originates dominantly from the process (either intentionally or inadvertently), this is usually termed a process-related signature, which can provide information on the possible hydrometallurgical and metallurgical production process used. Typically, these components are the materials or their residuals that are fed into the process line, such as organic residues of solvent extractants, precipitant in the final step of UOC production or alloying metals.

In contrast, if the measured parameter derives largely from the feed material (e.g. ores or secondary sources, such as phosphate fertilizer), the

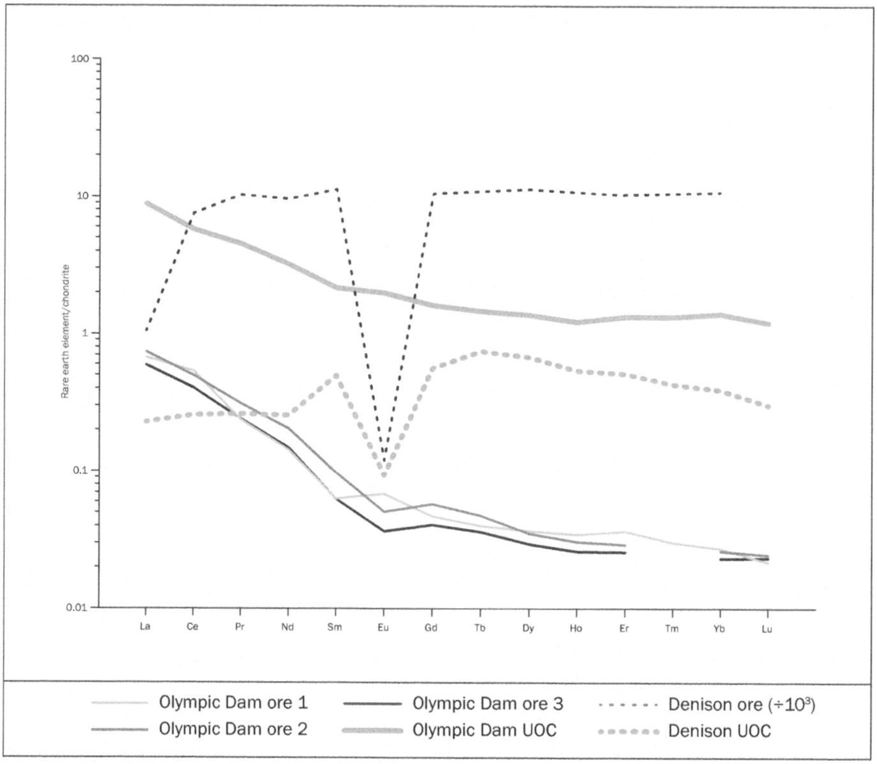

Figure 5.13. The rare-earth elemental pattern of various uranium ores and the respective uranium ore concentrates produced

Source: Varga, Z., Wallenius M. and Mayer, K., 'Origin assessment of uranium ore concentrates based on their rare-earth elemental impurity pattern', *Radiochimica Acta*, vol. 98, no. 12 (2010).

measured parameter is called a source-related signature. Typical examples of this group are the isotopic composition of uranium, strontium or neodymium and their rare-earth elemental composition. As an example, looking at the rare-earth elemental pattern of various uranium ores and the produced respective UOC (see figure 5.13) it is apparent that the ores and the respective UOCs have similar patterns, and the generally used production processes do not alter this signature. Thus, this parameter is indicative of the source (feed) material. The shape of the rare-earth elemental pattern is indicative of the geological deposit type from which the uranium was mined.[12] Applying this methodology, an unknown UOC material can be traced back to the original uranium ore, and the mine or

[12] Varga, Z., Wallenius M. and Mayer, K., 'Origin assessment of uranium ore concentrates based on their rare-earth elemental impurity pattern', *Radiochimica Acta*, vol. 98, no. 12 (2010).

geological location can therefore be identified or verified after the measurement of the rare-earth elemental pattern.

The concentration of metallic impurities in uranium samples has been investigated by several groups. Generally, the challenge associated with data interpretation arises from the broad range of concentrations encountered in different samples, from the sample (in)homogeneity and from the complexity of the data. A number of multivariate statistical techniques, such as discriminant analysis (DA), principal component analysis (PCA) and classification and regression tree (CART), have been used to identify distinguishing parameters (between groups) or common features (within groups).[13]

IV. Isotopic characteristics and signatures

Plutonium isotopes

Plutonium does not occur in nature (except for the longest living isotope, ^{244}Pu, which is found in ultratrace amounts in certain minerals), but it is produced in nuclear reactors by irradiating uranium with neutrons. Subsequent neutron capture leads to the build-up of different plutonium isotopes. The isotopic composition of the plutonium is largely determined by the number of neutrons to which the starting material (i.e. the fuel in the reactor) has been exposed and the energy distribution of the neutrons. In turn, the plutonium isotopic composition can be used as a nuclear forensic signature, providing clues to the type of reactor and the starting material from which the plutonium was produced.

According to its isotopic composition and depending on its application, plutonium can be categorized as weapon grade, reactor grade, or power or heat source. The first two categories can also have subcategories (see table 5.3).

To distinguish between weapon-grade plutonium (WGPu, also called low burn-up plutonium) and reactor-grade plutonium (RGPu, or high burn-up plutonium) the decisive parameter is the abundance of the isotope ^{240}Pu, which should not exceed 7 per cent for the weapon-grade material.[14] This is because the presence of ^{240}Pu limits applicability as weapon material since it has a relatively high spontaneous fission rate, raising the background neutron level and thus increasing the risk of predetonation. In addition, the abundance of ^{238}Pu in weapon-grade material should be kept at a minimum as this isotope produces significant amounts of decay heat, which is not compatible with the requirements for nuclear warheads.

[13] Robel, M., Kristo, M. J. and Heller, M. A., 'Nuclear forensic inferences using iterative multi-dimensional statistics', Institute of Nuclear Materials Management (INMM), *50th Annual Meeting of the Institute of Nuclear Materials Management 2009 (INMM 50)* (INMM: Deerfield, IL, 2009).

[14] For the definition of 'burn-up' see the glossary in this volume.

Table 5.3. Plutonium categories and subcategories

Category/ Subcategory	Isotopic composition (%)		
	Pu-238	Pu-239	Pu-240
Weapon grade (WGPu)	–	≥93	<7
Super grade (SGPu)	–	≥97	<3
Reactor grade (RGPu)	–	50–65	≥19
Fuel grade (FGPu)	–	–	7–19
Mixed oxide grade (MGPu)	–	–	>30
Power or heat source	~90	–	–

Sources: Carlson, J. et al., 'Plutonium isotopics: non-proliferation and safeguards issues', IAEA-SM-351/64, *Proceedings of the IAEA Symposium on International Safeguards*, Vienna, 13–17 Oct. 1997 (IAEA: Vienna, 1997); and US Department of Energy (DOE), *Plutonium: The First 50 Years*, DOE/DP-0137 (DOE: Washington, DC, Feb. 1996).

Fuel-grade plutonium (FGPu) is produced in some nuclear reactors that have a spent fuel burn-up lower than that resulting in RGPu, but higher than that resulting in WGPu. Such reactors include tritium-production and power reactors, where the fuel is discharged after only one year of irradiation.

RGPu is produced in commercial power reactors, such as LWRs and heavy-water reactors (HWRs) at irradiation times of about three years and a burn-up of more than 30 gigawatt-days per tonne (GWd/t). At discharge of the fuel, the plutonium typically contains 19 per cent or more of the isotope ^{240}Pu and around 50–65 per cent ^{239}Pu. As there are many types of power reactor, and differences in fuel composition, coolant and moderator system, and burn-up level, RGPu has a large variability of isotopic compositions. Consequently, the isotopic composition of plutonium (in combination with model calculations) can be exploited as a nuclear forensic signature.

Determination of the reactor type used for plutonium production

Just the isotopic composition of plutonium provides some clues about the reactor type in which the Pu was produced. WGPu is usually produced in so-called plutonium-production reactors. All production reactors are on-load refuelled—that is, they are refuelled while the reactor is operating—to allow for short fuel irradiation times. Reactor types that are typically considered as production reactors include heavy water- or graphite-moderated reactors, which use natural or slightly enriched uranium as a fuel. Another route for the production of WGPu makes use of the irradiation of ^{238}U with fast neutrons. Such conditions are met, for example, in the natural or depleted uranium fuel blanket of a liquid metal fast breeder reactor

(LMFBR). This leads to the production of plutonium with an isotope abundance of ^{240}Pu of about 4 per cent.[15]

Reactor-grade plutonium is produced in commercial power reactors, including LWRs, pressurized-water reactors (PWRs), water–water energetic reactors (*vodo-vodyanoi energetichesky reaktor*, VVER, a variant of a PWR originally developed in the Soviet Union), boiling-water reactors (BWR), HWRs and graphite-moderated reactors (e.g. the Soviet-designed high-power channel reactor, *reaktor bolshoy moshchnosty kanalny*, RBMK). Because of the difference in the initial ^{235}U enrichment of the fuel used in these reactors and the different neutron energy spectra (due to the difference in the moderation, e.g. heavy water, light water or graphite) the isotopic composition of plutonium in the spent fuel is quite different. Computer code-calculated isotopic correlations can be used to illustrate these differences and, consequently, identify the reactor type that produced the plutonium (see figure 5.14).[16]

Plutonium-238 is used in radioisotope thermoelectric generators (RTGs), which are a kind of battery that is used as a power source in satellites, space probes and unmanned remote facilities, such as the series of lighthouses built by the Soviet Union inside the Arctic Circle. In the past, small ^{238}Pu-powered RTGs were also used in implanted heart pacemakers to ensure a long battery life. In addition to RTGs, ^{238}Pu was used in smoke detectors produced in the Soviet Union. It is produced by neutron irradiation of ^{237}Np in a reactor, and the ^{237}Np needs to be separated from spent fuel during reprocessing.

Determination of the age of plutonium

The decay of plutonium isotopes can be used in two ways in nuclear forensics. First, the amount of decay products (i.e. daughter products) can be measured and, the age of the plutonium can be determined based on their relation to the plutonium isotopes. Second, the isotopic composition of freshly discharged reactor plutonium differs from, for instance, that of plutonium that has been stored for several years, due to the rapid decay of

[15] Carlson, J. et al., 'Plutonium isotopics: non-proliferation and safeguards issues', IAEA-SM-351/64, *Proceedings of the IAEA Symposium on International Safeguards*, Vienna, 13–17 Oct. 1997 (IAEA: Vienna, 1997).

[16] Luksic, A. T. et al., 'Isotopic measurements: interpretation and implications of plutonium data', Institute of Nuclear Materials Management (INMM), *51st Annual Meeting of the Institute of Nuclear Materials Management 2010 (INMM 51)* (INMM: Deerfield, IL, 2010); Wallenius, M., Peerani, P. and Koch, L., 'Origin determination of plutonium material in nuclear forensics', *Journal of Radioanalytical and Nuclear Chemistry*, vol. 246, no. 2 (2000); and Robel, M. and Kristo, M. J., 'Discrimination of source reactor type by multivariate statistical analysis of uranium and plutonium isotopic concentrations in unknown irradiated nuclear fuel material', *Journal of Environmental Radioactivity*, vol. 99, no. 11 (2008).

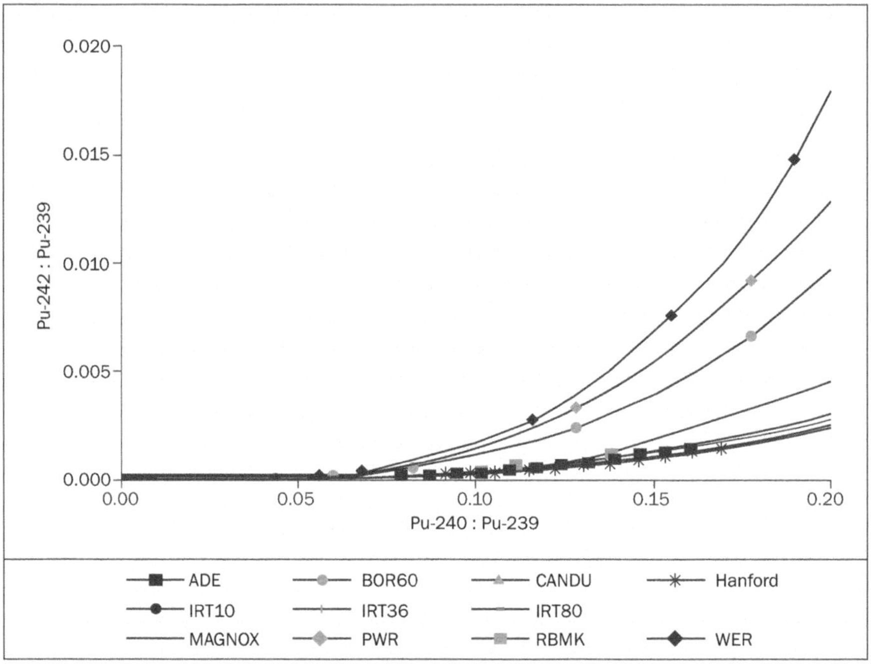

Figure 5.14. The plutonium-240 : plutonium-239 ratio versus the plutonium-242 : plutonium-239 ratio in different reactor types

Source: Luksic, A. T. et al., 'Isotopic measurements: interpretation and implications of plutonium data', Institute of Nuclear Materials Management (INMM), *51st Annual Meeting of the Institute of Nuclear Materials Management 2010 (INMM 51)* (INMM: Deerfield, IL, 2010).

the shortest living plutonium isotope ^{241}Pu (with a half life $T_{½}$ = 14.35 years). This difference can be used to determine the time of discharge.

Age since processing

Age determination in nuclear forensics defines the time that has elapsed since the radioactive material was most recently chemically separated (i.e. when the 'daughter' nuclides were removed from their 'parents'). Progenies (decay products) of the nuclear materials are continuously growing-in within the nuclear material due to the radioactive decay of the parent nuclides.[17] A chemical separation removes these decay products from the parent nuclides, that is, from the actual nuclear material. However, the parent nuclides continue to decay. The amount of decay products is proportional to the amount of parent nuclides and the time elapsed since the most recent chemical separation, if the removal of the progenies was complete (see figure 5.15).

[17] For the definition of 'growing-in' see the glossary in this volume.

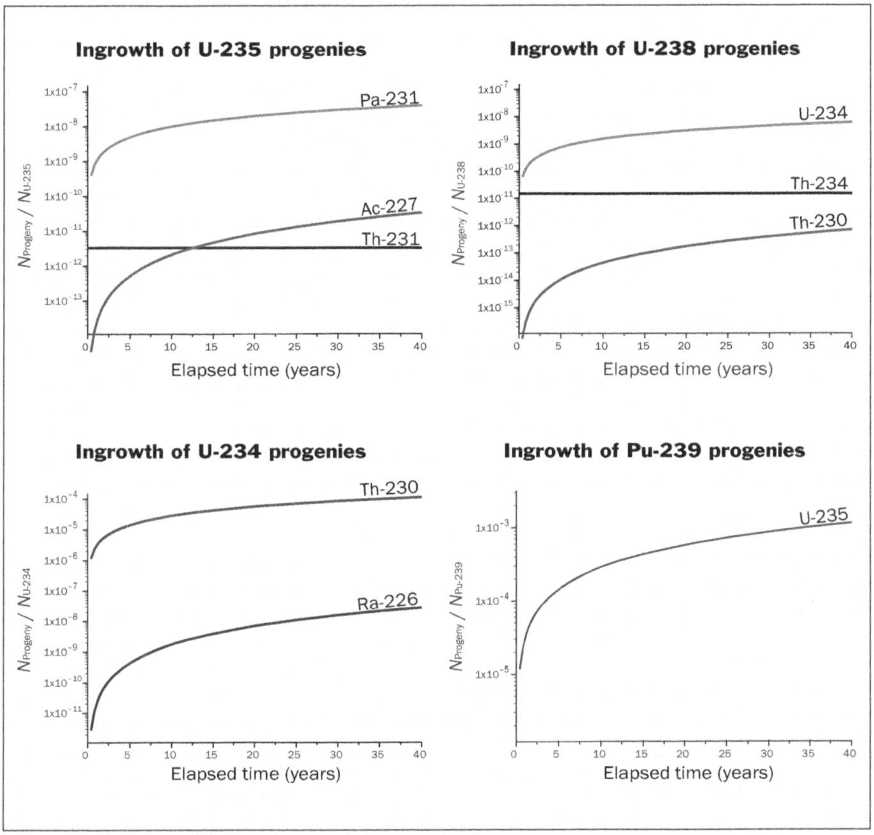

Figure 5.15. The variation of main progeny-to-parent nuclide atom ratios for the most important nuclear radionuclides as a function of time

Source: Unpublished result from the European Commission, Joint Research Centre (JRC), Institute for Transuranium Elements (ITU), Karlsruhe.

The variation of the decay product can be expressed by the following decay equation:

$$N_{\text{daughter}}(t) = \frac{\lambda_{\text{daughter}}}{\lambda_{\text{parent}} - \lambda_{\text{daughter}}} N^0_{\text{parent}} \left(e^{-\lambda_{\text{parent}} t} - e^{-\lambda_{\text{daughter}} t} \right) + N^0_{\text{daughter}}\, e^{-\lambda_{\text{daughter}} t},$$

where $N_{\text{daughter}}(t)$ is the number of daughter nuclides as a function of time; λ_{parent} and $\lambda_{\text{daughter}}$ are the decay constants of the parent and daughter nuclides, respectively; t is the elapsed time since the last separation of the radionuclides (age of the material); and N^0_{parent} and N^0_{daughter} are the number of initial parent and daughter nuclides after the separation (at $t = 0$), respectively. The model assumes that there is no further fractionation of the parent and daughter nuclides following production (i.e. it behaves as a closed system).[18]

The measurement of progenies in nuclear materials is widely used for nuclear forensic investigations, as their determination allows calculation of the date of production. The time elapsed since the most recent separation of the progenies can be calculated on the basis of the above decay equation. In several cases this equation can be simplified if the parent nuclide is long-lived (i.e. $N^0_{\text{parent}} = N_{\text{parent}}(t)$), if the separation was complete (i.e. $N^0_{\text{daughter}} = 0$) or if there is a big difference between the half-lives of the parent and the daughter nuclides.

In the case of plutonium, separation could have taken place, for example, when the material was reprocessed or, after a longer storage period, when the material was purified (i.e. the ingrown ^{241}Am was removed). The prerequisite for a correct result is that the daughters and granddaughters were removed completely when the chemical separation took place. If this is not the case, the determined result will be positively biased.[19] The age of plutonium can be determined, in principle, using five different parent–daughter relations and additional parent–granddaughter relations. The five 'direct' parent : daughter ratios are ^{238}Pu : ^{234}U, ^{239}Pu : ^{235}U, ^{240}Pu : ^{236}U, ^{241}Pu : ^{241}Am and ^{242}Pu : ^{238}U. However, in practice, only the first four ratios are used due to the long half-life of ^{242}Pu (3.75×10^5 years), which results in only a minute amount of ingrown ^{238}U. These four different parent : daughter ratios for age determination, also including two different elemental ratios (Pu : U and Pu : Am), make it fairly easy to detect inconsistencies. In particular, the above-mentioned incomplete separation of daughter nuclides can be perceived easily. If all four ratios indicate the same age, it is certain that the separation was complete when the material was processed and no residues of U and Am daughters were left.

Due to the two different parent : daughter elemental ratios available, different techniques to determine the age of the plutonium can also be used. Traditionally, the age of Pu has been determined by high-resolution gamma spectrometry (HRGS) from the ^{241}Pu : ^{241}Am ratio. In addition to its being a non-destructive technique, the method has advantages owing to its speed and because it requires no chemical preparation. The main disadvantage is that it is not possible to know whether any Am remains were left in the material (which is often the case) as only one parent : daughter ratio is determined. Therefore, in order to address this problem, various mass spectrometry techniques (i.e. thermal ionization mass spectrometry, TIMS, and inductively coupled plasma mass spectrometry, ICP-MS) are

[18] For the definition of 'fractionation' see the glossary in this volume.

[19] Mayer, K. et al., 'Development of analytical methodologies in response to recent challenges', Institute of Nuclear Materials Management (INMM), *42nd Annual Meeting of the Institute of Nuclear Materials Management 2001 (INMM 42)* (INMM: Deerfield, IL, 2001).

often applied to the three Pu : U parent : daughter ratios.[20] Due to the 'moderate' short half-lives of the plutonium isotopes ^{238}Pu, ^{239}Pu and ^{240}Pu (87.74 years, 24 110 years and 6563 years, respectively) age determination of plutonium can also be performed for small particle samples (μm-sized) since detectable amounts of daughter nuclides grow in and are present within a few years.[21] This is not the case for small uranium particles.[22]

Age since discharge from a reactor

With its half-life of 14.35 years, ^{241}Pu decays much faster that the neighbouring isotopes ^{240}Pu and ^{242}Pu (6563 years and 3.75×10^5 years, respectively). Thus, when irradiation in the reactor is stopped, its correlation with the other isotopes suffers due to the much faster decay. Therefore, by comparing the measured ^{241}Pu abundance to the theoretical one obtained by computer code calculations at the time of discharge, it is possible to calculate the time that has passed since irradiation was stopped.[23]

Uranium isotopes

Uranium is found in the earth's crust (on average) at a concentration level of a few parts per million (10^{-6} g g^{-1}). Natural uranium has three long-lived radioactive isotopes, ^{234}U, ^{235}U and ^{238}U, with the respective relative isotope abundances of 0.0054 per cent, 0.7204 per cent and 99.2742 per cent.[24] Ultratrace amounts of ^{236}U can also be found in natural uranium, but highly sophisticated measurement techniques, such as accelerator mass spectrometry (AMS), are required for their detection. The applicability of variations of the ^{236}U abundance for nuclear forensics in natural uranium samples (selected ores and UOC) has been studied.[25] Several researchers have also investigated small variations in the isotopic composition of natural uranium using TIMS.[26] These studies have shown that the

[20] Wallenius, M. and Mayer, K., 'Age determination of plutonium material in nuclear forensics by thermal ionisation mass spectrometry', *Fresenius' Journal of Analytical Chemistry*, vol. 366, no. 3 (Feb. 2000); Nygren, U., Ramebäck, H. and Nilsson, C., 'Age determination of plutonium using inductively coupled plasma mass spectrometry', *Journal of Radioanalytical and Nuclear Chemistry*, vol. 272, no. 1 (2007); and Zhang, H. T. et al., 'Age determination of plutonium material by alpha spectrometry and thermal ionization mass spectrometry', *Radiochimica Acta*, vol. 96, no. 6 (2008).

[21] Wallenius, M., Tamborini, G. and Koch, L., 'The "age" of plutonium particles', *Radiochimica Acta*, vol. 89 (2001).

[22] Glaser, A. and Bürger, S., 'Verification of a fissile material cutoff treaty: the case of enrichment facilities and the role of ultra-trace level isotope ratio analysis', *Journal of Radioanalytical and Nuclear Chemistry*, vol. 280, no. 1 (2009).

[23] Luksic et al. (note 16).

[24] *Karlsruher Nuclide Chart*, 8th edn (Joint Research Centre, Institute for Transuranium Elements: Karlsruhe, 2012).

[25] Srncik, M. et al., 'Investigation of the $^{236}U/^{238}U$ isotope abundance ratio in uranium ores and yellow cake samples', *Radiochimica Acta*, vol. 99, no. 6 (June 2011).

[26] Ovaskainen, R. et al., 'Unusual isotope abundances in natural uranium samples', *Proceedings of the 19th ESARDA Annual Symposium on Safeguards and Nuclear Material Management*, Montpellier,

Table 5.4. Uranium categories

Category	U-235 (%)
Depleted uranium (DU)	<0.7
Natural uranium (NU)	~0.7
Low-enriched uranium (LEU)	0.7–20
Slightly enriched uranium (SEU)	0.9–2
Reactor-grade uranium	3–5
Highly enriched uranium (HEU)	>20
Weapon-grade uranium	>90

variations in the isotopic composition of natural uranium, in particular the variability of the isotope abundance of ^{234}U, may serve as useful parameters in nuclear forensics. This variability can be explained by natural fractionation (e.g. due to preferential leaching of ^{234}U after alpha recoil, that is the decay of the parent nuclide ^{238}U) and due to the varying (weathering) conditions in different types of uranium deposit.[27] Such an isotopic signature of the ore is preserved through the production of UOC, uranium oxide, and UF_4 to UF_6. The signature is wiped out once the uranium is subjected to isotope enrichment.

Uranium-235 isotope abundance

Through enrichment, the abundance of the naturally occurring fissile uranium isotope, ^{235}U, can be increased in order to use uranium—for example, in nuclear power reactors or nuclear weapons. The product stream of the enrichment process results in uranium with a ^{235}U isotope abundance that is higher than in natural uranium. The tailings (i.e. remains) of the enrichment process are depleted in ^{235}U, typically showing abundances in ^{235}U between 0.2 and 0.3 per cent (see table 5.4). The ^{235}U isotope abundance in uranium is colloquially referred to as 'uranium enrichment' or the 'enrichment level'. The level of ^{235}U enrichment is a direct indicator of the intended use of the uranium material.

13–15 May 1997 (European Safeguards Research and Development Association: Ispra, 1997); Richter, S. et al., 'Isotopic "fingerprints" for natural uranium ore samples', *International Journal of Mass Spectrometry*, vol. 193, no. 1 (1999); and Buerger, S. et al., 'The range of variation of uranium isotope ratios in natural uranium samples and potential application to nuclear safeguards', IAEA-CN-184/256, *Symposium on International Safeguards: Preparing for Future Verification Challenges*, Vienna, 1–5 Nov. 2010 (IAEA: Vienna, 2010)

[27] Andersen, M. B. et al., 'Toward epsilon levels of measurement precision on $^{234}U/^{238}U$ by using MC-ICPMS', *International Journal of Mass Spectrometry*, vol. 237, nos 2–3 (Oct. 2004); and Brennecka, G. A. et al., 'Natural variations in uranium isotope ratios of uranium ore concentrates: understanding the $^{238}U/^{235}U$ fractionation mechanism', *Earth and Planetary Science Letters*, vol. 291, nos 1–4 (Mar. 2010).

Minor uranium isotopes as indicators of reprocessing

The main signature in uranium that indicates reprocessing or irradiation activities is the isotope ^{236}U. According to most chemistry textbooks, ^{236}U does not occur in nature. However, recent studies have shown that ^{236}U is present in nature, although in quantities so minute that it can only be detected by the use of advanced mass-spectrometry techniques such as AMS. It is considered an anthropogenic isotope, which is formed by neutron capture of ^{235}U. In recycled and re-enriched uranium the isotope abundance of ^{236}U may range up to 0.3 per cent. Other minor, long-lived isotopes of uranium that are produced by irradiation in a reactor are ^{232}U and ^{233}U, but they are present in trace amounts (typically $<10^{-8}$). Due to the relatively short half-life of ^{232}U ($T_{½}$ = 68.9 years), it can be detected easier than ^{233}U by using radiometric techniques, such as alpha or gamma spectrometry.[28]

Determination of the age of uranium

Determining the age of uranium-containing materials is a more complex task than in the case of plutonium materials for three reasons. First, only two parent : daughter ratios are potentially available to determine age: ^{234}U : ^{230}Th and ^{235}U : ^{231}Pa. Other candidate parent : daughter ratios, such as ^{236}U : ^{232}Th, cannot be used because ^{232}Th is often present in trace amounts in nominally pure uranium materials, thus leading to a biased result. In addition, the ^{238}U : ^{234}U ratio cannot be used as the daughter nuclide is present in the sample and is not separated from its parent nuclide during processing. Second, the long half-lives of the parent nuclides (2.455×10^5 years for ^{234}U and 7.038×10^8 years for ^{235}U) result in only small concentrations of daughter nuclides; thus, extremely sophisticated chemical separation methods and highly sensitive measurement techniques are required. Third, no long-lived spike isotope for protactinium can be used in isotope dilution analysis (the second longest-lived isotope of protactinium is ^{233}Pa, with a half-life of 27 days).

The most common approach uses the ^{230}Th : ^{234}U ratio variation. Determining the production date from the ^{230}Th : ^{234}U ratio is based on the decay of ^{234}U (which is relatively long-lived, with $T_{½}$ = 245 250 ± 490 years) to ^{230}Th ($T_{½}$ = 75 690 ± 230 years) and the disequilibrium between these two radionuclides. After the most recent chemical separation of ^{234}U in the course of the preparation of the nuclear material, the concentration of the

[28] Nguyen, C. T. and Zsigrai, J., 'Basic characterization of highly enriched uranium by gamma spectrometry', *Nuclear Instruments and Methods in Physics Research, Section B: Beam Interactions with Materials and Atoms*, vol. 246, no. 2 (May 2006); and Varga, Z. and Surányi, G., 'Detection of previous neutron irradiation and reprocessing of uranium materials for nuclear forensic purposes', *Applied Radiation and Isotopes*, vol. 67, no. 4 (Apr. 2009).

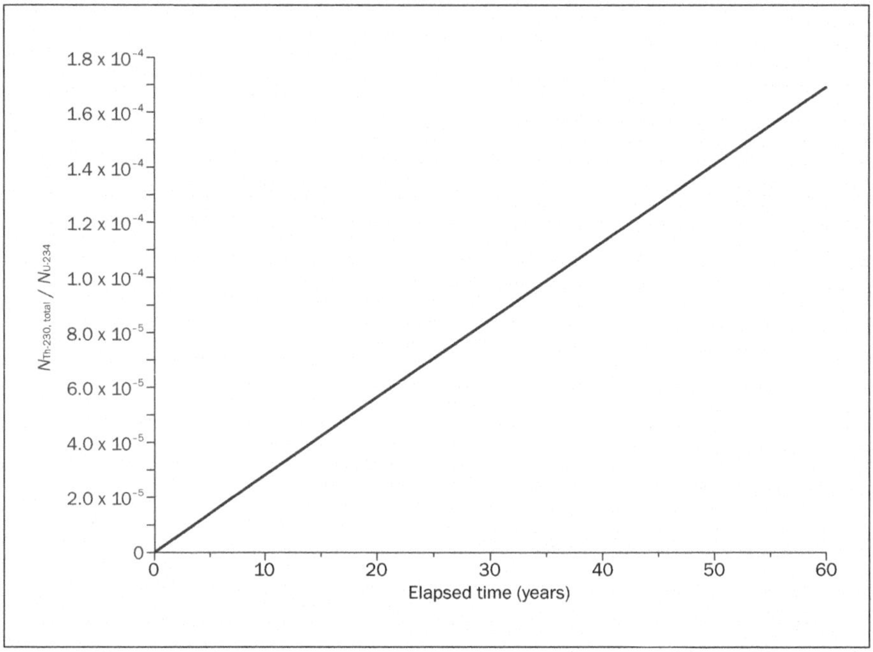

Figure 5.16. Variation of the thorium-230 : thorium-234 atom ratio in uranium oxide as a function of time

Source: Unpublished result from the European Commission, Joint Research Centre (JRC), Institute for Transuranium Elements (ITU), Karlsruhe.

^{230}Th daughter nuclide continuously increases in the uranium oxide material (see figure 5.16). The theoretical amount of ^{230}Th formed by the decay can be calculated by the use of equations for the radioactive decays, assuming that the initial concentration of the daughter nuclide is zero after the last chemical separation (i.e. the separation was complete) and using the approximation that the amount of ^{234}U in the sample is constant over the investigated time scale. These assumptions are usually valid for nuclear fuel samples. The variation of the ^{230}Th : ^{234}U atom ratio as a function of time can be calculated as

$$\frac{N_{\text{Th-230}}}{N_{\text{U-234}}} = \frac{\lambda_{\text{U-234}}}{\lambda_{\text{Th-230}} - \lambda_{\text{U-234}}} \left(1 - e^{(\lambda_{\text{U-234}} - \lambda_{\text{Th-230}})t}\right),$$

where $N_{\text{Th-230}} / N_{\text{U-234}}$ is the atom ratio in the sample; $\lambda_{\text{Th-230}}$ and $\lambda_{\text{U-234}}$ are the decay constants of ^{230}Th and ^{234}U, respectively; and t is the elapsed time since the last separation of the radionuclides (i.e. the age of the material).

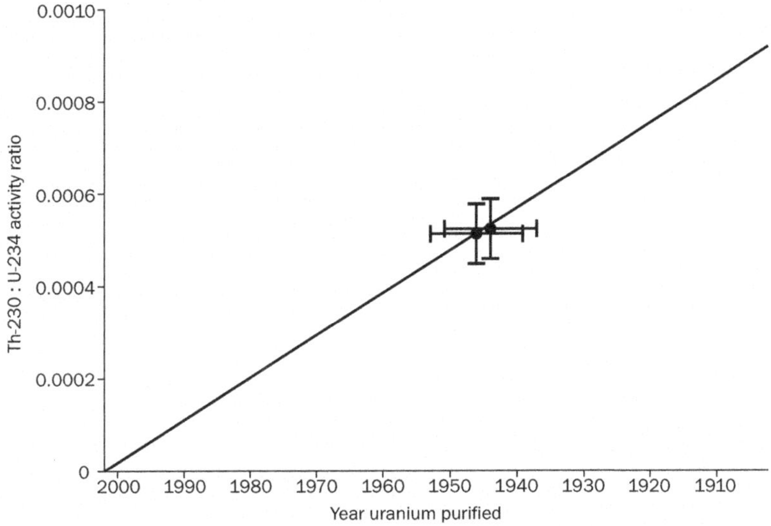

Age of uranium (2σ uncertainties): sample 1, 1946 ± 7 years
sample 2, 1945 ± 7 years

Figure 5.17. Age of uranium samples determined from the uranium-234 : thorium-230 ratio by thermal ionization mass spectrometry

Source: LaMont, S. P. and Hall, G., 'Uranium age determination by measuring the $^{230}Th/^{234}U$ ratio', *Journal of Radioanalytical and Nuclear Chemistry*, vol. 264, no. 2 (2005).

If the atom ratio of ^{230}Th and ^{234}U is experimentally determined, then the elapsed time and the production date can be calculated as

$$t = \frac{1}{\lambda_{U\text{-}234} - \lambda_{Th\text{-}230}} \ln\left(1 - \frac{N_{Th\text{-}230}}{N_{U\text{-}234}} \frac{\lambda_{Th\text{-}230} - \lambda_{U\text{-}234}}{\lambda_{U\text{-}234}}\right).$$

Despite these complexities, the age of uranium can be determined using several techniques, including mass spectrometry, alpha spectrometry and gamma spectrometry (see figure 5.17).[29]

[29] Wallenius, M. et al., 'Determination of the age of highly enriched uranium', *Analytical and Bioanalytical Chemistry*, vol. 374 (2002); LaMont, S. P. and Hall, G., 'Uranium age determination by measuring the $^{230}Th/^{234}U$ ratio', *Journal of Radioanalytical and Nuclear Chemistry*, vol. 264, no. 2 (2005); Varga, Z. and Surányi, G., 'Production date determination of uranium-oxide materials by inductively coupled plasma mass spectrometry', *Analytica Chimica Acta*, vol. 599, no. 1 (2007); Nguyen, C. T. and Zsigrai, J., 'Gamma-spectrometric uranium age-dating using intrinsic efficiency calibration', *Nuclear Instruments and Methods in Physics Research, Section B: Beam Interactions with Materials and Atoms*, vol. 243, no. 1 (Jan. 2006); and Varga, Z., Wallenius, M. and Mayer, K., 'Age determination of uranium samples by inductively coupled plasma mass spectrometry using direct

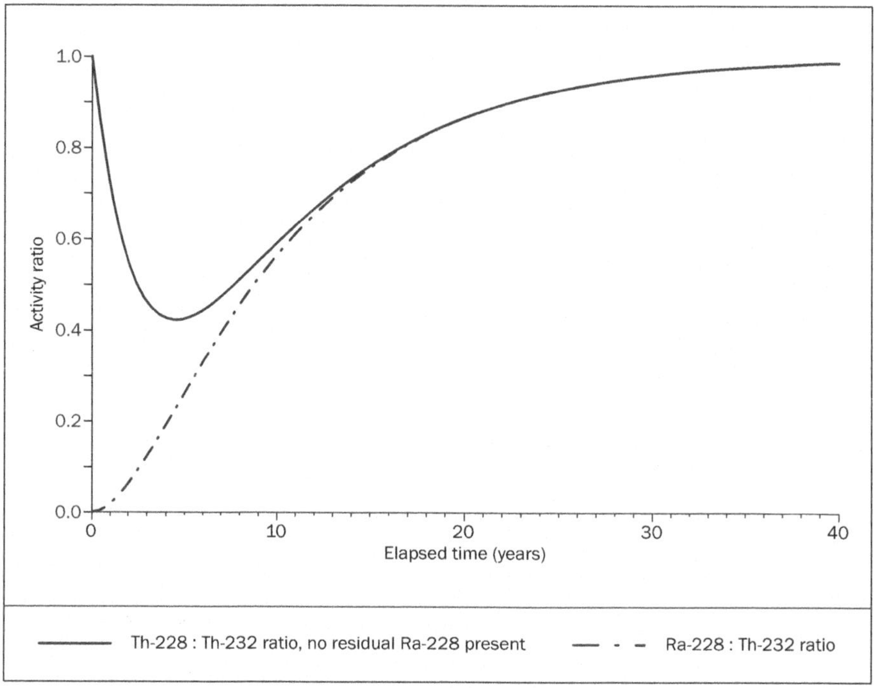

Figure 5.18. Variation of the thorium-228 : thorium-232 and radium-228 : thorium-232 ratios as a function of time

Source: Varga, Z. et al., 'Alternative method for the production date determination of impure uranium ore concentrate samples', *Journal of Radioanalytical and Nuclear Chemistry*, vol. 290, no. 2 (2011).

The prerequisite for a correct result is, obviously, that the chemical separation of the daughter nuclide had been completed when the uranium material was processed. However, this is not always the case, especially for intermediate products of uranium (e.g. UOC), where residual thorium is often found. To avoid this dilemma, age can be determined by using the $^{228}Th : ^{232}Th$ ratio, which is also applicable for impure uranium samples (see figure 5.18).[30]

Other important stable isotope ratios

In addition to the isotopic composition of the major fissile elements, the isotopic composition of other elements present can be used to assess origin.

measurement and spectral deconvolution', *Journal of Analytical Atomic Spectrometry*, vol. 25, no. 12 (2010).

[30] Varga, Z. et al., 'Alternative method for the production date determination of impure uranium ore concentrate samples', *Journal of Radioanalytical and Nuclear Chemistry*, vol. 290, no. 2 (2011).

Table 5.5. Typical examples of isotopes of impurities investigated in nuclear forensic analysis

	Isotopes investigated	Typical sources of variation
Stable isotopes	H-2 : H-1, C-13 : C-12, O-18 : O-16, N-15 : N-14, S-34 : S-32	Isotope fractionation in the feed and chemicals used
Decay products	Pb-206 : Pb-208, Sr-87 : Sr-86, Nd-143 : Nd-144, Th-230 : Th-232	Decay of a long-lived radionuclide
Cosmogenic isotopes	C-14 : C-12	Presence of organics with different C-14 : C-12 ratio
Activation products	U-236 : U-238, Pu-240 : Pu-239	Global fallout, blend/ contamination, natural reactors

Sources: Tamborini, G. et al., 'Oxygen isotopic measurements by secondary ion mass spectrometry in uranium oxide microparticles: a nuclear forensic diagnostic', *Analytical Chemistry*, vol. 74, no. 23 (2002); and Miller, D. L. et al., 'Analysis of concentrated uranium ores using stable isotopes and elemental concentrations', *Eos, Transactions, American Geophysical Union*, vol. 87, no. 52 (2006), Fall Meeting Supplement, Abstract V21A-0555.

Several elements in nature show isotopic variation because of, for instance, natural isotopic fractionation (e.g. by light elements, such as hydrogen, carbon or oxygen, where this effect is more pronounced) or due to the presence of a long-lived primordial isotope, such as naturally occurring ^{87}Rb ($T_{1/2}$ = 4.88 × 10^{10} years) or ^{147}Sm ($T_{1/2}$ = 1.06 × 10^{11} years). A number of isotopes can be used for nuclear forensic studies (see table 5.5 for a few examples). Depending on the source of the element (or isotopes) of interest, the isotopic signature of the impurities in a nuclear sample can be indicative of either the process and the production location or the feed material. In the former case, for example, the isotopic composition of hydrogen and oxygen reflects that of the process water used, and thus largely the location of the facility. In the latter case, since strontium derives mainly from the feed material, its isotopic composition will give information on the starting material.

By measuring these isotopes it is possible to effectively verify the assumed origin of a material (i.e. measurement serves as a verification tool). Moreover, more information can be deduced by supplementing the measured data with geological and industrial information (e.g. the lead isotopic composition can permit an assessment of the uranium's deposit type and age).[31]

Radionuclides, both short and long-lived nuclides (e.g. ^{236}U, plutonium or ^{241}Am), that are present at minor or trace level play an important role in nuclear forensics. These nuclides are introduced into the starting material if previously irradiated uranium has been recycled and is used as the base

[31] Varga, Z. et al., 'Application of lead and strontium isotope ratio measurements for the origin assessment of uranium ore concentrates', *Analytical Chemistry*, vol. 81, no. 20 (2009).

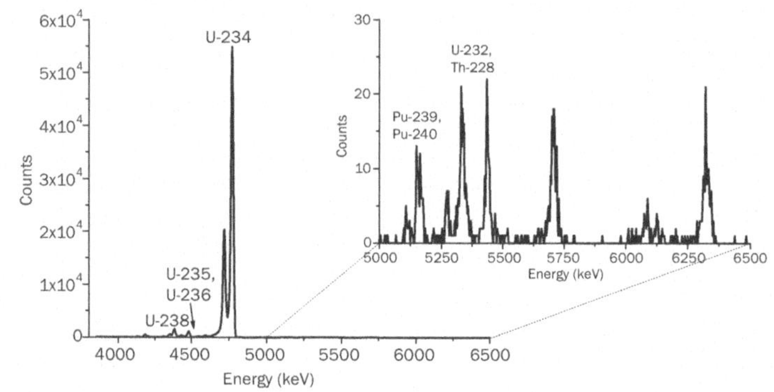

Figure 5.19. Alpha spectrum of a confiscated nuclear material sample with the presence of trace-level actinides

Source: Unpublished data from the European Commission, Joint Research Centre (JRC), Institute for Transuranium Elements (ITU), Karlsruhe.

material or if irradiated and recycled material is mixed with unirradiated material for the purpose of fuel production. Contamination during the fabrication process may be another source of radionuclides at trace level. Fission and activation products in the nuclear material may also be present if the sample has been irradiated after production (e.g. spent fuel samples). As the isotopic composition of activation and fission products varies significantly depending on burn-up and reactor design, this can be used to determine origin.[32]

The alpha spectrum of a confiscated sample provides a good example (see figure 5.19). In addition to the other uranium isotopes with higher activity (^{234}U, ^{235}U, ^{236}U and ^{238}U), traces of ^{232}U and Pu were detectable in the material, which implies that the sample contained a component that had previously been irradiated. The respective amounts and isotopic compositions are indicative of the formation condition (i.e. irradiation and reactor type, and neutron flux). It is noteworthy that the trace-level ^{232}U present in the spectrum was also found to be in secular equilibrium with its ^{228}Th daughter nuclide ($T_{½}$ = 1.91 years), showing that the investigated material was produced long before the measurement (i.e. its age was more than 10 years).

Lead

Three of lead's four stable isotopes are final products of the uranium and thorium decay series: $^{238}\text{U} \rightarrow {}^{206}\text{Pb}$, $^{235}\text{U} \rightarrow {}^{207}\text{Pb}$ and $^{232}\text{Th} \rightarrow {}^{208}\text{Pb}$.

[32] Varga and Surányi (note 28); and Wallenius, M. et al., 'Nuclear forensic investigations with a focus on plutonium', *Journal of Alloys and Compounds*, 2007, vols 444–45 (Oct. 2007).

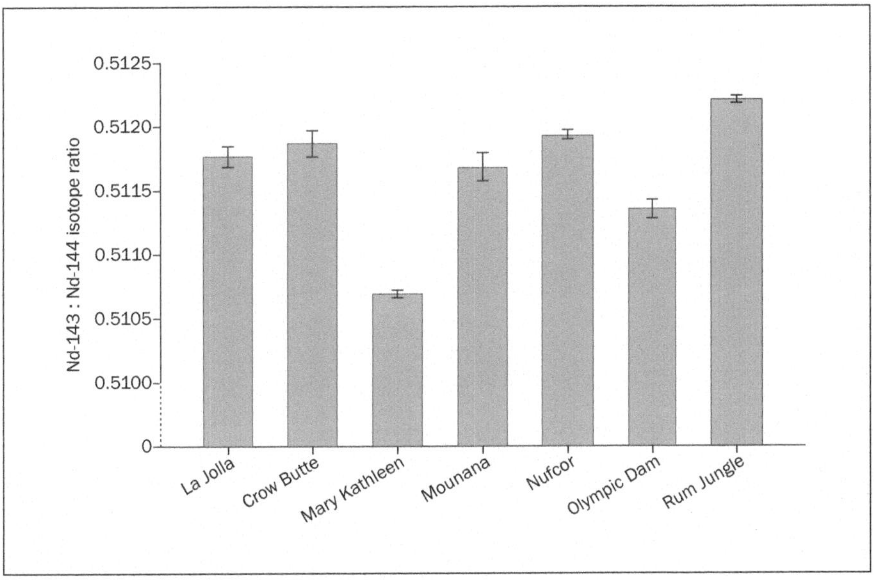

Figure 5.20. The neodymium-143 : neodymium-144 ratio in uranium ore concentrate samples

Source: Unpublished data from the European Commission, Joint Research Centre (JRC), Institute for Transuranium Elements (ITU), Karlsruhe.

The lead isotopes are found in varying ratios, depending on the uranium : thorium ratio of the uranium deposit and its age. The differences can be orders of magnitude, and so they are relatively easy to measure (e.g. by ICP-MS).[33] However, as lead is found commonly everywhere, the possibility of cross-contamination is rather high, which might diminish the value of this signature, especially for materials that have gone through several process steps. Lead isotopes are commonly used for dating in geology, and this common Pb–Pb dating method can also be used in nuclear forensics, for example for UOC samples, to estimate the age of the deposit from which the uranium originates.[34]

Strontium

Strontium has four stable isotopes: ^{84}Sr, ^{86}Sr, ^{87}Sr and ^{88}Sr. The ^{88}Sr : ^{86}Sr ratio is constant in nature, but the ^{87}Sr : ^{86}Sr ratio shows small variations. This is caused by ^{87}Rb, which decays with a long half-life (4.8×10^{10} years)

[33] Varga et al. (note 31); and Švedkauskaite-LeGore, J. et al., 'Investigation of the isotopic composition of lead and of trace elements concentrations in natural uranium materials as a signature in nuclear forensics', *Radiochimica Acta*, vol. 95 (2007).

[34] Varga et al. (note 31).

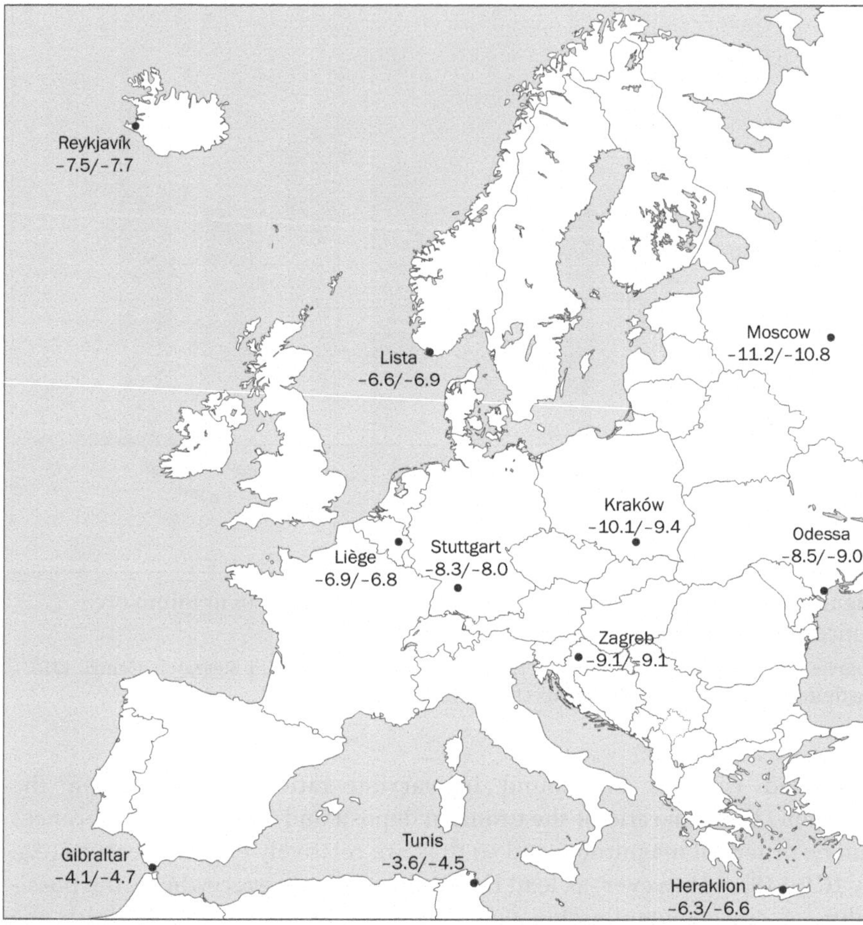

Figure 5.21. Observed rainwater delta-oxygen-18 values in Europe

The $\delta^{18}O$ value for each location expresses the recorded $^{18}O:^{16}O$ isotopic ratio normalized to a standard. In each case, the first figure is the long-term arithmetic mean and the second figure is the weighted mean.

Source: Pajo, L., *UO2 Fuel Pellet Impurities, Pellet Surface Roughness and* $n(^{18}O)/n(^{16}O)$ *Ratios, Applied to Nuclear Forensic Science*, Doctoral dissertation (University of Helsinki: Helsinki, 2001), pp. 14–15.

to ^{87}Sr. Depending on the Rb : Sr ratio in the uranium rock and the age of the deposit, a variety of $^{87}Sr:^{86}Sr$ ratios have been observed.[35]

Neodymium

Neodymium has five stable isotopes: ^{142}Nd, ^{143}Nd, ^{145}Nd, ^{146}Nd and ^{148}Nd. The isotope ratio $^{146}Nd:^{144}Nd$ is considered to be constant in nature,

[35] Varga et al. (note 31). See also figure 3.4 in chapter 3 in this volume.

similar to the $^{88}Sr:^{86}Sr$ ratio, and it can be used for normalization of the $^{143}Nd:^{144}Nd$ ratio determination, which shows variations due to the decay of ^{147}Sm ($T_{1/2} = 1.06 \times 10^{11}$ years) to ^{143}Nd (see figure 5.20). Neodymium, like all other rare-earth elements, behaves similarly to uranium in chemical processes; thus, it is an excellent signature for origin determination (i.e. the type of uranium deposit) of natural uranium samples.

Oxygen

The three stable isotopes of oxygen, ^{16}O, ^{17}O and ^{18}O, are found in the abundances 99.757 per cent, 0.038 per cent and 0.205 per cent, respectively.[36] However, due to the natural fractionation of the isotopes, relative variations of up to 5 per cent can be measured in the $^{18}O:^{16}O$ ratio. The natural fractionation is caused by, for example, the temperature and distance to the sea in the region. Thus, smaller $^{18}O:^{16}O$ ratios are found in colder areas than in warmer ones (see figure 5.21). As water is involved in most uranium processing, the $^{18}O:^{16}O$ ratio of water from a region will be reflected in the uranium oxide product. Therefore, this signature can indicate the geographic location of the uranium production facility.[37]

[36] *Karlsruher Nuclide Chart* (note 24).

[37] Pajo, L. et al., 'A novel isotope analysis of oxygen in uranium oxides: comparison of secondary ion mass spectrometry, glow discharge mass spectrometry and thermal ionization mass spectrometry', *Spectrochimica Acta, Part B: Atomic Spectroscopy*, vol. 56, no. 5 (May 2001).

6. Radionuclide signatures for post-nuclear explosion environments

LARS-ERIK DE GEER

As explained in earlier chapters, some characteristics of nuclear or other radioactive materials change with every use or at every stage of the nuclear fuel cycle. Analysis of informative combinations of material characteristics—nuclear forensic signatures—allows understanding the history of the materials, as well as their purpose. For material within the nuclear fuel cycle, relevant signatures are discussed in chapter 5.

Nuclear explosions disintegrate the materials in the nuclear explosive device and so erase most forensic signatures based on physical or chemical properties. However, the atoms of these materials would either survive the explosion intact or transform via a limited and predictable number of nuclear reactions. The resulting radioactive substances can be collected from the debris at the site of the explosion, from the fallout (as particles) or from the atmosphere (as aerosols and gases). They can then be subjected to the standard nuclear forensic approach: sample collection, material characterization, and nuclear forensic interpretation leading to a determination of the material's history (see chapter 2). Nuclear forensic analysis, in this case, could result in verification of the nuclear nature of the explosion, as well as, perhaps, understanding of some features of the nuclear explosive device.

Such 'post-explosion nuclear forensics' is used under various names in at least five contexts. First, individual states can assess their own nuclear weapon performance this way. For example, radiochemical analysis of weapon debris (combined with weapon design information) provides the most accurate method of measuring a weapon's actual yield after a nuclear test.[1] Second, states have used this approach for the long-range detection and analysis of foreign nuclear weapon tests since 1949, and continue to do so.[2] Third, post-explosion nuclear forensics is currently being developed as an investigative tool to be used in case of a nuclear or dirty bomb explosion by terrorists. Fourth, the same approach has been used for many years to verify compliance with international treaties limiting nuclear weapon tests,

[1] US National Research Council, Committee on the Evaluation of Quantification of Margins and Uncertainties Methodology for Assessing and Certifying the Reliability of the Nuclear Stockpile, *Review of the DOE National Security Labs' Use of Archival Nuclear Test Data: Letter Report* (National Academies Press: Washington, DC, 2009), pp. 11–12.

[2] See chapters 7 and 8 in this volume.

including the 1996 Comprehensive Nuclear-Test-Ban Treaty (CTBT).[3] Finally, understanding the radioactive contents of fallout from nuclear weapon tests is important for radiation protection of the public.

The literature discussing signatures pertinent to post-explosion nuclear forensics systematically or in any detail is extremely scarce. This is due to the technical complexity of the issue, as well as to the fact that the first three of the above-mentioned contexts are connected with nuclear weapon development or nuclear intelligence.

This chapter is based on the work of the author first at the Swedish Defence Research Establishment (Försvarets forskningsanstalt, FOA), and later at the Preparatory Commission for the Comprehensive Nuclear-Test-Ban Treaty Organization (CTBTO). The Swedish nuclear weapon programme, which used to be led by FOA on a rather advanced level, was closed down in 1968 when the country joined the 1968 Non-Proliferation Treaty (NPT).[4] Sweden continued its programme of long-range detection of foreign nuclear tests, but it was oriented mainly towards radiation protection, a much less restricted context.[5] FOA's efforts focused on collection of airborne particles and noble gases, as well as their characterization by means of gamma spectrometry. Sweden then played a leading role in designing the CTBTO's International Monitoring System (IMS), including its radionuclide component.[6] Radionuclide monitoring technologies used by the IMS are primarily based on detection of gamma rays. For this reason this chapter focuses on radioactive isotopes detectable by gamma spectrometry.

Currently, 2391 different radioactive ground and metastable states—that is, 2391 radionuclides—with half-lives ranging from 1 second to 100 million years are known.[7] The clearly spelled-out methodology of choosing 'relevant radionuclides' reduces the number of nuclides to be addressed, helps to make their selection unbiased and easily explainable, and ensures that a nuclide having or able to have an impact on a certain problem is not forgotten or overlooked. The following sections describe how relevant radionuclides were chosen in the contexts of calculating aggregated radiation doses to humans (section I); calculating doses to humans from under-

[3] Comprehensive Nuclear-Test-Ban Treaty (CTBT), opened for signature 24 Sep. 1996, not in force, <http://treaties.un.org/Pages/CTCTreaties.aspx?id=26>.

[4] Jonter, T., 'The Swedish plans to acquire nuclear weapons, 1945–1968: an analysis of the technical preparations', *Science and Global Security*, vol. 18, no. 2 (2010). Treaty on the Non-Proliferation of Nuclear Weapons (Non-Proliferation Treaty, NPT), opened for signature 1 July 1968, entered into force 5 Mar. 1970, IAEA Information Circular INFCIRC/140, 22 Apr. 1970, <http://www.iaea.org/Publications/Documents/Treaties/npt.html>.

[5] See chapter 8 in this volume.

[6] Dahlman, O., Mykkeltveit, S. and Haak, H., *Nuclear Test Ban: Converting Political Visions to Reality* (Springer: Dordrecht, 2009); and Dahlman, O. et al., *Detect and Deter: Can Countries Verify the Nuclear Test Ban?* (Springer: Dordrecht, 2011).

[7] US National Nuclear Data Center (NNDC), Brookhaven National Laboratory, 'Evaluated nuclear structure file, ENSDF file', 14 July 2014, <http://www.nndc.bnl.gov/nudat2>.

Table 6.1. Radionuclides relevant for estimating worldwide average effective dose commitments from past nuclear weapon testing

Radionuclide	Half-life	Worldwide average effective dose commitment (μSv)	Worldwide average truncated effective dose commitment, 1945–99 (μSv)
C-14	5 700 years	2 500	140
Cs-137	30.07 years	460	320
Sr-90	28.90 years	120	110
Zr-95	64.02 days	84	84
I-131	8.02070 days	68	68
Ce-144	284.893 days	60	60
Ru-106	373.59 days	60	60
Ba-140	12.752 days	28	28
H-3	12.32 years	24	24
Mn-54	312.12 days	19	19
Pu-239	24 110 years	13	13
Ru-103	39.26 days	13	13
Sb-125	2.7582 years	12	12
Am-241	432.6 years	11	11
Pu-240	6 561 years	8.9	8.9
Fe-55	2.737 years	6.6	6.6
Sr-89	50.53 days	4.5	4.5
Y-91	58.51 days	4.1	4.1
Pu-241	14.290 years	4.0	4.0
Ce-141	32.501 days	1.9	1.9
Pu-238	87.7 years	1.0	1.0
Total		**3 500**	**990**

Source: United Nations Scientific Committee on the Effects of Atomic Radiation (UNSCEAR), *Sources and Effects of Ionizing Radiation*, vol. 1, *Sources*, 2000 Report to the General Assembly, with scientific annexes, E.00.IX.3 (United Nations: New York, 2000).

ground nuclear testing (section II); and verifying the CTBT by analysis of fallout particles (section III) and noble gas (section IV), or during on-site inspections (section V). Together, these contexts provide a reasonably complete set of nuclides relevant for post-explosion environments.

I. Radiation doses to humans

The concept of relevant radionuclides began to be used in the early 1980s when the United Nations Scientific Committee on the Effects of Atomic Radiation (UNSCEAR) sought to re-evaluate the health risk to humans (dose commitment) of nuclear weapon testing, especially tests in the atmosphere.[8] Previously, doses from fallout had been estimated for major

[8] A person will receive a radiation dose from external irradiation only while in a contaminated environment. In contrast, if radioactive material is absorbed into the body, the person is 'committed' to receiving the dose, because some of that material will be distributed in tissues and will keep

contributors, such as ^{14}C, ^{137}Cs and ^{90}Sr, and for certain nuclides, such as ^{136}Cs and ^{85}Kr, which were included even though the doses caused by them were very small. At the same time, several other nuclides that give small doses, although higher than those of ^{136}Cs and ^{85}Kr, were not addressed.

In a 2000 report, UNSCEAR estimates the worldwide average effective dose commitment of relevant radionuclides from nuclear testing, where 'relevant' means a dose of 1 microsievert (μSv) or more (see table 6.1).[9]

As the quite long-lived nuclide ^{14}C totally dominates the dose commitment, a time limit is often set at the end of the 20th century—yielding what is often just referred to as the 'truncated effective dose commitment'. The total truncated effective dose commitment from past nuclear testing was about 1 mSv, which is about 40 per cent of the average dose from natural sources in a single year, or less than 1 per cent of the natural dose for the 55-year period 1945–99.

II. Nuclides from an abandoned underground nuclear test site

The next time that the need to define relevant radionuclides arose was in connection with a study initiated in 1996 by the French Government and administered by the International Atomic Energy Agency (IAEA), to analyse the radiological situation at Mururoa and Fangataufa, two atolls in the Pacific Ocean where France had conducted 147 underground nuclear tests between 1975 and 1996.[10] A working group was tasked with estimating the amount of radionuclides produced by the 147 underground nuclear explosions that could have a long-range effect on the dose received by humans.

The half-life of a relevant nuclide (i.e. one that would have the potential to deliver a dose from nuclides buried underground) was determined to be between 1 year and 10 billion years. This quite generous span included only 124 nuclides. The lower limit is low enough to include all geological and hydrological processes needed to transfer the nuclides to humans, and the upper limit is of the order of time remaining in the existence of the earth and its solar system. Among the 124 nuclides, six different classes were defined that potentially could have been created in an underground nuclear explosion: (*a*) residues of fuel materials, (*b*) non-fission reaction products of fuel materials, (*c*) fission products, (*d*) activation products of non-fuel

irradiating them, possibly for many years. The total effective dose delivered in such a case over a lifetime (70 years for infants, 50 years for adults) is called the committed effective dose. It characterizes total irradiation received by the whole body over the lifetime. In contrast, the dose commitment integrates the dose rate to a specified population for all time (or up to a defined limit when it is referred to as a truncated dose commitment).

[9] United Nations Scientific Committee on the Effects of Atomic Radiation (UNSCEAR), *Sources and Effects of Ionizing Radiation*, vol. 1, *Sources*, 2000 Report to the General Assembly, with scientific annexes, E.00.IX.3 (United Nations: New York, 2000).

[10] International Advisory Committee, *The Radiological Situation at the Atolls of Mururoa and Fangataufa: Main Report* (International Atomic Energy Agency: Vienna, 1998).

Table 6.2. The 36 radionuclides relevant for estimating underground inventories at the Mururoa and Fangataufa nuclear test sites

Figures are terabecquerels (TBq) as of 1 May 1996 (5 months after the last French nuclear test, at Fangataufa lagoon on 27 Jan. 1996).

			Location (TBq)				
Nuclide	Half-life (years)	Class[a]	Lava	Rubble	Gas	Water	Total
H-3	12.33	*a, b*	–	–	6 000	274 000	280 000
Cs-137	30.07	*c*	4 300	10 500	–	–	14 800
Pm-147	2.6234	*c*	10 500	500	–	–	11 000
Sr-90	28.78	*c*	4 300	6 500	–	–	10 800
Fe-55	2.73	*d, e*	7 200	400	–	–	7 600
Ru-106	1.0229	*c*	5 100	2 200	–	–	7 300
Pu-241	14.35	*a, b*	6 700	100	–	–	6 800
Co-60	5.2708	*d*	2 300	300	–	–	2 600
Pu-239	24 110	*a, b*	1 080	20	–	–	1 100
Kr-85	10.772	*c*	–	100	800	100	1 000
Sb-125	2.7582	*c*	510	220	–	–	730
Sm-151	90	*c*	480	20	–	–	500
Eu-155	4.7611	*c*	450	20	–	–	470
Ni-63	100.1	*d*	430	20	–	–	450
Am-241	432.2	*a, b*	370	10	–	–	380
Eu-152	13.537	*e*	310	20	–	–	330
Pu-240	6 564	*a, b*	295	5	–	–	300
Pu-238	87.7	*b*	195	5	–	–	200
Eu-154	8.593	*c, e*	47	3	–	–	50
C-14	5 730	*d*	–	3	22	3	28
Ni-59	76 000	*d*	3.6	0.2	–	–	3.8
Cd-113m	14.1	*c*	2.3	1.0	–	–	3.3
Tc-99	211 100	*c*	2.0	0.5	–	–	3.0
Cl-36	301 000	*d, e*	0.9	0.7	–	0.1	1.7
Ca-41	103 000	*e*	0.9	0.4	–	–	1.3
Cs-134	2.0648	*c*	0.19	0.75	–	–	0.94
Sn-121m	55	*c*	0.22	0.14	–	–	0.36
Zr-93	1 530 000	*c*	0.30	0.02	–	–	0.32
Cs-135	2 300 000	*c*	0.06	0.21	–	–	0.27
Np-237	21 444 000	*a, b*	0.23	0.02	–	–	0.25
Pd-107	6 500 000	*c*	0.15	0.06	–	–	0.21
Sn-126	~100 000	*c*	0.13	0.05	–	–	0.18
U-236	23 420 000	*a, b*	0.12	0.02	–	–	0.14
Se-79	≤650 000	*c*	0.008	0.003	–	–	0.011
Pu-242	373 300	*b*	0.0090	0.0002	–	–	0.0092
I-129	15 700 000	*c*	0.0031	0.0024	–	0.0006	0.0061

[a] The 6 classes are (*a*) residues of fuel materials, (*b*) non-fission reaction products of fuel materials, (*c*) fission products, (*d*) activation products of non-fuel bomb materials, and (*e*) activation products in stemming (filling) materials and rocks surrounding the explosion. Fluence detectors were used in atmospheric tests but, as there are no openly reported observations of such nuclides in underground testing, no relevant nuclide is listed in this category.

Source: International Advisory Committee, *The Radiological Situation at the Atolls of Mururoa and Fangataufa: Main Report* (International Atomic Energy Agency: Vienna, 1998), table 23.

bomb materials, (*e*) activation products in stemming (filling) materials and rocks surrounding the explosion, and (*f*) activation products deriving from neutron fluence detectors.[11]

Analysing which nuclides could be produced and deliver dose to humans, the 124 nuclides were reduced to 36 considered to be relevant for an abandoned underground nuclear test site (see table 6.2).

III. Particulate radionuclides relevant for verification of the CTBT

'Particulate' radionuclides are attached to, or part of, small solid particles that can easily be collected by blowing large volumes of air through a filter.[12] In the framework of verification of the CTBT, the CTBTO is developing a worldwide network of 321 monitoring stations—the IMS—to detect signals from nuclear explosions, whether clandestine or not. At 80 of the stations, particulate radionuclides are or will be sampled and analysed to determine whether the radionuclides typically created by nuclear explosions are present in the air around the station. In late October 2014, 63 of these 80 stations were certified and in use in order to gain experience and preparedness for entry into force of the CTBT. In addition, 3 stations had been installed, 10 were under construction and 4 were still at the planning stage (with 1 of these in South Asia waiting for treaty signature).

More than 500 cubic metres of air pass the filters at these stations each hour and after one day's sampling and another day of waiting for some extraneous radon daughters to decay, the filter is (manually or automatically) exposed to a gamma ray-sensitive detector inside a lead shield at the station. Counting takes 24 hours, after which the resulting spectrum is immediately sent via satellite link to the CTBTO's International Data Centre (IDC) in Vienna, Austria, where it is, again immediately and automatically, analysed and the raw data and results are sent to the CTBT states signatories. A day or so later a second analysis of the information is sent to them after the spectrum has been interactively checked and reviewed by an analyst at the IDC. When the IMS becomes fully operational, national data centres will receive 80 particulate spectra and reviewed analyses per day.

To help these data centres navigate in the stream of analyses a five-level categorization scheme has been designed that indicates the treaty relevance of each measurement. It ranges from the typical background spec-

[11] For a definition of 'neutron fluence' see the glossary in this volume. Fluence detectors were used in atmospheric tests but, as there are no openly reported observations of such nuclides in underground testing, no relevant nuclide is listed in this category.

[12] Matthews, M. and Schulze, J., 'The radionuclide monitoring system of the comprehensive Nuclear-Test-Ban Treaty Organisation: from sample to product', *Kerntechnik*, vol. 66, no. 3 (May 2001), pp. 102–20; and Matthews, K. M. and De Geer, L.-E., 'Processing of data from a global atmospheric monitoring network for CTBT verification purposes', *Journal of Radioanalytical and Nuclear Chemistry*, vol. 263, no. 1 (Jan. 2005), pp. 235–40.

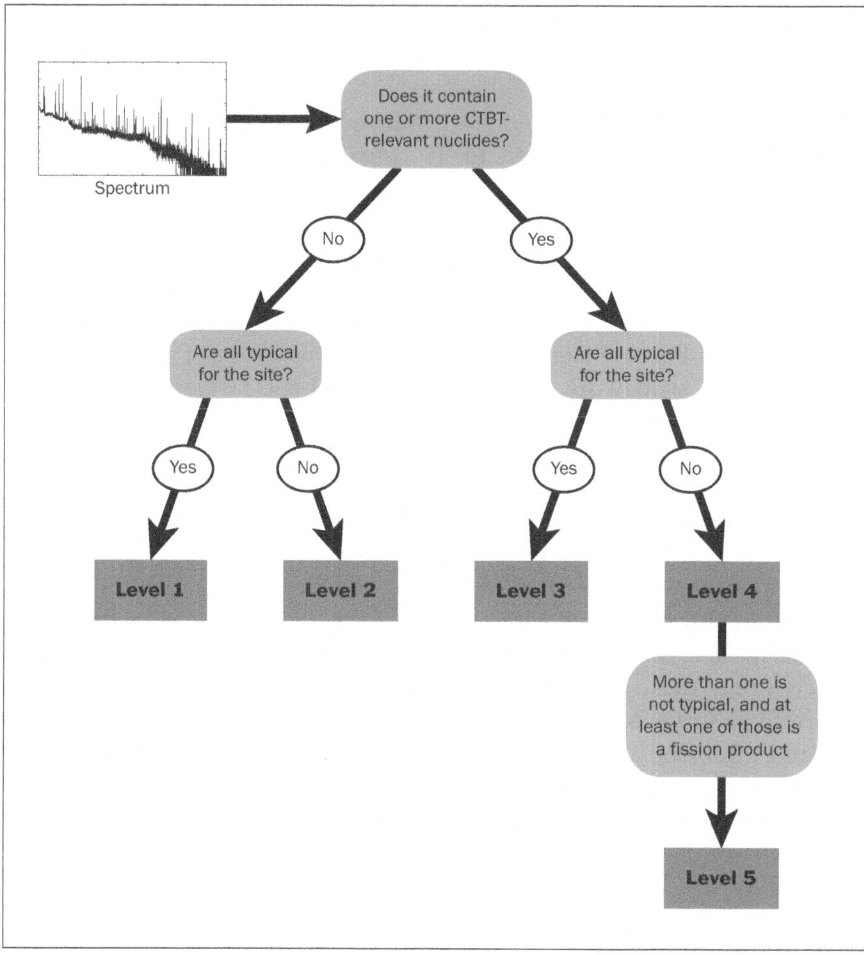

Figure 6.1. The CTBTO's five-level categorization scheme

trum (Level 1), to a spectrum with multiple atypical relevant nuclides with at least one relevant fission product (Level 5; see figure 6.1).

To define CTBT-relevant radionuclides, a number of quantitative criteria were set so that only nuclides within the realm of the IMS particulate network are considered.[13] In addition, the CTBTO conducted a fairly extensive and conservative review to reduce the risk of disregarding an important nuclide and to provide material of interest for the more specialized

[13] De Geer, L.-E., *CTBT Relevant Radionuclides*, Technical Report PTS/IDC-1999/02 (Preparatory Commission for the Comprehensive Nuclear-Test-Ban Treaty Organization: Vienna, Apr. 1999); and De Geer, L.-E., 'Comprehensive Nuclear-Test-Ban Treaty: relevant radionuclides', *Kerntechnik*, vol. 66, no. 3 (May 2001), pp. 113–20. The first of these provides more details about the selection process.

analyses that might be requested from the radionuclide laboratories. There are 16 well-equipped laboratories worldwide that can scrutinize Level 5 samples or other samples that might have generated special interest. The Mururoa study's selection technique was used but with a couple of new categories and a different selection of parameters.

It must be stressed that it is not possible to state that a certain nuclide absolutely cannot be produced in a nuclear weapon explosion. For any existing nuclide, a more or less odd nuclear test arrangement can be imagined (using rare materials or even adding special experimental devices to the charge itself) that would produce it. However, if such a nuclide were to be detected from a bomb test, the probability is high that a set of other more typical nuclides would be seen simultaneously. Thus, sharp division lines are not possible but, taking a conservative approach, a list of CTBT-relevant radionuclides to be used in the practical categorization and screening process can certainly be drawn up.

In a nuclear weapon test explosion, radioactive debris can be formed in a number of ways, including the most obvious one, fission, and more esoteric reactions (e.g. those used for diagnostics). In this section, the following 10 categories are distinguished: residues of fuel materials (Category 1), non-fission reaction products in fuel materials (Category 2), fission products (Category 3), activation products of non-fuel bomb materials (Category 4), activation products in stemming materials and rocks surrounding an underground explosion (Category 5), activation products in the ground below a near-surface atmospheric explosions (Category 6), activation products in seawater around an underwater or a near-sea surface explosion (Category 7), activation products in air around an atmospheric explosion (Category 8), activation products deriving from neutron fluence detectors (Category 9), and added tracers (Category 10). These 10 criteria were thus modelled on the 6 used for underground explosions in the Mururoa study, but with the additions of three categories (categories 6–8) in order to also cover underwater and atmospheric tests categories and one (Category 10) that could also have been a principal category for the French underground tests.

For the study of CTBT relevance, the lower half-life criterion had to be set at a limit that barely allowed for transport and counting by the IMS. A lower half-life limit of six hours was considered reasonable as it corresponded to a significant number of half-lives during transport, sampling, radon progeny decay and counting. It was also consistent with the long experience of various national CTBT-like systems. The upper half-life limit could actually be reduced to, for example, 1000 years as longer half-lives imply very low specific activities and signals generally below the capability of the IMS detection system. Again, past experience was well in concert with such a limit, but to ensure that no potentially interesting

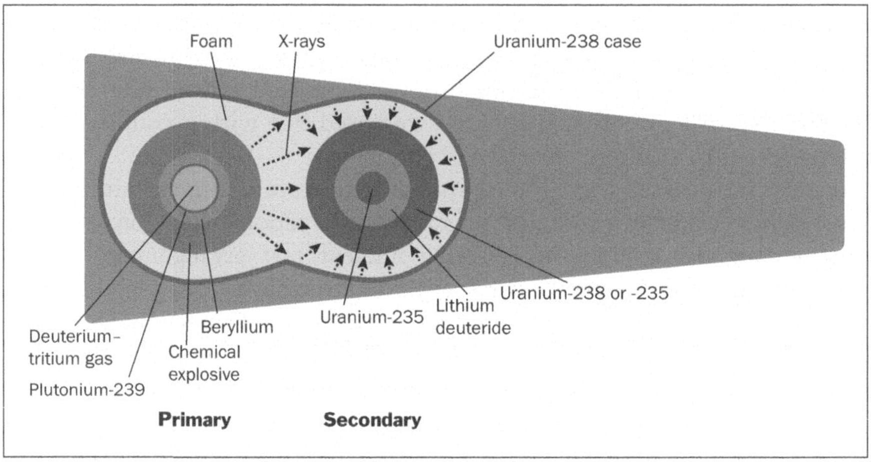

Figure 6.2. A schematic picture of a thermonuclear weapon

Notes: To the left is the primary ('normal') fission device, with concentric layers consisting of (from the inside out) a mixture of heavy hydrogen isotopes for boosting, fissile fuel for fission, a neutron-reflecting tamper material and high explosives. To the right is the secondary device, with concentric layers consisting of (again from the inside out) a fissile material ('spark plug'), fusion fuel and a pusher. The pusher absorbs the high flux of X-rays from the primary and squeezes the secondary to the extreme density needed for fusion.

Source: Adapted from International Panel on Fissile Materials.

radionuclide is overlooked and that basic data is recorded for nuclides that can at least be questioned by CTBT states signatories and later might be detected in laboratory analyses, that reduction is left to the final selection stage. For the initial scan of nuclides, an upper half-life limit of 1 billion years was adopted.

A traditional problem in studies where nuclides are selected based on fixed criteria is how to treat short-lived daughter nuclides. The IMS strictly abides to the six-hour half-life cut-off. That means that a few well-known nuclides, such as ^{137m}Ba ($T_{1/2}$ = 2.552 minutes) and ^{132}I ($T_{1/2}$ = 2.295 hours), are only implicitly treated if the best analysis gamma happens to be emitted in the decay of these daughters.

A schematic picture of a thermonuclear bomb (see figure 6.2) is useful for the following discussion of the 10 categories.

Category 1. Residues of fuel materials

The fuel is composed of isotopes of uranium, plutonium, lithium and hydrogen, together with traces of their decay products. If normal weapon characteristics are assumed, the following compounds and isotopes should be considered: tritium (^{3}H), lithium deuteride (^{6}LiD, ^{7}LiD), weapon-grade uranium (^{234}U, ^{235}U and ^{238}U), natural or depleted uranium (^{235}U and ^{238}U), weapon-grade plutonium (^{239}Pu, ^{240}Pu, ^{241}Pu, ^{241}Am), and other uranium isotopes (^{232}U, ^{233}U).

Tritium is used for boosting a fission device. A mixture of tritium and deuterium gas in the centre of an imploding fission assembly gives a small thermonuclear flame, and the resulting neutrons substantially increase the burn-up of the fissile fuel. Some tritium may also be used in the main thermonuclear stage of fusion devices.

Lithium deuteride is the main fuel of the thermonuclear stage. It is the source of tritium (through reactions with neutrons) and deuterium during the explosion, but itself contains only stable nuclides.

Weapon-grade uranium contains more than 90 per cent ^{235}U (by weight), which is enriched from natural uranium containing about 0.711 per cent ^{235}U (by weight). Normally the enrichment process also increases the content of ^{234}U, but due to the very low natural abundance of ^{234}U (0.0055 per cent by weight), less than 1 per cent of weapon-grade uranium consists of this nuclide.[14]

Depleted uranium, which like natural uranium may form the tamper in a fission weapon or the pusher of a secondary device in a thermonuclear weapon is readily obtained from the tailings of the enrichment process for production of weapon-grade uranium, and therefore contains mainly ^{238}U. Normally, between one-third and one-half of the ^{235}U in the natural uranium feed (0.2–0.3 per cent by weight) is left in the depleted uranium tailings.[15] The function of a tamper in a fission device is to act as a heavy 'hammer' in the compression process and also to help reflect back neutrons that are on their way out of the reaction zone. The pusher, the heavy material wrapped around the fusion fuel, is there to absorb the X-ray energy from the fission primary or 'trigger' and by ablation to squeeze the fusion fuel to the necessary extreme densities.

Plutonium is considered to be of weapon grade if it, in addition to its main ^{239}Pu, contains less than 7 per cent ^{240}Pu. Some ^{241}Pu will also be present. A typical isotope mix in weapon-grade plutonium is 93.5 per cent ^{239}Pu, 6.5 per cent ^{240}Pu and 0.5 per cent ^{241}Pu.[16] A low content of ^{240}Pu is desirable in a reliable weapon to minimize stray neutrons from spontaneous fission, which might prematurely start the fission chains during the compression phase and produce a fizzle. To achieve this, production reactors are run on a low burn-up cycle, which also results in low production of ^{238}Pu, ^{242}Pu, ^{243}Pu and ^{244}Pu. These isotopes and their progeny are therefore disregarded here. ^{240}Pu, however, decays to ^{236}U, which must

[14] Albright, D., Berkhout, F. and Walker, W., SIPRI, *Plutonium and Highly Enriched Uranium 1996: World Inventories, Capabilities and Policies* (Oxford University Press: Oxford, 1997).

[15] Albright, Berkhout and Walker (note 14).

[16] Albright, Berkhout and Walker (note 14).

be considered further. ^{241}Pu, with its rather short half-life of 14.35 years, decays to ^{241}Am; therefore, depending on the age of the fuel, it will contain some varying amount of this isotope. In the long term, this decay chain will also produce the very long-lived nuclides ^{237}Np, ^{233}U and ^{229}Th, and their chain of short-lived daughters down to stable ^{209}Bi.

Uranium-233 is a fissile isotope, which can be made in reactors by irradiating thorium. It was early considered as a fuel material for nuclear weapons and there are authoritative publications stating that ^{233}U weapons have been tested.[17] However, irradiating thorium (^{232}Th is the isotope of thorium available in nature) will also inevitably produce the fairly short-lived isotope ^{232}U ($T_{1/2}$ = 70 years), which through alpha decay feeds into the natural thorium decay chain. This makes ^{233}U an impractical weapon fuel as it will emit hard and penetrating gamma rays, like the 2.6 mega-electronvolt (MeV) gamma from the decay of ^{208}Tl. However, since the possibility that a state will attempt to test a ^{233}U weapon cannot be excluded, special attention should be paid to sudden increases in the $^{224}Ra : ^{228}Ac$ ratio. Actinium-228 (with principal gamma ray 911.2 keV with an intensity of 25.8 per cent) is a measure of the natural decay chain, while radium-224 (with principal gamma ray 241.0 keV with an intensity of 4.1 per cent) would also incorporate direct feed from the decay of ^{232}U. In normal IMS spectra the background radium-224 is not detected because of a combination of the facts that it is not an airborne radon progeny nuclide and has a fairly low intensity gamma ray.

The neutron injector used to start the fission process may involve other radionuclides. Polonium-210 is known to have been used, together with beryllium, for this purpose. However, it emits extremely few gamma rays and cannot be expected to be picked up by the IMS. More modern injectors involve tritium in a miniature accelerator but in very small amounts compared to those in other parts of the device.

In addition to the increase in the thorium natural decay chain from ^{228}Th and down, which might result from a test of a ^{233}U weapon, only one of the radionuclides mentioned above has a reasonably short half-life (1000 years or less) and emits gamma rays with an intensity above 1 per cent: ^{241}Am, which emits 59.5 keV gamma rays with an intensity of 35.9 per cent. To fully follow the goal identified at the outset—to list nuclides irrespective of any upper half-life limit—would require analysing the detailed decay chains of ^{237}Np ($T_{1/2}$ = 2 144 000 years), ^{233}U ($T_{1/2}$ = 159 200 years) and ^{229}Th ($T_{1/2}$ = 7340 years) just to gather information on radionuclides which, no doubt, would later be disregarded because of their extremely low specific activity.

[17] Hansen, C., *Swords of Armageddon*, vol. 1 (Chukelea Publications: Sunnyvale, CA, 2007), p. 275.

Recently, there has been debate about whether or not long-lived isotopes of neptunium (^{237}Np, $T_{1/2}$ = 2 144 000 years) and americium (mainly ^{241}Am, with $T_{1/2}$ = 432.2 years, but also including ^{242m}Am, with $T_{1/2}$ = 141 years, and ^{243}Am, with $T_{1/2}$ = 7370 years) should also be considered as nuclear weapon materials. All of these nuclides, however, feed decay chains that have been already considered and only bring three additional isotopes, ^{242m}Am, ^{243}Am and ^{242}Cm into focus. These nuclides, in turn, only emit low-intensity gamma rays of energies lower than 50 keV.

Category 2. Non-fission reaction products of fuel materials

Reactions in thermonuclear fuel produce tritium, all of which would ideally be consumed in a fusion explosion. In actual fusion explosions, however, a substantial proportion is left. About 240 exabecquerels (671 kilograms) of tritium has been estimated to have remained after about 273 megatons (Mt) of fusion during atmospheric weapon tests worldwide (*c*. 2.5 kg of tritium per Mt).[18]

The secondary stage of a thermonuclear explosion produces extremely intense neutron fluences, which, in addition to fission, will give multiple neutron capture in the pusher. As many as 19 consecutive neutron captures have been reported from large thermonuclear tests with natural or depleted uranium components. The resulting high mass uranium isotopes are very short-lived and decay rapidly, by beta decay, to longer-lived isotopes of heavier elements. In the first two-stage fusion explosion, in 1952 (code-named Mike, 10.4 Mt), this resulted in the discovery of the new elements fermium and einsteinium. This does not, however, mean that those resulting isotopes of heavier elements are very abundant. In the Mike explosion, the relative abundances of mass chains 239, 240, 241 and 242 were 1 to 0.363 to 0.039 to 0.019, respectively. For global fallout—the cumulative effect of atmospheric testing—the corresponding relative abundances were 1 to 0.18 to 0.013 to 0.0043, respectively.[19] These ratios were set largely by high-yield thermonuclear devices with very high neutron fluences. Note also that it is reasonable to assume that, especially in underground tests, natural or depleted uranium is not often used as pusher material, as it creates high yields and fission products without always contributing to an understanding of the processes involved. The abundance curve thus decreases rapidly with mass, and the neutron capture chain can

[18] United Nations Scientific Committee on the Effects of Atomic Radiation (UNSCEAR), *Sources and Effects of Ionizing Radiation,* 1993 Report to the General Assembly, with scientific annexes, E.94.IX.2 (United Nations: New York, Oct. 1993); and De Geer, L.-E., 'Slutnotan över kärnladdningsproven' [The final cost of the nuclear explosion tests], FOA-tidningen no. 5, Swedish Defence Research Establishment, Dec. 1996.

[19] De Geer (note 18); and Krey, P. H. et al., 'Mass isotopic composition of global fall-out plutonium in soil', *Transuranium Nuclides in the Environment* (IAEA: Vienna, 1976), p. 671.

be disregarded beyond mass 241 for ^{238}U. That leaves ^{239}U, ^{240}U and ^{241}U, which decay to the corresponding neptunium isotopes and further to ^{239}Pu, ^{240}Pu and ^{241}Pu, respectively, over the course of only days, hours and minutes.

As mentioned above, natural uranium may also be used as material in tampers of fission weapons. Fission weapons do not produce sufficiently high neutron fluences for significant multiple neutron capture to occur. One step capture will, however, occur in the tamper, and ^{239}U will be produced, which subsequently decays via ^{239}Np to ^{239}Pu. Plutonium in the core of a fission device will also capture single neutrons to produce ^{240}Pu (with decay product ^{236}U), ^{241}Pu (with decay products ^{241}Am, ^{237}Np, ^{233}U, ^{229}Th etc.) and ^{242}Pu. Americium-242m, which is a neutron capture product of ^{241}Am in aged plutonium fuels, is disregarded because its cross-section is extremely small (in the order of microbarns, i.e. 10^{-34} m^2).[20] Highly enriched uranium as a neutron target in fission weapons, in primaries or in secondaries, will produce ^{236}U.

Fuel materials very close to the thermonuclear burn region are exposed to very high-energy neutrons (up to 14.1 MeV) and can thus be subject to (n,2n) reactions. This process results in the element being unchanged but with the mass decreased by one unit. In a full thermonuclear weapon many (n,2n) reactions occur in the ^{238}U pusher, if there is one. More ^{237}U atoms are often made this way than there are ^{137}Cs or ^{90}Sr atoms produced by fission. In the debris from a Chinese 4-Mt thermonuclear explosion in 1976, for example, three times more ^{237}U was observed than ^{137}Cs.[21] Uranium-237 is fairly short-lived ($T_{1/2}$ = 6.75 days) and decays to ^{237}Np, with its decay products ^{233}U and ^{229}Th and further progeny. The high-energy tail of the fission neutron spectrum in a fission device, and the high-energy neutrons in a booster, also cause a limited number of (n,2n) reactions in the fissile fuel, producing ^{238}Pu or ^{234}U or both.

Analysis of the fallout from the unboosted all-plutonium Nagasaki bomb showed a $^{238}Pu : ^{239}Pu$ atom ratio of 0.00023—sufficiently low to disregard ^{238}Pu at this stage.[22] Furthermore, most full-size thermonuclear explosions also show a very low $^{238}Pu : ^{239}Pu$ ratio due to all the ^{239}Pu deriving from neutron capture in the pusher (formed by decay long after all nuclear reactions have stopped). As an example, an atom ratio of 0.000045 was

[20] For a definition of 'cross-section' see the glossary in this volume.

[21] De Geer, L.-E. et al., *Particulate Radioactivity, Mainly from Nuclear Explosions, in Air and Precipitation in Sweden Mid-year 1975 to Mid-year 1977*, FOA report C 40089-T2(A1) (Försvarets Forskningsanstalt: Sundbyberg, Nov. 1978), <http://www.iaea.org/inis/collection/NCLCollectionStore/_Public/11/543/11543720.pdf>. Also published in *Environmental Quarterly*, Report EML-349 (Environmental Measurements Laboratory: New York, 1979).

[22] Yamamoto, M. et al., 'Pu isotopes, ^{241}Am and ^{137}Cs in soils from the atomic bombed areas in Nagasaki and Hiroshima', *Journal of Radiation Research*, vol. 26, no. 2 (1985), p. 211.

found in the debris from the above-mentioned Chinese test.[23] In a boosted fission bomb, however, fluences of high-energy neutrons impinge on the fissile material and the $^{238}Pu : ^{239}Pu$ ratio gets higher. A recent analysis of material from a 140-kiloton underground explosion that in 1965 created Lake Chagan at the Soviet test site at Semipalatinsk, Kazakhstan, showed a $^{238}Pu : ^{239}Pu$ atom ratio of 0.0020, about 10 times higher than the ratio in the Nagasaki non-boosted fission bomb and about 40 times higher than the ratio in the Chinese fusion device.[24] This was a so-called clean explosion with a very low fission : fusion ratio (5 per cent), indicating the absence of a uranium pusher and the use of an efficient primary (i.e. a booster).[25]

Among the mentioned fuel-activation products, only ^{239}Np and ^{237}U are readily detectable in a gamma spectrum of bomb debris. Moreover, their ratio is a convenient measure of the thermonuclear character of the test. An atom ratio around 1 was normally observed in past atmospheric test of thermonuclear weapons, while fission bombs displayed $^{239}Np : ^{237}U$ ratios above 30.

Category 3. Fission products

Fission produces hundreds of short-lived, neutron-rich nuclides that normally decay by beta emission through the isobaric chains to higher Z elements, longer-lived and finally stable nuclides.[26] The detailed mass distribution of the products, the production probabilities and the yield function depend on the nucleus split, and on the energy of the initiating neutron. In a 'simple' fission weapon, the effect of fission-spectrum neutrons on ^{235}U or ^{239}Pu should be considered. In a 'booster', where there is a small thermonuclear flame in the centre of the exploding fission device (the effect of the fusion-derived neutrons is mainly to boost the number of fission events), only a small proportion of the fission events are induced by high-energy neutrons, and therefore high-energy yield curves for the fissile materials in the booster can be disregarded. In full-scale thermonuclear weapons the effects of high-energy neutrons on ^{238}U must be considered as well as on ^{235}U or ^{239}Pu used in the primary or, more importantly, if also used in the secondary stage.

[23] De Geer et al. (note 21). See also appendix 8A in this volume.

[24] Yamamoto M., Tsukatani T. and Katayama Y., 'Residual radioactivity in the soil of the Semipalatinsk nuclear test site in the former USSR', *Health Physics*, vol. 71 (Aug. 1996), p. 142.

[25] Russian Ministry of Atomic Energy and Russian Ministry of Defence, *USSR Nuclear Weapons Tests and Peaceful Nuclear Explosions, 1949 through 1990* (All-Russian Research Institute of Experimental Physics, Russian Federal Nuclear Center (VNIIEF): Sarov, 1996).

[26] Z denotes the number of protons in an atom, and each Z corresponds to a given element. In the beta decay process, Z increases 1 step and the number of neutrons, N, decreases 1 step.

Maximum cumulative yield values are around 6 per cent in all yield distributions.[27] Caesium-137, for example, is produced by 6.58 per cent of all fissioning ^{239}Pu atoms exposed to a fission-neutron spectrum flux.[28] For present purposes, all fission products produced with yields higher than 0.1 per cent, in at least one of the fission types, should be considered. This is a conservative level in the sense that no fission product detected in global fallout in the past will be missed.[29]

Category 4. Activation products of non-fuel bomb materials

Clearly, the non-radioactive materials used to construct a nuclear test device cannot be known exactly. Steel, aluminium, brass, plastics, beryllium and high explosives are, however, obvious components. Titanium has also reportedly been used. From these constituents, the following elements are there to be activated: hydrogen, beryllium, carbon, nitrogen, oxygen, aluminium, chlorine, titanium, chromium, manganese, iron, cobalt, nickel, copper and zinc. In some devices alternative tamper materials, such as lead and tungsten, are used to reduce the residual activity. Lead, along with boron, is also used in nuclear tests as a radiation shield. Gallium is a phase-stabilizing component in plutonium and would therefore clearly be an activation target. The same applies to niobium in uranium. Minor target elements would be gold and silver, which are used to cover surfaces or in electronic packages. Caesium might also be found in small quantities in electronic components (it is mainly included for historical reasons, as it was an issue in the debate on the French underground testing in the Pacific). In debris from one of the atmospheric tests in China, activation products of antimony were detected.[30] Antimony is used to increase the strength of lead and can therefore be regarded as an expected nuclear device material.

Most activation will occur by slowing down neutrons causing (n,γ) capture. Higher-energy neutrons, typically in the MeV range, will also be capable of inducing (n,p), (n,α) and (n,2n) reactions. In a few cases, involving fairly light target nuclides, the cross section for (n,p) reactions can be significant also for lower neutron energies. For high-energy neutrons a lot of reactions that take the target nucleus up to 10 nucleons away

[27] The cumulative yield of a nuclide member in an isobaric decay chain is the time-integrated number of nuclides that will be formed of that nuclide expressed in relation (mostly %) to the number of fissions.

[28] For a definition of 'neutron flux' see the glossary in this volume.

[29] The yield data used was taken from the compilation England, T. R. and Rider, B. F., 'Evaluation and compilation of fission product yields', Los Alamos National Laboratory Report LA-UR-94-3106, Oct. 1994, <http://ie.lbl.gov/fission/endf349.pdf>. Other compilations can be found at Organisation for Economic Co-operation and Development, Nuclear Energy Agency, Janis 4 browser, <http://www.oecd-nea.org/janis/>.

[30] See chapter 7 in this volume.

(e.g. (n,t2α)) are possible and also reported in the literature, but the cross sections for these reactions are generally so low that they can be disregarded here.

Category 5. Activation products in stemming (filling) materials and rocks surrounding an underground explosion

Underground tests will cause activation of nuclides in the surrounding ground. These activations depend heavily on the composition of the ground at the site of the test as well as on the emplacement and design of the test. For present purposes, the selection of target nuclides to be considered is based on the average abundances in the earth's crust and sea.[31] The values vary from 4×10^{-13} parts per million (ppm) to 46.1 per cent in the crust and from 6×10^{-16} ppm to 85.7 per cent in seawater. A cut-off limit of 0.1 per cent was chosen for both crust and seawater isotopes to be considered as targets.

Having chosen the target nuclides in the environment, the analysis is carried out in the same way as for non-fuel bomb materials. It is clear that the environment will be irradiated with a much softer neutron spectrum than the exploding device itself and that therefore (n,p), (n,α) and (n,2n) activation reactions, which normally have a cut-off energy in the range of 1–10 MeV, will be less common in the materials deriving from the outside than in materials deriving from the device itself.[32] It is difficult, however, to quantify this in a general way, as it is so heavily dependent on the device layout and on how the test is set up. Because of the softer neutron spectrum the thermal cross-section has been taken as a production indicator.

The ground contains a large number of trace elements in the 1–1000 ppm (by weight) range, but their abundance is too low to give significant amounts of neutron capture products, unless the cross sections are extremely large. Some elements, most notably rare earths, do exhibit such high neutron capture cross sections (several thousand barn as compared to more 'normal' values of millibarns up to a few or even tens of barn) and should therefore be considered. There are seven cases where the capture cross section is higher than 1000 barns and the resulting radionuclide has a half-life of more than six hours. If crust and isotopic abundances are also taken into account, then ^{152}Eu ($T_{1/2}$ = 13.542 years) and ^{152m1}Eu ($T_{1/2}$ = 9.274 hours) are the most prominent products created in this way. This is also supported by evidence from material

[31] Values are taken from Firestone, R. B. and Shirley, V. S. (eds), *Table of Isotopes*, 8th edn (Wiley: New York, 1996), table 2, p. A5.

[32] 'Neutron spectrum' here refers to a distribution of neutrons by energies. 'Softer' spectrum refers to a distribution where there are more neutrons with lower energies.

collected at ground zero of both the first United States nuclear weapon tests, in 1945, and the first Soviet test, in 1949.[33]

Category 6. Activation products in the ground below a near-surface atmospheric explosion

There is, in principle, no difference in the target nuclides in the ground that are available for activation by an atmospheric nuclear explosion or by an underground one. Depending on the altitude of the explosion, the neutron spectrum will be more or less soft, but no 'new' radionuclides will be omitted from the list of relevant radionuclides because of this.

One difference is, of course, that for underground explosions the crust abundance values were used to choose the target nuclides, while for atmospheric tests similar data for the biosphere should be added. A big difference in the choice of target nuclides is, however, not expected, although some biological elements, such as carbon and nitrogen, should probably be considered—even though at least these two would not be the source of any CTBT-relevant nuclide. ^{14}C would be produced, but it is a long-lived nuclide ($T_{1/2}$ = 5730 years) with no gamma ray emission.

Category 7. Activation products in seawater around an underwater or near-sea surface explosion

Seawater was already included in the analysis of underground explosions, in part because seawater in some environments is part of the underground, but more so because the major elemental components of seawater are also major elemental components of the earth's crust. No new nuclide is therefore expected under this category.

Category 8. Activation products in air around an atmospheric explosion

No CTBT-relevant nuclide is expected to be produced by neutron activation of the air constituents. However, the relevance to the CTBT is here sharply in contrast to dose relevance as ^{14}C, the long-lived beta-only-emitter mentioned above, is the major contributor (70 per cent) to the worldwide dose commitment from all past nuclear testing. ^{14}C is produced by (n,p) reactions in the nitrogen (^{14}N) of the air.

[33] De Geer, L.-E., 'Analyses by gamma spectroscopy of samples taken at ground zero of the first US and USSR nuclear test detonations', Unpublished memorandum from the Swedish National Defence Research Establishment, 1996.

Category 9. Activation products deriving from neutron fluence detectors

In many atmospheric tests, it has been common to include materials in vital parts of the device that allow the fluence of high-energy neutrons to be deduced, by measuring single and multiple (n,2n) products. Stable nuclides suitable for this purpose have produced radionuclides with long enough half-lives (hours to days) to allow for analysis in global fallout in the past. Examples, detected in debris from atmospheric tests, include yttrium and gold (used at least twice by China) and iridium (used by France). A Chinese thermonuclear test in November 1976 used an yttrium detector that produced about 0.022 atoms of ^{88}Y ($T_{1/2}$ = 106.65 days) per fission, enough to let ^{88}Y be easily detected in the air of the northern hemisphere for more than a year after the event.[34]

A scan of the Chart of Nuclides reveals only a few stable nuclides that will produce (n,2n) and (n,2n)^2 products that are fairly easily detected by the testing party. If reasonable gamma ray activity is assumed for the products, the possible detectors are ^{75}As, ^{85}Rb, ^{89}Y, ^{90}Zr, ^{103}Rh, ^{107}Ag, ^{169}Tm, ^{191}Ir and ^{197}Au. For these nuclides the maximum (n,2n) cross sections below 14 MeV neutron energy are all in the range of 0.3–2 barns. Unavoidable side products created through the same reactions that were considered above for other materials, that is (n,γ), (n,p) and (n,α), have to be considered here.

Category 10. Added tracers

In nuclear weapon tests, as in many other nuclear experiments, added tracers might prove useful for diagnostics. They provide a convenient way to calculate the absolute production of isotopes of the same element and provide information on the fluences and production rates in different parts of the device. In contrast to the high-energy neutron fluence detectors, a tracer should be chosen and added in such a way that it is not itself produced or altered by some reaction induced by the experiment (i.e. the explosion).

Tracers have been used to calculate the fusion yield of experiments, but these are gaseous tracers and will not be collected by the IMS particulate system. However, other nuclides in the transuranium region of the isotopic map will be collected by the particulate system and are used by the tester to better understand the reactions in the heavier elements. A few that have been mentioned in the literature are ^{233}U, ^{237}Np, ^{238}Pu, ^{242}Pu, ^{241}Am, ^{243}Am

[34] De Geer et al. (note 21). See also appendix 8A in this volume.

and ^{244}Cm.[35] Of these, based on specific activity as well as gamma ray energies and intensities, only ^{241}Am has a realistic possibility of being detected in the CTBT system. Other nuclides have been suggested to be added as tracers in past weapon testing programmes, such as, for example, ^{57}Co in Chinese tests.[36] In the current author's 1999 study, ^{133}Ba in the 1945 Trinity test was interpreted as a tracer.[37] In reality, it was probably activated barium from the weapon's high explosives (baratol) and would instead belong in category 4 (activation products of non-fuel bomb materials).

Making a final list of CTBT-relevant particulate nuclides[38]

As pointed out above, the selection process has been quite conservative, not fully taking into account decay mode, half-life or production cross section. The reason was twofold: to make sure that not a single nuclide of interest would slip through, and to give a background for identifying radionuclides that would be relevant for the radionuclide laboratories in their more in-depth analyses, also employing other techniques than gamma spectroscopy.

Many nuclides appear several times in the above 10 categories. Because of this and also as a result of applying more stringent selection rules for the routine gamma ray analysis scheme, a final list of CTBT- (or, more stringently, CTBT- and IDC-) relevant particulate nuclides will become much shorter. To make a realistic list it is reasonable to apply the following constraints.

1. The half-life should be between 6 hours and 1000 years.
2. A primary gamma ray should exist.
3. The primary gamma ray should have an energy above 50 keV.
4. The primary gamma ray intensity should be larger than 0.1 per cent.
5. The relevant production cross sections in non-fuel bomb materials and fluence detectors should be larger than 0.1 barn.
6. For target elements in the environment, at least one of the abundances in the earth's crust and the sea should be larger than 0.1 per cent.
7. In the case of radiative capture in ground materials, the product of the abundance in the earth's crust or the sea (maximum in ppm), the isotopic abundance (normalized to 1) and the cross section (in barns) should be

[35] Smith, D. K., *Challenges in Defining a Radiological and Hydrologic Source Term for Underground Nuclear Test Centers, Nevada Test Site, Nye County, Nevada*, Preprint UCRL-JC-120389 (Lawrence Livermore National Laboratory: Livermore, CA, June 1995.

[36] De Geer et al. (note 21).

[37] De Geer (note 13).

[38] Note that the word 'final' here refers to being final at the time of writing. Of course, data might become available in the future from different countries.

larger than 100. For other reactions in the ground, the corresponding product should be larger than 1000.

8. No gases should be considered.

9. No (n,2n)^2, two-step high-energy neutron fluence detector reaction products should be included.

In the first constraint, the lower half-life criterion is based on the notion that a nuclide cannot realistically be counted on an IMS detector before three days have passed after a nuclear test. Three days allow for just one day in atmospheric transport and the inherent average delay in the IMS system of two days. During three days more than 99.97 per cent of a radionuclide with a half-life of less than six hours would have decayed away. The upper half-life criterion is set to exclude radionuclides with very low specific activity. It is chosen here as 1000 years (nuclides 33.2 times less active than ^{137}Cs), but it is worth noting that even increasing it to almost 100 000 years would not add a single nuclide to the list of relevant nuclides.

The second constraint—the very clear criterion that there should be at least one gamma ray emitted by the nuclide—if applied as the first exclusion criterion, excludes 27 no-γ nuclides. In the third constraint, the lower primary gamma ray energy limit of 50 keV is set below the pragmatic IMS requirement that detectors should be well-calibrated: between 88 and 1836 keV. The reason is mainly to include ^{241}Am with its primary gamma ray energy of just 59.5 keV. Otherwise the limit could be set to 88 keV, or even 100 keV, without omitting more than two nuclides (i.e. ^{109}Pd and ^{170}Tl) from the list of relevant nuclides.

The fourth constraint, that the intensity, or the gamma–beta(alpha) branching ratio, should be larger than 0.1 per cent is based mainly on past experience. The probability that a gamma ray with an absolute intensity of less than 0.1 per cent should be instrumental in the first detection of a nuclear operation has been shown by national networks to be extremely minute. That does not mean, however, that such weak gamma rays would not add valuable information in the careful analysis after a spectrum has been flagged as being of interest.

The neutron activation of non-fuel bomb materials must necessarily be hard to estimate as it has to be based on educated guesses about the structure and functioning of the device itself. The materials to be expected are defined above, where it is also noted that, although no material can be fully excluded, it would not be conceivable to find an activation product from such a material without also detecting some more prominent remnant of the process. A lower limit of production is set, in the fifth constraint, by requesting that the relevant production cross section be higher than 0.1 barn. The same limit is conservatively applied for all activation processes in order to compensate for the fact that there is no exact knowledge

Table 6.3. The 42 particulate fission products relevant to the International Monitoring System of the CTBT

				Cumulative fission yield (%)					
				Fission induced by fast neutrons			Fission induced by high-energy neutrons		
Fission product	Half-life	Primary γ (keV)	Primary intensity (%)	U-235	U-238	Pu-239	U-235	U-238	Pu-239
Sr-91	9.65 h	1 024.3	*33.5*	*5.73*	*4.04*	*2.51*	*4.81*	*3.87*	*2.22*
Y-91[a]	58.51 d	1 204.8	*0.26*	*5.73*	*4.04*	*2.52*	*4.82*	*3.87*	*2.24*
Y-93	10.18 h	266.9	*7.4*	*6.25*	*4.91*	*3.82*	*5.19*	*4.53*	*3.22*
Zr-95[a]	64.032 d	756.7	*54.38*	*6.43*	*5.14*	*4.67*	*5.17*	*4.89*	*3.92*
Nb-95[a]	34.991 d	765.8	*99.808*	*6.43*	*5.14*	*4.67*	*5.18*	*4.89*	*3.93*
Zr-97[a]	16.749 h	743.4	*93.09*	*6.00*	*5.56*	*5.27*	*5.14*	*5.28*	*4.40*
Mo-99[a]	65.976 h	140.5[b]	*97.9*	*5.94*	*6.17*	*5.98*	*5.14*	*5.71*	*4.75*
Tc-99m	6.0067 h	140.5	*89*	*5.23*	*5.43*	*5.26*	*4.52*	*5.02*	*4.18*
Ru-103[a]	39.247 d	497.1	*91.0*	*3.24*	*6.28*	*6.83*	*3.21*	*4.62*	*5.21*
Rh-105[a]	35.36 h	319.1	*19.1*	*1.20*	*4.05*	*5.36*	*1.87*	*3.22*	*4.27*
Ru-106[a]	371.8 d	621.9	*9.93*	*0.53*	*2.49*	*4.36*	*1.61*	*2.45*	*3.54*
Ag-111[a]	7.45 d	342.1	*6.7*	*0.04*	*0.07*	*0.36*	*1.08*	*0.99*	*1.54*
Pd-112	21.03 h	617.5[b]	*50.5*	*0.04*	*0.06*	*0.19*	*1.08*	*1.03*	*1.38*
Cd-115m	44.56 d	933.8	*2*	*0.00*	*0.00*	*0.01*	*0.46*	*0.07*	*0.12*
Cd-115[a]	53.46 h	336.2	*50.2*	*0.03*	*0.03*	*0.07*	*0.64*	*0.79*	*1.16*
Sn-125[a]	9.64 d	1 067.1	*10*	*0.04*	*0.03*	*0.12*	*0.92*	*0.64*	*1.18*
Sb-125[a]	2.75856 y	427.9	*29.6*	*0.07*	*0.05*	*0.18*	*1.46*	*1.20*	*1.95*
Sb-126[a]	12.35 d	695.0	*99.6*	*0.01*	*0.01*	*0.04*	*0.34*	*0.19*	*0.62*
Sb-127[a]	3.85 d	685.7	*36.8*	*0.31*	*0.14*	*0.50*	*2.16*	*1.49*	*2.14*
Sb-128	9.01 h	743.2	*100*	*0.01*	*0.00*	*0.04*	*0.42*	*0.14*	*0.68*
Te-129m[a]	33.6 d	695.9	*3.1*	*0.14*	*0.17*	*0.24*	*0.71*	*0.36*	*0.93*
I-130	12.36 h	536.1	*99*	*0.00*	*0.00*	*0.00*	*0.03*	*0.00*	*0.13*
Te-131m	33.25 h	773.7	*36.8*	*0.43*	*0.26*	*0.92*	*1.34*	*0.42*	*1.88*
I-131[a]	8.0252 d	364.5	*81.5*	*3.22*	*3.29*	*3.88*	*4.10*	*3.99*	*4.35*
Te-132[a]	3.204 d	772.6[b]	*77.9*	*4.66*	*5.13*	*5.15*	*4.09*	*4.65*	*3.30*
I-133[a]	20.83 h	529.9	*87.0*	*6.72*	*6.76*	*6.91*	*5.36*	*6.00*	*4.48*
I-135	6.58 h	1 260.4	*28.7*	*6.30*	*6.94*	*6.08*	*4.22*	*5.50*	*3.96*
Cs-136[a]	13.16 d	1 048.1	*80*	*0.01*	*0.00*	*0.12*	*0.23*	*0.02*	*0.75*
Cs-137[a]	30.08 y	661.7[b]	*85.1*	*6.22*	*6.05*	*6.58*	*4.93*	*5.15*	*4.45*
Ba-140[a]	12.7527 d	537.3	*24.39*	*5.98*	*5.82*	*5.32*	*4.50*	*4.61*	*3.70*
La-140[a]	1.67855 d	1 596.2	*95.40*	*5.98*	*5.82*	*5.33*	*4.53*	*4.61*	*3.84*
Ce-141[a]	32.508 d	145.4	*48.29*	*5.95*	*5.34*	*5.15*	*4.49*	*4.38*	*3.56*
Ce-143[a]	33.039 h	293.3	*42.8*	*5.73*	*4.62*	*4.34*	*3.82*	*3.91*	*2.80*
Ce-144[a]	284.91 d	133.5	*11.09*	*5.27*	*4.55*	*3.67*	*3.17*	*3.72*	*2.68*
Nd-147[a]	10.98 d	531.0	*13.4*	*2.14*	*2.59*	*1.99*	*1.62*	*2.09*	*1.71*
Pm-149	53.08 h	286.0	*3.1*	*1.04*	*1.63*	*1.24*	*0.81*	*1.46*	*1.06*
Pm-151	28.40 h	340.1	*22.5*	*0.41*	*0.80*	*0.78*	*0.36*	*0.80*	*0.73*
Sm-153	46.28 h	103.2	*29.25*	*0.17*	*0.41*	*0.43*	*0.20*	*0.39*	*0.46*
Eu-155[a]	4.753 y	105.3	*21.1*	*0.04*	*0.14*	*0.21*	*0.08*	*0.16*	*0.23*
Sm-156	9.4 h	203.8	*21*	*0.02*	*0.08*	*0.15*	*0.05*	*0.11*	*0.17*
Eu-156[a]	15.19 d	1 153.8	*11.5*	*0.02*	*0.08*	*0.15*	*0.06*	*0.11*	*0.21*
Eu-157	15.18 h	370.5	*11.0*	*0.01*	*0.04*	*0.11*	*0.04*	*0.08*	*0.11*

CTBT = Comprehensive Nuclear-Test-Ban Treaty; d = day; h = hour; y = year.

[a] These nuclides have been detected in the past by a national system similar to the CTBTO International Monitoring System (the Swedish).

[b] The primary gamma ray is emitted in the decay of a short-lived daughter.

Source: De Geer, L.-E., *CTBT Relevant Radionuclides*, Comprehensive Nuclear-Test-Ban Treaty Organization (CTBTO) Provisional Technical Secretariat (PTS) Technical Report PTS/IDC-1999/02 (CTBTO: Vienna, Apr. 1999); De Geer, L.-E., 'Comprehensive Nuclear-Test-Ban Treaty: relevant radionuclides', *Kerntechnik*, vol. 66, no. 3 (May 2001), pp. 113–20; National Nuclear Data Center, NuDat 2 database extraction, Brookhaven National Laboratory, 6 Jan. 2015, <http://www.nndc.bnl.gov/nudat2/>; and England, T. R. and Rider, B. F., 'Evaluation and compilation of fission product yields', Los Alamos National Laboratory Report LA-UR-94-3106, Oct. 1994, <http://ie.lbl.gov/fission/endf349.pdf>. Other yield data compilations can be found at Organisation for Economic Co-operation and Development, Nuclear Energy Agency, Janis 4 browser, <http://www.oecd-nea.org/janis/>.

of the layout and function of all possible devices. Reactions where cross sections were not found at all are disregarded (assuming that no important cross section would be unknown). The 0.1 barn cut-off is also applied to the side reaction products of the fluence detectors.

For activation products in the environment, which are assumed to be homogeneous, the product of abundance and cross section is used for the relevance test in the sixth and seventh constraints. In this way the different categories represented by (n,p), (n,α), (n,2n) and (n,γ) reactions can be treated in a balanced way. For high-energy reactions, the same product of crustal or sea abundance, isotopic abundance and relevant cross section is used, but the cut-off is set higher (conservatively just a factor of 10) to make allowance for the lower neutron fluence in the environment at higher energies.

The eighth constraint disregards gases because the filter material used will not capture them and the ninth constraint disregards the two-step neutron fluence detector reactions because they are second-order effects.

Applying these rules to the nuclides previously selected in categories 1–10 results in a list of 92 CTBT-relevant radionuclides, 47 of them fission products and 45 neutron activated nuclides, fuel residues or added tracers.

At a CTBT workshop in Melbourne, Australia, in January 2000, where this list was discussed, eight nuclides were suggested to be removed from the list for various reasons: ^{241}Am apparently due to some oversensitivty about transuranium isotopes and the fission products ^{109}Pd, ^{123}Sn, ^{125m}Te, ^{127m}Te, ^{127}Te and activation products ^{64}Cu and ^{170}Tm due to their quite low primary gamma intensities (<4 per cent). That left a list of 42 fission products and 42 non-fission CTBT-relevant radionuclides (see tables 6.3 and 6.4). This list was recommended by the CTBTO's Working Group B for testing and development and was accepted by the 11th session of the CTBTO Preparatory Commission, and in early 2015 it is still in use. One further nuclide, ^{92m}Nb, detected in the last Chinese atmospheric test in 1980, was later interpreted as a (n,2n) product of niobium used in corrosion-

Table 6.4. The 42 particulate non-fission products relevant to the International Monitoring System of the CTBT and 3 particulate non-fission nuclides not sanctioned as relevant by the CTBT Preparatory Commission

				Production mode[a]												
						Device, 4				Environment, 5–8						
Nuclide	Half-life	Primary γ (keV)	Primary intensity (%)	1	2	γ	p	α	2n	γ	p	α	2n	γ*	9	10
Na-24[b]	14.997 h	1 368.6	*99.9936*	–	–	–	–	x	–	x	x	x	–	–	–	–
K-42	12.360 h	1 524.6	*18.08*	–	–	–	–	–	–	x	–	–	–	–	–	–
Sc-46	83.79 d	889.3	*99.9840*	–	–	–	x	–	–	–	–	–	–	–	–	–
Sc-47	3.3492 d	159.4	*68.3*	–	–	–	x	–	–	–	–	–	–	–	–	–
Cr-51[b]	27.7010 d	320.1	*9.910*	–	–	x	–	–	x	–	–	–	–	–	–	–
Mn-54[b]	312.05 d	834.8	*99.9760*	–	–	–	x	–	x	–	x	–	–	–	–	–
Co-57[b]	271.74 d	122.1	*85.60*	–	–	–	–	–	–	–	–	–	–	–	–	x
Co-58[b]	70.86 d	810.8	*99.450*	–	–	–	x	–	x	–	–	–	–	–	–	–
Fe-59	44.495 d	1 099.2	*56.5*	–	–	x	–	–	–	x	–	–	–	–	–	–
Co-60[b]	1 925.28 d	1 332.5	*99.9826*	–	–	x	x	–	–	–	–	–	–	–	–	–
Zn-65[b]	243.93 d	1 115.5	*50.04*	–	–	x	–	–	x	–	–	–	–	–	–	–
Zn-69m	13.756 h	438.6	*94.85*	–	–	x	–	–	x	–	–	–	–	–	–	–
Ga-72	14.10 h	834.1	*95.45*	–	–	x	–	–	–	–	–	–	–	–	x	–
As-74	17.77 d	595.8	*59*	–	–	–	–	–	–	–	–	–	–	–	x	–
As-76	26.24 h	559.1	*45.0*	–	–	–	–	–	–	–	–	–	–	–	x	–
Rb-84	32.82 d	881.6	*68.9*	–	–	–	–	–	–	–	–	–	–	–	x	–
Rb-86	18.642 d	1 077.0	*8.64*	–	–	–	–	–	–	–	–	–	–	–	x	–
Y-88[b]	106.627 d	1 836.1	*99.2*	–	–	–	–	–	–	–	–	–	–	–	x	–
Zr-89	78.41 h	909.2	*99.04*	–	–	–	–	–	–	–	–	–	–	–	x	–
Rh-102	207.3 d	475.1	*46*	–	–	–	–	–	–	–	–	–	–	–	x	–
Ag-106m	8.28 d	717.3	*28.9*	–	–	–	–	–	x	–	–	–	–	–	x	–
Ag-108m	438 y	722.9	*90.8*	–	–	x	–	–	–	–	–	–	–	–	x	–
Ag-110m	249.83 d	657.8	*95.61*	–	–	x	–	–	–	–	–	–	–	–	–	–
Sb-120	5.76 d	1 171.7	*100*	–	–	–	–	–	x	–	–	–	–	–	–	–
Sb-122[b]	2.7238 d	564.2	*70.67*	–	–	x	–	–	x	–	–	–	–	–	–	–
Sb-124[b]	60.20 d	602.7	*97.8*	–	–	x	–	–	–	–	–	–	–	–	–	–
Cs-132	6.480 d	667.7	*97.59*	–	–	–	–	–	x	–	–	–	–	–	–	–
Ba-133[b]	10.551 y	356.0	*62.05*	–	–	x	–	–	–	–	–	–	–	–	–	–
Cs-134	2.0652 y	604.7	*97.62*	–	–	x	–	–	–	–	–	–	–	–	–	–
Eu-152m	9.3116 h	841.6	*14.2*	–	–	–	–	–	–	–	–	–	–	x	–	–
Eu-152[b]	13.517 y	1 408	*20.87*	–	–	–	–	–	–	–	–	–	–	x	–	–
Tm-168	93.1 d	816.0	*50.95*	–	–	–	–	–	–	–	–	–	–	–	x	–
W-187	24.000 h	685.8	*33.2*	–	–	x	–	–	–	–	–	–	–	–	–	–
Ir-190	11.78 d	186.7	*52*	–	–	–	–	–	–	–	–	–	–	–	x	–
Ir-192[b]	73.829 d	316.5	*82.86*	–	–	–	–	–	–	–	–	–	–	–	x	–
Au-196	6.1669 d	355.7	*87*	–	–	–	–	–	x	–	–	–	–	–	x	–
Au-196m	9.6 h	147.8	*43.5*	–	–	–	–	–	x	–	–	–	–	–	–	–
Au-198[b]	2.6947 d	411.8	*95.62*	–	–	x	–	–	–	–	–	–	–	–	x	–
Pb-203	51.92 h	279.2	*80.9*	–	–	–	–	–	x	–	–	–	–	–	–	–
Ra-224	3.66 d	241.0	*4.10*	x	–	–	–	–	–	–	–	–	–	–	–	–
U-237[b]	6.75 d	208.0	*21.2*	–	x	–	–	–	–	–	–	–	–	–	–	–
Np-239[b]	2.356 d	277.6	*14.44*	–	x	–	–	–	–	–	–	–	–	–	–	–
Particulate non-fission nuclides not sanctioned by the CTBT Preparatory Commission																
Nb-92m	10.15 d	934.4	*99.15*	–	–	–	–	–	x	–	–	–	–	–	–	–
Ba-131	11.50 d	496.3	*48.0*	–	–	x	–	–	–	–	–	–	–	–	–	–
Am-241	432.6 y	59.5	*35.9*	x	x	–	–	–	–	–	–	–	–	–	–	x

CTBT = Comprehensive Nuclear-Test-Ban Treaty; d = day; h = hour; y = year.

[a] Production mode refers to the following 9 categories: residues of fuel materials (Category 1), non-fission reaction products in fuel materials (Category 2), activation products of non-fuel bomb materials (Category 4), activation products in stemming materials and rocks surrounding an underground explosion (Category 5), activation products in the ground below a near-surface atmospheric explosions (Category 6), activation products in seawater around an underwater or a near-sea surface explosion (Category 7), activation products in air around an atmospheric explosion (Category 8), activation products deriving from neutron fluence detectors (Category 9), and added tracers (Category 10). Note that Category 3, fission products, is covered by table 6.3. In categories 4–8, γ, p, α, 2n and γ^* signify the emitted particle in the neutron activation reaction, where γ^* stands for the special very high cross-section neutron activation products in the environment.

[b] These nuclides have been detected in the past by a national system similar to the CTBTO International Monitoring System (the Swedish).

Source: De Geer, L.-E., *CTBT Relevant Radionuclides*, Comprehensive Nuclear-Test-Ban Treaty Organization (CTBTO) Provisional Technical Secretariat (PTS) Technical Report PTS/IDC-1999/02 (CTBTO: Vienna, Apr. 1999), pp. 25–26.

resistant uranium–niobium alloys. It is now included here together with ^{241}Am and ^{131}Ba (see table 6.4).[39]

During the discussion among CTBT states signatories before the decision about the composition of the list there was a heated debate on whether such a long list should be used. There was apprehension among nuclear weapon states and some other states that a long list would produce too many false alarms and frivolous accusations of treaty violations. The argument was made that if there is a violation at least one or a few, if any, of the most prominent radionuclides would be detected and that there was no need for a comprehensive list. Sweden argued from its own experience that such a scenario would not necessarily be the case. In the 1968 US test 'Schooner', for example, no more than one activated tungsten isotope was detected.[40] Furthermore, it is philosophically difficult to consider clear nuclear explosion products as non-relevant for nuclear-explosion detection. Experience at the IDC since 2000 has also shown that there have not been excessive false detections or accusations. Although a few nuclides have repeatedly created Level 4 or 5 samples—most notably ^{99m}Tc, ^{24}Na and ^{60}Co—they could be dealt with in a separate manner. The half-life of ^{99m}Tc is 6.0067 hours (thus very close to the limit of 6 hours) and could easily be excluded by increasing the lower half-life limit by just 25 seconds.[41] The other two nuclides are often present due to cosmic radiation or terrestrial background and, as such, should be possible to neutralize.

39 Brown, D. W. et al., 'Aging and deformation of uranium-niobium alloys', *Los Alamos Science*, no. 30 (2006).

40 Persson, G., *Radioactive Tungsten in the Atmosphere Following Project Schooner*, FOA report C 4460-28 (Försvarets Forskningsanstalt: Sundbyberg, 1971).

41 Technetium-99m appears repeatedly at stations (e.g. station ARP01 in Buenos Aires, Argentina) that are close to laboratories that work with this isotope as a radiopharmaceutical.

Table 6.5. The four noble gas fission products relevant to the International Monitoring System of the CTBT

				Cumulative fission yield (%)					
				Fission induced by fast neutrons			Fission induced by high-energy neutrons		
Fission product	Half-life	Primary γ (keV)	Primary intensity (%)	U-235	U-238	Pu-239	U-235	U-238	Pu-239
Xe-131m	11.84 d	163.9	*1.95*	*0.05*	*0.05*	*0.05*	*0.06*	*0.06*	*0.07*
Xe-133m	2.198 d	233.2	*10.12*	*0.19*	*0.19*	*0.24*	*0.29*	*0.18*	*0.42*
Xe-133	5.2475 d	81.0	*36.9*	*6.72*	*6.76*	*6.97*	*5.53*	*6.02*	*4.86*
Xe-135	9.14 h	249.8	*90.00*	*6.60*	*6.97*	*7.54*	*5.67*	*5.84*	*6.18*

CTBT = Comprehensive Nuclear-Test-Ban Treaty; d = day; h = hour.

Source: De Geer, L.-E., *CTBT Relevant Radionuclides*, Comprehensive Nuclear-Test-Ban Treaty Organization (CTBTO) Provisional Technical Secretariat (PTS) Technical Report PTS/IDC-1999/02 (CTBTO: Vienna, Apr. 1999); De Geer, L.-E., 'Comprehensive Nuclear-Test-Ban Treaty: relevant radionuclides', *Kerntechnik*, vol. 66, no. 3 (May 2001), pp. 113–20; National Nuclear Data Center, NuDat 2 database extraction, Brookhaven National Laboratory, 6 Jan. 2015, <http://www.nndc.bnl.gov/nudat2/>; and England, T. R. and Rider, B. F., 'Evaluation and compilation of fission product yields', Los Alamos National Laboratory Report LA-UR-94-3106, Oct. 1994, <http://ie.lbl.gov/fission/endf349.pdf>. Other yield data compilations can be found at Organisation for Economic Co-operation and Development, Nuclear Energy Agency, Janis 4 browser, <http://www.oecd-nea.org/janis/>.

The IDC categorization scheme and its related relevant radionuclides are, as noted above, basically tools to help CTBT states signatories see which radionuclide samples are more interesting than others. However, after entry into force of the CTBT, any national data centre could, in its own right, detect any gamma-active radionuclide and use it to back up a request to the CTBTO's Executive Council for an on-site inspection. There are two formal implications of a Level 5 categorization. First, the sample should, as a rule, be split into two pieces that are sent to 2 randomly selected laboratories among the 16 IMS laboratories in order to corroborate the results, or even possibly detect something that was not noted in the monitoring station spectrum. The latter is quite possible as at the time the sample reaches the laboratories the radon daughter background will have been significantly reduced. Laboratories might also use larger detectors and longer counting times that will further increase sensitivity for not too short-lived nuclides. Second, the meteorological analysis of a Level 5 sample will be repeated at a number of the Regional Specialized Meteorological Centres of the World Meteorological Organization (WMO). Every sample is analysed at the Vienna centre for its field of regard (FOR), which indicates the geographical regions within which an emission that could have caused a potential detection should be sought. In the case of a Level 5 sample there is also a need to better understand the uncertainties of the meteorological analysis and, hopefully (as it could well be a very controversial analysis), reach a kind of global consensus on the results.

IV. Noble gas radionuclides relevant to the IMS of the CTBT

When the CTBT was designed in 1996 to encompass 80 particulate radionuclide stations it was also decided that half of them should be equipped with xenon-analysis capabilities. The latter was basically a compromise between states that trusted the technology and states that did not. After the experience of the North Korean tests it must be hard to oppose equipping all 80 radionuclide stations with xenon capabilities.[42] This option is laid out in the CTBT's Protocol and is to be decided at the first annual meeting of the Conference of the States Parties after the treaty's entry into force.

Noble gases are of particular interest in detecting emissions from underground nuclear explosions as they are chemically inert and therefore quite hard to contain underground. When extreme pressures are generated in the explosion cavity a substantial possibility/risk exists that some noble gases will escape within seconds to the atmosphere and then blow with the winds to one or several stations in a verification network. Unless special care is taken, noble gases can also escape a few days after the explosion when the testing team gain access to the experimental chamber.

However, the inertness of these gases means that they are also much more complicated to collect from the air than aerosols or particulates. In recent decades, however, effective technologies and equipment have been developed, mainly in France, Russia, Sweden and the USA. As of late October 2014, the IMS network comprises 30 certified noble gas stations, with an additional 2 just installed, 5 under construction and 3 being planned.

Among the noble gases only xenon has a set of isotopes suitable for detection: ^{131m}Xe ($T_{½}$ = 11.84 days), ^{133m}Xe ($T_{½}$ = 2.198 days), ^{133}Xe ($T_{½}$ = 5.2475 days) and ^{135}Xe ($T_{½}$ = 9.14 hours). These isotopes have detectable radiation, half-lives that are long enough to survive atmospheric transport and short enough not to build up a background of old emissions, as is the case with the 10.76 year half-life of ^{85}Kr. They are also produced with high yields in fission (see table 6.5). One problem with xenon isotopes is that a number of medical isotope factories (in Australia, Belgium, Canada, South Africa and other countries) that, in the process of producing ^{99m}Tc 'cows', emit substantial amounts of xenon radioisotopes in a mix quite similar to that produced by nuclear weapon explosions. To deal with what can be a significant background problem for some stations, a three-step categor-

[42] On the North Korean tests see chapter 8 in this volume.

Table 6.6. The particulate and gaseous radionuclides being considered for search during an on-site-inspection

Radionuclide	Half-life	Production
Ar-37	34.95 days	(n,α) product in environment
Cr-51	27.7025 days	(n,γ) and (n,2n) product in steel
Mn-54	312.12 days	(n,p) product in steel
Co-58	70.86 days	(n,p) product in steel
Fe-59	44.495 days	(n,p) product in steel
Zn-65	243.66 days	(n,γ) and (n,2n) product in steel
Zr-95	64.02 days	Fission product
Nb-95	34.975 days	Fission product
Mo-99	65.94 hours	Fission product
Ru-103	39.26 days	Fission product
Ru-106	373.59 days	Fission product
Cd-115m	44.6 days	Fission product
I-131	8.02070 days	Fission product
Xe-131m	11.84 days	Fission product
Te-132	3.204 days	Fission product
Xe-133m	2.198 days	Fission product
Xe-133	5.2475 days	Fission product
Xe-135	9.14 hours	Fission product
Ba-140	12.752 days	Fission product
Ce-141	32.501 days	Fission product
Ce-144	284.893 days	Fission product
Nd-147	10.98 days	Fission product
U-237	6.75 days	(n,2n) product in uranium-238
Np-239	2.3565 days	(n,v) product in uranium-238

ization scheme is applied for xenon: (*a*) a sample in the first step has no detection of any of the four isotopes, (*b*) a sample in the second step has detections that are typical for that station, and (*c*) a sample in the third step has at least one of the four isotopes in concentrations exceeding three times the difference between the lower limit of the upper quartile and the upper limit of the lower quartile in the distribution of detections during the preceding year. In addition, flags are developed that indicate properties of isotopic ratios, station state of health and prevailing atmospheric transport sensitivity for sources known to sometimes influence the station.

V. Particulate and gaseous nuclides relevant for on-site inspection

The concept of IDC-relevant radionuclides discussed above is fairly straightforward and should not be politically sensitive as it is basically an advisory tool. For the CTBT on-site-inspection (OSI) regime a similar need exists to define relevant radionuclides, but that is a much more politically critical matter as only OSI-relevant radionuclides, and nothing else, are supposed to be searched for and analysed by an OSI search team. Operations are intended to focus only on OSI-relevant radionuclides, and some states advocate that all measurement results must be partly blinded so that inspection teams cannot extract information other than that related to the relevant nuclides.

The OSI regime has been a subject for heated discussions at the CTBTO headquarters in Vienna for many years. A particular issue has been the definition of the set of radionuclides that an OSI-team would look for in an inspection. A 2010 internal CTBTO paper identified 16 particulate and 5 gaseous radionuclides as 'agreed on'. Five other particulates were identified as requiring further consideration.

At an OSI experimental advanced course in Paris, on 12–20 November 2001, the current author suggested the creation of an OSI-relevant particulate gamma detectable radionuclide list. That list was based on the IDC list (see sections III and IV) but changed the half-life range to 2 days–2 years, the primary gamma ray intensity requirement to >5 per cent, and the relevant production cross section in bomb materials from ≥0.1 barn to ≥0.5 barn. These constraints result in a list that is almost identical to the list being discussed in the CTBTO's Working Group B, except for two obvious bomb products, uranium-237 and neptunium-239—nuclides that have often totally dominated atmospheric test residues in the past. However, that list also includes radioxenon and another noble gas isotope, ^{37}Ar ($T_{1/2}$ = 35.04 days), which might be produced (and has been detected in the past from underground explosions) by neutrons impinging on calcium underground via the reaction $^{40}Ca(n,\alpha)^{37}Ar$.

The list includes ^{135}Xe despite its short half-life (see table 6.6 for the combined list of particulate and gaseous nuclides that should be relevant for OSI). The reason for the cautious approach to ^{237}U and ^{239}Np is probably an indication of the same apprehension about discussing uranium and transuranium data by decision makers in the nuclear weapon states that was observed when ^{241}Am was excluded from the IDC list (see section III). They are, however, kept in the list here to fulfil the principle of applying fairly general selection criteria.

It is important to bear in mind the different objectives of the IDS- and OSI-relevant radionuclides lists, which has not always been understood. For example, ^{137}Cs is on the IDC list but is not on the OSI list. That can appear to be a loophole, but in fact it is not: it is important not to include on the OSI list nuclides that are too long-lived as that could cause, maybe (possibly frivolous) accusations of treaty violations when such nuclides are detected at former test sites. That was exactly the reason why two years was chosen as an upper half-life limit in the 2001 Paris list.

Part II

Nuclear forensics in practice

7. The origins of nuclear forensic analysis I: the United States and the Soviet Union

VITALY FEDCHENKO AND ROBERT KELLEY

A lot of the ideas and techniques that are now part and parcel of nuclear forensics originated in 1940s and 1950s within national nuclear weapon programmes or in response to the programmes of other countries. This chapter describes that process during those years for the two first nuclear weapon states: the United States and the Soviet Union. Both states followed a generally similar path (with some differences depending on local situation, such as availability of particular measurement techniques), using their own nuclear materials and weapon-production experience to advance long-range monitoring of nuclear explosions and other nuclear intelligence techniques. Such ideas and techniques were later adopted by the rest of the international community to verify international agreements and enforce nuclear security (see chapter 9).

Section I of this chapter describes the US origin of nuclear forensics in the 1940s, first to investigate Germany's nuclear capabilities, then to measure the USA's own tests and then to analyse the first Soviet test. Section II describes the first instances of nuclear forensic analysis in the Soviet Union, first to assess its own tests in the 1940s and then to detect US tests in the 1950s. The following chapter, chapter 8, gives a longer history of the development of the capacity to detect nuclear explosions around the world of a non-nuclear state—Sweden.

I. The origin of nuclear forensics in the United States

The need for what has later come to be called nuclear forensic analysis was probably first recognized and formulated by Brigadier General Leslie R. Groves, the head of the USA's Manhattan Project.[1] When Groves became

[1] For general history of the Manhattan Project (formally known as the Manhattan Engineer District) see Rhodes, R., *The Making of the Atomic Bomb* (Simon & Schuster: New York, 1986). For history of its technical dimensions see Hoddeson, L. et al., *Critical Assembly: A Technical History of Los Alamos during the Oppenheimer Years, 1943–1945* (Cambridge University Press: Cambridge, 1993); and Groueff, S., *Manhattan Project: The Untold Story of the Making of the Atomic Bomb* (Little, Brown and Company: Boston, 1967). For a summary of scientific, technological and engineering results of the Manhattan Project see Smyth, H. D., *Atomic Energy for Military Purposes: The Official Report on the Development of the Atomic Bomb under the Auspices of the United States Government, 1940–1945* (Princeton University Press: Princeton, NJ, 1945); Hawkins, D., Truslow, E. C. and Smith, R. C., *Project Y: The Los Alamos Story* (Tomash Publishers: Los Angeles, CA, 1961); and Serber, R.,

the leader of the project to develop an atomic bomb, in September 1942, he was briefed on the British and US intelligence agencies' perceptions of two aspects of the 'German nuclear threat'. First, although the chance of Germany producing a nuclear weapon was perceived as 'remote' at the time, the general attitude in the US Government did not allow the threat to be completely dismissed.[2] In 1944 Columbia University's Harold Urey summarized the commonly held view that additional information should be sought and a '10% chance is too much to neglect'.[3]

Second, Groves took seriously the possibility of Germany producing what was called a 'radioactivity bomb' or 'an ordinary bomb containing radioactive material'. The US scientists concluded that the creation of such a weapon would be precluded by the challenges of handling the enormous amount of radioactive material needed and the protection measures required.[4] However, Groves believed that 'the safety problems that deterred U.S. scientists would not inhibit the Germans because the Nazis would use the technicians and workers drawn from "inferior" groups within the populations they controlled'.[5]

The first steps in bulk environmental sample analysis

Monitoring Germany's nuclear reactor operation was key to detecting its development of either a nuclear weapon or radioactivity bomb. In order to succeed, the notion of 'radiological intelligence' was introduced. In its first incarnation—analysing environmental samples in order to monitor Germany's reactors—it focused on analysis of air and two liquids: water and wine.

Air

In the autumn of 1943 Groves summoned Luis W. Alvarez, a future Nobel laureate in physics, to ask him 'how we could find out whether the Germans were operating nuclear reactors' and gave him one week to develop a technical method to achieve that aim.[6] Alvarez came up with a method

The Los Alamos Primer: The First Lectures on How to Build an Atomic Bomb (University of California Press: Berkeley, CA, 1992).

[2] Ziegler, C. A. and Jacobson, D., *Spying without Spies: Origins of America's Secret Nuclear Surveillance System* (Praeger: Westport, CT, 1995), p. 3.

[3] Richelson, J. T., *Spying on the Bomb: American Nuclear Intelligence from Nazi Germany to Iran and North Korea* (W. W. Norton & Company: New York, 2006), p. 45.

[4] The large amount of radioactivity required is explained by the fact that, at the time, a 'radioactivity bomb' was perceived as a war-fighting tool, as opposed to the 'radiological dispersal devices' or 'dirty bombs' that in current discussions on terrorism are viewed as 'weapons of mass disruption'. Bernstein, B. J., 'Radiological warfare: the path not taken', *Bulletin of the Atomic Scientists*, vol. 41, no. 7 (Aug. 1985), p. 44.

[5] Ziegler and Jacobson (note 2), p. 3.

[6] Alvarez, L. W., *Alvarez: Adventures of a Physicist* (Basic Books: New York, 1987), p. 120.

involving the detection of radioactive gases that reactors emit during normal operation, and he decided that it would be particularly easy to detect a certain radioactive isotope of xenon: ^{133}Xe.

That isotope is generated at a high rate during fission of ^{235}U, ^{238}U and ^{239}Pu, which means that any reactor produces it in significant quantities. Since xenon is a noble gas, it escapes a reactor in detectable quantities instead of chemically reacting with other elements. The half-life of ^{133}Xe is 5.247 days, which means that it does not appear in the atmosphere naturally. It is also relatively easy to separate from the nitrogen and oxygen in the air, since these gases have significantly different boiling points. It produces distinctive gamma and beta radiation. All these qualities make ^{133}Xe suitable for detection as a 'signature' of an operating nuclear facility.[7]

By the summer of 1944, with the help of General Electric, Alvarez had developed a xenon-detection system consisting of air-sampling equipment and a ground-based laboratory for sample analysis. Alvarez described the air-sampling apparatus in his memoirs:

> The system passed an air sample through activated charcoal. That trapped any xenon and radon but not nitrogen and oxygen. Radon, also a noble gas, is present in the atmosphere at a lower concentration than xenon and also has a much higher boiling point (as its name implies, it's a decay product of radium, which is in turn a decay product of uranium, which occurs naturally throughout the world in low concentration but great volume). After an overflight, we could heat the activated charcoal to boil off both the radon and the xenon into a stream of helium gas. Passing that gas stream again through activated charcoal at ice temperature would freeze out the radon but allow the xenon to get through. More activated charcoal, this time at dry-ice temperature, would absorb the xenon. With the helium pumped out, this last filter would be heated to drive off pure xenon. The resulting highly concentrated sample could then be examined for radioactive xenon, the presence of which would indicate an operating German nuclear reactor.[8]

The equipment built by Alvarez was first tested in the USA and then deployed in Germany. In the autumn of 1944 a few Douglas A-26 Invader aircraft made a number of flights over locations that were considered potentially related to the German nuclear programme. These locations were pinpointed by analysis of open source information, aerial photographs and reports from British and US agencies. No ^{133}Xe was found.[9]

The idea of sampling noble gases was also thought of by Anthony Turkevitch, a scientist at the Manhattan Project's Metallurgical Laboratory in Chicago, Illinois. It cannot be established exactly when this idea occurred

[7] Saey, P. R. J., 'Ultra-low-level measurements of argon, krypton and radioxenon for treaty verifycation purposes', *ESARDA Bulletin*, no. 36 (July 2007), p. 44; and Kalinowski, M. B. et al., 'Environmental sample analysis', eds R. Avenhaus et al., *Verifying Treaty Compliance: Limiting Weapons of Mass Destruction and Monitoring Kyoto Protocol Provisions* (Springer: Heidelberg, 2006), pp. 376–77.

[8] Alvarez (note 6), pp. 120–21.

[9] Ziegler and Jacobson (note 2), pp. 7–8.

to Turkevitch, but in his memoir Stanisław Ulam, a prominent employee at the Los Alamos Laboratory, New Mexico, stated that '[The idea] to detect the presence of nuclear work anywhere in the world ... by examining air samples from the atmosphere for the presence of certain gases which came from uranium fission ... came from Tony Turkevitch .... I remember his mentioning such a plan in my presence in Los Alamos during the war.'[10] Ulam himself arrived at Los Alamos on 4 February 1944.[11]

Water and wine

In September 1943 the Manhattan Project's scientific director, J. Robert Oppenheimer, was asked by the project's head of foreign intelligence, Robert Furman, for advice on how to monitor the German nuclear programme. Among other measures, Oppenheimer suggested that an operating reactor could be detected by investigating 'the radioactivity of rivers some miles below any suspicious and secret plant'. Oppenheimer suggested that a few cubic centimetres of water collected downstream of a nuclear reactor and analysed in the US laboratory would be sufficient to determine if the reactor was in operation.[12]

Following up Oppenheimer's suggestion, Furman identified Lake Constance (Bodensee) and the upper reaches of the Rhine River, accessible from Switzerland, as locations where water sampling could be done without infiltrating German territory.[13] Even though that part of the Rhine is upstream from most of Germany, small rivers entering the Rhine and Lake Constance from the north could bring some traces of radioactivity with them, if, for instance, their water had been used as a reactor coolant. The German experimental reactor at Haigerloch was located about 100 kilometres north of the sampling locations identified by Furman.

In May 1944 Furman also consulted Alvarez, who provided instructions on procedures for collection of water samples, suggesting that water should be put into 'containers, with a label to indicate the geographical position and the date (including time of day) at which the sample was collected'. He also instructed that river water should be collected from the fast-flowing part, and 'lake water should be taken as near shore and as close to small rivers which empty into the lake on the German side'. 'Men in fishing boats, equipped with water bottles as part of a lunch basket' were supposed to collect samples without raising suspicion.[14] It is not clear if water samples were actually collected at those locations.[15]

[10] Ulam, S. M., *Adventures of a Mathematician* (Charles Scribner's Sons: New York, 1976), p. 210.

[11] Ulam, S.M., 'Stan Ulam: Vita—excerpts from *Adventures of a Mathematician*', *Los Alamos Science*, no. 15 (1987), p. 14.

[12] Richelson (note 3), p. 34.

[13] Richelson (note 3), p. 41.

[14] Richelson (note 3), p. 41.

[15] Ziegler and Jacobson (note 2), pp. 7–8.

In October 1944 the Alsos mission team, which followed the Allied forces across Europe to investigate German scientific developments, did collect water samples from the lower Rhine, near the Dutch town of Nijmegen. The sampling procedure recommended by Alvarez was followed diligently, and samples were promptly flown to the USA for analysis, but no trace of radioactivity was found in the water.[16]

A member of the Alsos mission that collected water at Nijmegen, Captain Robert Blake, brought samples to the mission's headquarters at the Hotel Royal Monceau in Paris for packaging and shipment to Washington, DC. Another officer, Russell A. Fischer, who had just returned to Paris from Marseilles, had brought some Roussillon wine with him. On a whim, Furman, who was also in Paris at the time, decided to add a bottle of the wine to the water samples as a joke, inscribing the label 'Test this for activity, too'. His intention was that the recipient would 'test' the wine by drinking it, but 'Washington did not get the joke'. A few days later Alsos received an urgent cable stating 'Water negative, wine positive, send more'. The explanations that wine and mineral water can naturally contain trace radioactivity did not have any effect, so Fischer was sent back to southern France 'on a ten-day mission to collect a sampling of French wines—two of each, one for Washington and a "file copy" for the office in Paris'.[17]

Environmental sampling for verification of the US nuclear explosions

During the planning for the first US nuclear test in 1945, code-named Trinity, the decision was made to collect samples of soil to calculate the explosion yield. The responsibility for this work was given to Herbert Anderson, a Columbia University graduate student who came to Los Alamos in November 1944, and Nathan Sugarman, from the Manhattan Project's Metallurgical Laboratory.[18]

It was recognized that, in the event of a successful test, the 'best samples would probably be found in the ground beneath the blast', and that it would be desirable to collect such samples as soon as possible after the event. After considering various ways to collect samples, such as by helicopter or airship, the decision was made to use two T4 Sherman tanks. One tank had compartments for the driver and the sample collector, with breathing equipment, that were lined with 2 inches (5 centimetres) of lead (weighing 11 tonnes in total). The sample collector was able to gather earth samples 'through a hole in the floor of his compartment with a vacuum cleaner or

[16] Ziegler and Jacobson (note 2), p. 8.

[17] Powers, T., *Heisenberg's War: The Secret History of the German Bomb* (Da Capo Press: Cambridge, MA, 2000), p. 362.

[18] Knight, J. D. and Sattizahn, J. E., 'Tracking the isotopes', *Los Alamos Science*, no. 8 (summer 1983), p. 6; and Hoddeson et al. (note 1), p. 358.

by driving a hollow pipe into the ground'. The other tank was equipped to fire rockets 'fitted with sampling noses and with cables for retrieval' into the centre of the crater from a distance of 500 yards (460 metres).[19]

The first sample collection at 'ground zero' was done about 90 minutes after the test in a shielded tank driven by Sergeant William Smith. Anderson, who collected the samples, became the first to discover trinitite—the glass produced from the desert sand by the Trinity explosion. Later in the day, the tanks made five trips through the crater to collect samples, which were brought back to Los Alamos for processing and radiochemical analysis.[20]

Anderson developed the method of sample analysis, which was based on first determining the efficiency of the nuclear device by measuring the ratios of fission products to the remaining plutonium, and then calculating the yield from the determined efficiency and the known amount of plutonium in the device.[21] The radiochemical analysis worked, and after one week Anderson reported to Oppenheimer that 'the blast had an 18 percent efficiency and was equivalent to 20 000 tons of TNT' (i.e. 20 kilotons).[22] The determination of efficiency and yield by Anderson's method was 'probably accurate to ±10 per cent'.[23]

The idea to detect nuclear explosions from a distance by collecting airborne debris emanating from them was not put forward until shortly after the Trinity test. It stemmed from the 1888 report by the Royal Society of London on the effects of the 1883 eruption of the Krakatoa volcano in what is now Indonesia.[24] Turkevitch and John Magee of the Manhattan Project's Metallurgical Laboratory read that report 'as an indication of the type of long range effects' that might be expected from a nuclear explosion. In order to collect the dust from a nuclear explosion, an aeroplane was outfitted with filters.[25] The filters consisted of the 'AirMat' soft tissue paper used in aircraft air intake filters, wrapped around the inside of a 'perforated sheet iron' cylinder, 24 inches (60 cm) in diameter and 24 inches (60 cm) long. A contemporary Los Alamos report described the debris-collection apparatus in the following way:

[19] Knight and Sattizahn (note 18), p. 7; Szasz, F. M., *The Day the Sun Rose Twice: The Story of the Trinity Site Nuclear Explosion, July 16, 1945* (University of New Mexico Press: Albuquerque, NM, 1984), pp. 116–17; and Bainbridge, K. T., 'A foul and awesome display', *Bulletin of the Atomic Scientists*, vol. 31, no. 5 (May 1975), pp. 44, 46.

[20] Szasz (note 19), p. 117.

[21] Bainbridge (note 19), p. 41.

[22] Szasz (note 19), p. 117.

[23] Knight and Sattizahn (note 18), p. 7.

[24] Symons, G. J. (ed.), *The Eruption of Krakatoa and Subsequent Phenomena: Report of the Krakatoa Committee of the Royal Society* (Trübner & Company: London, 1888).

[25] A. Turkevitch, letter to D. Jacobson, 27 Mar. 1990, quoted in Ziegler and Jacobson (note 2), p. 38.

To hold the paper in place it was placed between two layers of copper screen wire before being fitted into the iron cylinder. The ends of the cylinder had brass disks clamped over them. This device was supported by two wooden cross bars in the forward bomb-bay of a B-29 airplane. The air to be filtered came in through a scoop mounted above the bomb-bay and was led down to the filter through a section of six-inch-diameter pipe. This terminated at the center of one of the brass end-disks of the filter.[26]

By 10 August 1945, the first time a B-29 bomber from the US Army's Second Air Force was fitted with such a filter, the explosions over Hiroshima and Nagasaki, Japan, had both already taken place. A total of five sampling flights were flown in the USA between Bakersfield in California, Wendover Field in western Utah and Seattle in Washington. The chosen flight routes added an additional complication, since it was possible that the filters would pick up some radioactivity from plutonium-producing facilities at Hanford, Washington. Thus, although low levels of artificial radioactivity were indeed found on the filters, its unambiguous attribution to a specific explosion was not possible. The flights proved the feasibility of the techniques of radioactive debris collection with airborne filters and their subsequent analysis.[27] However, long-range debris collection and analysis were still in their infancy, and their effectiveness and reliability remained far from proved.

Further experience of post-explosion debris collection and analysis was gathered during the 1946 Operation Crossroads at Bikini Atoll, in the US-administered territory of the Marshall Islands. It consisted of two tests: an airburst, Able, conducted on 30 June, and an underwater explosion, Baker, on 24 July. Both were explosions of a slightly modified version of the Fat Man plutonium device exploded over Nagasaki.[28] Two kinds of post-explosion debris sample were gathered in the vicinity of the test site: from the atmosphere by remotely controlled, unmanned B-17 bombers and F-6F fighters and from the sea by remotely operated boats.[29] Collection of debris samples from the Baker test was anticipated to be especially difficult since no debris collection from an underwater explosion had been attempted before.[30]

The short-lived isotopes in the debris, such as ^{97}Zr and ^{99}Mo from fission and ^{239}Np from neutron capture of ^{238}U, were measured nearby in a laboratory set up on Kwajalein Atoll. The longer-lived fission products, such as ^{89}Sr, ^{95}Zr, ^{140}Ba and ^{144}Ce, were measured at Los Alamos, as was the ratio of

[26] Blair, J. M. et al., 'Detection of nuclear-explosion dust in the atmosphere', Report no. LA-418, Los Alamos National Laboratory, 2 Oct. 1945, <http://www.fas.org/sgp/othergov/doe/lanl/docs1/00423503.pdf>, pp. 2–3.

[27] Blair et al. (note 26).

[28] Hansen, C., *Swords of Armageddon*, vol. 7 (Chukelea Publications: Sunnyvale, CA, 2007), p. 86.

[29] Knight and Sattizahn (note 18), p. 8.

[30] Hawkins, Truslow and Smith (note 1), p. 275.

^{239}Pu to total plutonium in the debris. This ratio provides information about the fraction of a weapon's ^{239}Pu content spent in non-fission neutron reactions, such as (n,2n) and (n,γ). This information is useful for more precise calculation of weapon efficiency and yield.[31] The analysis of the fission-product measurements after Operation Crossroads demonstrated for the first time the phenomenon of fractionation.[32] Despite anticipated difficulties, the radiochemical method of efficiency and yield determination worked well for both tests, and its results were confirmed by other methods.[33]

The long-range debris collection and analysis experiments after Operation Crossroads were less sophisticated. The debris samples were indeed collected by aircraft equipped with filters 'at points thousands of miles from the explosions', but the quantity of fission products in the samples was too small for the radiochemical analysis procedures available at the time. Only general radioactivity measurements were conducted.[34] Thus, by the end of 1947 the long-range detection and attribution of nuclear explosions by analysis of radioactive debris was a promising, but certainly not proved, idea. The technology for sample attribution (i.e. unambiguous identification of the cause of radioactivity in the samples) was still missing.

On 14 December 1947 a section of the US Air Force specializing in the long-range detection of nuclear explosions was created and designated AFMSW-1.[35] Dr Ellis Johnson was recruited to head the technical development of long-range detection techniques. By that time the decision had already been taken to conduct three weapon tests in the Pacific Ocean in the spring of 1948 as part of Operation Sandstone. Its main purpose was to test new design principles of nuclear weapons: levitated and composite uranium–plutonium cores.[36] Additionally, a programme of long-range detection research related to Sandstone was planned and given the code name Fitzwilliam on 31 December 1947.[37]

During preparation for Operation Fitzwilliam, Johnson chose a small private company, Tracerlab, to provide a large part of the equipment, together with the training and personnel to maintain it. Tracerlab was also given a contract to develop the radiochemical techniques needed for attribution of samples from Operation Fitzwilliam. A scientist affiliated with Tracerlab, Dr Lloyd R. Zumwalt, stated that:

[31] Knight and Sattizahn (note 18), p. 8.

[32] For a definition of 'fractionation' see the glossary in this volume. See also chapter 2 in this volume.

[33] Hawkins, Truslow and Smith (note 1), p. 275.

[34] Ziegler and Jacobson (note 2), p. 111.

[35] The letters AF stand for Air Force, M for Deputy Chief of Staff for Materiel and SW-1 for Special Weapons Group, Section One.

[36] Hansen (note 28), p. 86.

[37] Ziegler and Jacobson (note 2), pp. 95, 113.

I had to use the coming A-bomb test in the Pacific as the means of trying to see if long range detection was possible . . . scientists at Los Alamos had developed a technique of sampling clouds very close in and subjecting the samples to radiochemical analysis to get information on the efficiency of the given atomic explosion. It was hoped that perhaps sufficient radioactivity would be collected at some considerable distance in order to derive similar information. . . . It was clearly stated that we were looking for Russian explosions.[38]

Tracerlab prepared its main laboratory in Boston, Massachusetts, as well as field stations in Guam, Hawaii and Kwajalein Atoll for analysis of filters with debris samples. The company introduced a novel autoradiography procedure for finding radioactive particles on filters: filters were placed in contact with sheets of X-ray film, which was developed after some exposure period. The particles' radioactivity produced dark spots in the developed film, thus creating a 'map' of the radioactive particles in the filter. Portions of filters with particles were cut out and dissolved, leaving bare particles for further radiochemical analysis.[39]

Samples were to be obtained by aircraft, ships and ground-based stations equipped with precipitation collectors to gather rainwater and 'high-speed blowers and filters' as part of a massive collection operation.[40] Samples from all three tests of Operation Sandstone—X-Ray, Yoke and Zebra—were collected by various means. It was found that ground-based 'blowers' with filters remained useful in terms of collecting informative debris up to 2000 miles (3200 km) away from the blast. Rainwater collection was found useful at a range of 9000 miles (14 500 km; see below), although its reliability was limited since this technique was highly dependent on the weather. Finally, debris collection by aircraft-mounted filters yielded 'samples of sufficient strength to allow radiochemical analysis' at a range of up to 12 000 miles (19 300 km) from the test site.[41] By obtaining these results, Operation Fitzwilliam proved the feasibility of long-range detection of ground- and airbursts and, ultimately, gave impetus to two parallel projects carried out by the US Air Force (the Interim Surveillance Research Net) and the US Navy (Project Rainbarrel). Additionally, the USA had cooperated with the United Kingdom since 1948 to complement its radioactive debris detection capabilities with British collection aircraft based in Gibraltar, Northern Ireland and Scotland.[42] These three components constituted a system that detected the first Soviet nuclear test in 1949.

[38] Ziegler and Jacobson (note 2), p. 125.
[39] Ziegler and Jacobson (note 2), p. 137.
[40] Ziegler and Jacobson (note 2), pp. 134–35.
[41] Ziegler and Jacobson (note 2), pp. 134–35.
[42] Goodman, M. S., *Spying on the Nuclear Bear: Anglo-American Intelligence and the Soviet Bomb* (Stanford University Press: Stanford, CA, 2007), pp. 44–46.

The Interim Surveillance Research Net

Operation Fitzwilliam was terminated on 6 June 1948. It was realized, however, that most of the radiological stations located in the northern Pacific used in that operation could prove useful for detection of any future Soviet nuclear test because the prevailing west winds would carry radioactive debris out of the Soviet Union in an eastward direction. Of 25 ground-based debris collection stations, only two were moved to the Atlantic Ocean to better prepare the network for detection of the Soviet test. The debris-collection aircraft and Tracerlab's laboratories were also maintained in operational condition in preparation for detection of a Soviet test. In July 1948 AFMSW-1 was transformed into AFOAT-1, with the 'O' indicating the addition of operational functions to existing research activities.[43] AFOAT-1 sponsored research at Tracerlab that focused on remedying the shortcomings of the methods of radiochemical analysis that were found during Operation Fitzwilliam.[44]

By 1949 the Interim Surveillance Research Net consisted of a number of WB-29 aircraft equipped for collection of both particulate debris and noble gases and ground stations with equipment for measuring air radioactivity and collecting precipitating debris. The aerial collection was done by regular flights of four very long range (VLR) weather reconnaissance squadrons, stationed in Alaska, California and Guam in the Pacific and Bermuda in the Atlantic. The ground stations were placed in an arc covering the eastern and western rims of the Pacific, with some additional stations in the Atlantic. Tracerlab's laboratories in Berkeley, California, and Boston, Massachusetts, handled the samples from the network, using the procedures and methodology of radiochemical analysis and radioassay of fission products that had been significantly improved during the year after Operation Fitzwilliam in cooperation with the Radiochemistry Group of Los Alamos Laboratory.[45]

Project Rainbarrel

The US Navy established the Naval Radiological Defense Laboratory after the 1946 Operation Crossroads, equipped several of its stations to monitor atmospheric radioactivity and, in general, remained interested in long-range detection of nuclear tests.[46] In the spring of 1947, after a request from the US Atomic Energy Commission, the Naval Research Laboratory (NRL) also began developing 'apparatus for detecting, collecting and measuring

[43] The letters AF stand for Air Force, O for Deputy Chief of Staff for Operations and AT-1 = Atomic Energy Office, Section One.

[44] Ziegler and Jacobson (note 2), pp. 178.

[45] Ziegler and Jacobson (note 2), pp. 186–87.

[46] Ziegler and Jacobson (note 2), p. 88.

airborne radioactive debris', which later participated in Operation Fitzwilliam.[47]

In early 1948, during tests of a large prototype gamma-ray detector, the NRL scientists observed the effect of rain 'scavenging' radioactivity from the atmosphere and, thus, concentrating it in rainwater.[48] In June 1948, two months after the Sandstone tests on Eniwetok Atoll in the Marshall Islands, the NRL scientists learned that the US Virgin Islands in the Caribbean relied on collected rain as a source of potable water and so had large volumes of it in 'concrete cisterns'.[49] On 6 June 1948 a suitable amount of water sufficiently old to contain radioactivity from the Sandstone tests was located on St Thomas Island, 14 500 km away from Eniwetok Atoll. After concentrating radioactive debris from 2500 gallons (6464 litres) of water, the rare-earth isotopes ^{91}Y, ^{141}Ce and ^{144}Ce were separated at the NRL, and it was shown that their 'ratios were in the correct proportions to be fallout from the Sandstone test'.[50] It is these results—obtained by the US Navy, and not the US Air Force's own attempts at rainwater collection and analysis—that were referenced in the report after Operation Fitzwilliam as proof of the possibility of registering a nuclear explosion by rainwater collection at a distance of 9000 miles (14 500 km).

Following on this success, the NRL collected pond water samples at Shemya in the Aleutian Islands, at Kodiak Island, both Alaska, and at Truk Lagoon (now known as Chuuk Lagoon) in the US-administered territory of Micronesia in the central Pacific. Apart from water, some 'mossy growth' was collected at Truk Lagoon, which was one of the first times that vegetation was collected for nuclear forensic purposes.[51]

Detection and analysis of debris from the first Soviet test

The first Soviet nuclear weapon test, designated RDS-1 in the Soviet Union, and referred to as Joe-1 in the West, was conducted on 29 August 1949 at the Semipalatinsk test site, Kazakhstan. By the spring of 1949 AFOAT-1 had coordinated its routine airborne particle-collection flights and rainwater collection with the British, as discussed above. All these efforts proved useful in September 1949.

[47] Friedman, H., Lockhart, L. B. and Blifford, I. H., 'Detecting the Soviet bomb: Joe-1 in a rain barrel', *Physics Today*, vol. 49, no. 11 (Nov. 1996), p. 38.
[48] Friedman, Lockhart and Blifford (note 47), p. 38.
[49] Friedman, Lockhart and Blifford (note 47), p. 40; and Ziegler and Jacobson (note 2), p. 190.
[50] Friedman, Lockhart and Blifford (note 47), p. 40.
[51] Friedman, Lockhart and Blifford (note 47), p. 40.

Joe-1 debris in airborne filters

Flights conducted along two routes—from Fairbanks, Alaska, to the North Pole and from Fairbanks to Yokota, Japan—in order to analyse air masses travelling eastward from Soviet territory proved to have a special significance. On 3 September 1949 the WB-29 aircraft returning to Fairbanks from Japan collected the first traces of radioactive particles, which were presumably carried to the Pacific from the Soviet nuclear test site in an air mass. During the following days an all-out effort was made to collect as many samples as possible. In the USA alone, between 3 and 16 September 1949, 92 sample-collection flights were conducted, resulting in the collection of over 500 radioactive samples.[52] Since the air mass containing particles had moved over the territory of North America to the northern regions of the Atlantic, the USA also enlisted the help of the British atomic energy authorities and the British Royal Air Force, which launched an independent sample-collection and radiochemical analysis effort.[53]

The abundance of sampling results permitted meteorologists to delineate a cloud shape and conclude that it was 'typical of a tracer whose initial distribution is in a vertical column', such as a volcanic eruption or a ground burst nuclear explosion.[54] Analysis by Tracerlab and the Radiochemistry Group of the Los Alamos Laboratory of the particles collected on airborne filters revealed the presence of fission products: ^{140}Ba and ^{99}Mo, as well as ^{111}Ag, ^{140}La, ^{91}Y, ^{95}Zr, ^{144}Ce, ^{144}Pr and other isotopes.[55] The calculations, based on measured fission product ratios, showed that all fission products were of the same age, providing the first 'hint' that their probable origin was a bomb explosion, rather than a nuclear reactor accident. The fission product yield curve was also more consistent with the fission of plutonium than of highly enriched uranium (HEU), so the US scientists guessed that the Soviet nuclear weapon was plutonium-based and, therefore, an implosion-type bomb.[56] In addition, the Tracerlab scientists tested the particles for traces of ^{239}Np, an isotope produced when ^{238}U absorbs a neutron.[57] The test allowed the conclusion to be drawn that the Joe-1 bomb

[52] Northrup, D. L. and Rock, D. H., 'The detection of Joe-1', *Studies in Intelligence*, vol. 10, no. 3 (fall 1966), p. 30.

[53] Goodman (note 42), pp. 46–49; and Ziegler and Jacobson (note 2), pp. 204–11.

[54] Machta, L., 'Finding the site of the first Soviet nuclear test in 1949', *Bulletin of the American Meteorological Society*, vol. 73, no. 11 (Nov. 1992), p. 1805.

[55] 'Al Ghiorso recalls a moment in time', *Actinide Research Quarterly*, vol. 13, no. 1/2, (spring/summer 2007), p. 13; Machta (note 54), p. 1800; and Northrup and Rock (note 52), p. 32.

[56] The fission product yield curve, sometimes referred to as the 'Mae West curve' due to its characteristic 2-peak form, is a graph of the mass or mole yield of fission products against their atomic number. Its shape depends on the fissile isotope and the energy of the neutrons inducing fission. See e.g. Saey (note 7), p. 43.

[57] Northrup and Rock (note 52), p. 31.

probably had a layer of natural uranium as a tamper and neutron reflector.[58]

At that time, neither seismic nor acoustic signals from the explosion were of any help in pinpointing the location of the Soviet test. That information also had to be obtained by radiochemical analysis of debris. Analysing the known meteorological data, meteorologists made backward projections of the trajectories of the air masses.[59] The calculated age of the radioactive isotopes in the samples gave an estimate of the time of the event: sometime between 03:00 UTC on 26 August and 03:00 UTC on 29 August 1949. This provided the cut-off time at which to stop the backward projection of air mass trajectories and thus defined, accurately but not precisely, the area where the test had been conducted.[60]

The real time of the event was apparently 02:00 UTC on 29 August 1949.[61] This was almost outside of the estimated time interval provided by radiochemical analysis. It has been suggested that the source of error could have been the fact that the fission time (i.e. the explosion time) was calculated from the ratio of ^{140}Ba and ^{99}Mo in the sample. It was assumed that the ratio was the same as it had been immediately after fissioning with an allowance for radioactive decay. It was not recognized at the time that most of the ^{140}Ba in the sample had not appeared directly from fission, but rather through the decay of the short-lived ^{140}Xe parent. This would cause significant fractionation of ^{140}Ba and ^{99}Mo (^{140}Ba would have a different condensation history than ^{99}Mo), and negate the original assumption, causing the error. Reportedly, the subsequent US nuclear test experience led to discontinuing the use of ^{140}Ba for yield-determination purposes.[62]

Bomb debris in a rainwater barrel

The Tracerlab and Los Alamos Laboratory analyses of airborne filters provided a relatively good indication that a nuclear device similar in design to the one exploded in the Trinity test had been detonated on a tower on the territory of the Soviet Union at the end of August 1949. However, a significant part of the US administration, including the Secretary of Defense, Louis A. Johnson, found it politically difficult to accept the fact that the Soviet Union had produced a nuclear weapon earlier than expected.[63] It was necessary to 'increase the power of analyses' by comple-

[58] For definitions of 'tamper' and 'neutron reflector' see the glossary in this volume.

[59] Machta (note 54), pp. 1798–99.

[60] Machta (note 54), pp.1801–1804; and Ziegler and Jacobson (note 2), pp. 204–11. On the terms accurate and precise see appendix 2A n this volume.

[61] Mikhailov, V. N. (ed.), [Nuclear tests in the USSR], vol. 1 (RFYaTs-VNIIEF: Sarov, 1997), p. 188 (in Russian).

[62] Machta (note 54), p. 1805.

[63] Ziegler and Jacobson (note 2), p. 208.

menting the measurements of the fission products with analysis of the plutonium itself.[64]

One of the bomb debris detector types set up by the NRL was simply rainwater barrels collecting several hundred square metres of runoff from the roof of buildings in Kodiak, Alaska, and Washington, DC. The collection barrels were collocated with filter paper-type air samplers continuously scanned for gamma activity. If a gamma signal were detected from the filter papers, the operators would add to the reservoir 0.2 grams of aluminium sulphate per gallon of water (*c.* 0.05 grams per litre) to produce aluminium flocculent. This procedure would precipitate matter suspended in the water and make it available for further chemical treatment and analysis. Precipitating solids by this method would essentially concentrate a sample by up to about seven orders of magnitude from a few hundred litres of water. In the periods 6–12 September and 13–17 September 1949, large gamma signals were detected and precipitation of the water took place.[65]

The NRL collected the samples and shipped them to Los Alamos, where they were chemically treated and purified for further analysis by alpha spectroscopy. Available samples turned out to be too small for the only 'alpha energy analyser' available at the time in Los Alamos. The samples had to be shipped to the team headed by Glenn Seaborg at the University of California, Berkeley, which was known for discovering new transuranium elements by analysis of alpha energies at extremely low disintegration rates. At Berkeley, the task of analysing plutonium in Joe-1 debris was given to Albert Ghiorso.[66]

First, the question was asked whether the plutonium in the samples had originated from a reactor accident or a nuclear explosion. Ghiorso was able to show that the plutonium in the available samples contained about 4 per cent of the isotope ^{238}Pu.[67] One of the 'production pathways' for ^{238}Pu is from ^{239}Pu in an (n,2n) reaction on fast neutrons: $^{239}Pu(n,2n) \rightarrow {}^{238}Pu$.[68] Since fast neutrons are abundant in a nuclear explosion but are quickly moderated in a plutonium production reactor, Ghiorso could conclude that the content of ^{238}Pu present was more consistent with a nuclear test than a reactor explosion. Moreover, he was aware of the 'typical signature of the alpha spectrum of plutonium after an atomic-bomb explosion' due to his experience with the measurements he had taken after the Trinity test. He

64 'Al Ghiorso recalls a moment in time' (note 55), p. 14.

65 Friedman, Lockhart and Blifford (note 47).

66 'Al Ghiorso recalls a moment in time' (note 55), p. 16; Northrup and Rock (note 52), p. 33; and Reed, T. and Kramish, A., 'Trinity at Dubna', *Physics Today*, vol. 49, no. 11 (Nov. 1996), p. 34.

67 'Al Ghiorso recalls a moment in time' (note 55), p. 17.

68 For further discussion see Luksic, A. T. et al., 'Isotopic measurements: interpretation and implications of plutonium data', Institute of Nuclear Materials Management (INMM), *51st Annual Meeting of the Institute of Nuclear Materials Management 2010 (INMM 51)* (INMM: Deerfield, IL, 2011).

stated that the amount of ^{238}Pu formed in the explosion is 'approximately proportional to the efficiency of the bomb', which means that the ratio of the alpha peak at 5.5 megaelectronvolts (corresponding to ^{238}Pu) to the peak at 5.16 MeV (corresponding to ^{239}Pu and ^{240}Pu) 'gives a good idea of the efficiency of the explosion'. In the case of the debris from Joe-1, Ghiorso calculated that ratio to be 4.8 per cent, and stated that this ratio was 'roughly half that produced in the Trinity Test'.[69] The efficiency of the Trinity device determined by Ghiorso is not clear. However, Herbert Anderson placed it at about 18 per cent (see above). Assuming that Ghiorso's result for Trinity was about the same, then his result for the efficiency of Joe-1 would have been about 9 per cent.

Second, Ghiorso devised a way to confirm that Joe-1 used reactor-bred plutonium as the explosive (as opposed to HEU explosive breeding some plutonium in the explosion from the tamper material). He proposed to look at the ratios of ^{241}Pu to ^{239}Pu.[70] Plutonium-241 is formed by a double neutron (non-fission) capture on plutonium-239: $^{239}\text{Pu} + \text{n} \rightarrow {}^{240}\text{Pu} + \text{n} \rightarrow {}^{241}\text{Pu}$. The intermediate isotope, ^{240}Pu, has a very low fission cross section at thermal energies but fissions reasonably well in a fast (*c.* 1 MeV and more) neutron environment. Hence, when ^{240}Pu is bombarded by neutrons in a reactor, there will be few fissions and almost all interactions will be captures, producing a relatively significant amount of ^{241}Pu. However, when ^{240}Pu is bombarded by neutrons in a nuclear bomb explosion, it will preferably fission, and therefore much less ^{241}Pu will be produced. Thus, the resulting ratio of ^{241}Pu to ^{239}Pu will be much higher in the case of the reactor. The relatively high ratio of ^{241}Pu to ^{239}Pu, measured by Ghiorso, served as yet another indication that the plutonium had been created in a reactor and therefore Joe-1 was a plutonium bomb. This was a relatively easy conclusion because very little plutonium is produced in an HEU nuclear explosion.

Third, and more importantly, Ghiorso realized that measuring the $^{241}\text{Pu} : {}^{239}\text{Pu}$ ratio could also provide insight into the operation mode of the reactor that produced plutonium used in a weapon. The amount of ^{241}Pu depends on how long the reactor had been operating before the fuel was removed from it to be processed. The ratio of ^{241}Pu to ^{239}Pu could therefore be used to extract information concerning the fuel irradiation time in a reactor. Assuming that the Soviet Union had produced its plutonium in reactors similar in neutron spectrum to the graphite-moderated thermal reactors at Hanford, and checking his measurements and calculations against samples of the US plutonium with various irradiation times,

[69] 'Al Ghiorso recalls a moment in time' (note 55), pp. 17, 19.
[70] 'Al Ghiorso recalls a moment in time' (note 55), p. 21.

Ghiorso concluded that the reactor that discharged the plutonium used in the Joe-1 test had operated for about one year.[71]

Finally, although Ghiorso probably could not look at the amount of ^{241}Am in the debris because the samples available to him were highly chemically purified, such analysis could have been very useful.[72] Americium-241 is a decay product from the beta decay of ^{241}Pu. When plutonium is reprocessed from a plutonium-production reactor, most of the ^{241}Am that is present is chemically removed. Hence, the ratio of ^{241}Am to ^{241}Pu found in a weapon's plutonium before detonation can serve as a chronometer, measuring the time since the separation and purification of plutonium in reprocessing. Only a tiny amount of ^{241}Am would have been produced in the bomb blast because its parent, ^{241}Pu, is itself produced in small amounts and there is no time for that parent to decay to ^{241}Am. Almost all of the ^{241}Am and ^{241}Pu in the debris would have come from the pre-detonation plutonium with their ratio retaining significance as a chronometer, measuring the time since the most recent chemical purification.

To sum up, the radiochemical analysis of debris from the first Soviet test was not just a matter of collecting the debris, but also of careful analysis of it by the USA. Such analysis revealed the approximate date and location of the test and indicated that the Soviet Union had exploded, most likely on a tower, a plutonium-based implosion device that had a natural uranium tamper, with plutonium that probably came from fuel irradiated in a reactor for about a year. The analysis also indicated that the efficiency of the Soviet device was half that of Trinity. It was not possible to discern the yield from the radiochemical data alone since the total amount of plutonium put into the device was not known to the USA.[73]

As demonstrated by many later publications, most of Ghiorso's estimates were correct. However, his estimates of the efficiency and reactor fuel irradiation time of Joe-1 are questionable. There is a well-substantiated claim that the reactor fuel was irradiated for about 100 days, and not 1 year.[74] The efficiency of Joe-1 is not available from known open sources, but it is widely accepted it was a similar device to that exploded in the Trinity test, including in terms of the plutonium content and achieved yield. As noted above, that device's efficiency was calculated by one method to be around 18 per cent, a significant difference from the estimate of 9 per cent.

[71] 'Al Ghiorso recalls a moment in time' (note 55), p. 21.

[72] 'Al Ghiorso recalls a moment in time' (note 55), p. 21.

[73] That data was later obtained by re-examining the records of another method—sonic detection. Ziegler and Jacobson (note 2), p. 210.

[74] Diakov, A., 'The history of plutonium production in Russia', *Science & Global Security*, vol. 19, no. 1 (2011), p.33; Ryabev, L.D., ed., *Atomnyjproekt SSSR: Dokumenty i Materialy* [USSR Atomic Project: Documents and Materials], 'Igor Kurchatov's Suggestion on Increasing Plutonium Production at A Reactor, 9 April 1949,' Document 244, Vol. II, Atomic Bomb, 1945–1954, Book 4, p. 638 (Moscow-Sarov, Nauka-Fizmatlit, 2003).

II. The first instances of nuclear forensic analysis in the Soviet Union

Analysis of debris from Soviet domestic tests

In early 1947, over two years before the first nuclear weapon test in the Soviet Union, the Radium Institute of the Soviet Academy of Sciences (Radievyi Institut Akademii Nauk, RIAN) received an assignment from Joseph Stalin, the Soviet leader, to develop a radiochemical method to determine nuclear weapon efficiency (i.e. the fraction of fissile material actually fissioned in the explosion).[75] This work, assigned to the group headed by Georgii M. Tolmachev, initially focused on developing a method to determine the efficiency and yield of the first Soviet nuclear test, on 29 August 1949.[76]

The RIAN methodology relied on the determination of activity ratios to ^{239}Pu of fission products.[77] Since the fission products by definition originate from fissioned plutonium, their ratios to remaining plutonium allow the calculation of the fraction of plutonium that has fissioned in the explosion. The total amount of plutonium in the weapon is known to the designers, so they can readily calculate how much of it has fissioned and determine the total energy of fission (i.e. the weapon's yield). The concentration of fission products in a sample, such as ^{99}Mo, ^{95}Zr, ^{95}Nb, ^{137}Cs and ^{144}Ce, was measured by their specific beta and gamma activity.[78] The concentration of ^{239}Pu was measured by a 'radiometric device for measuring α-activity against a high background of β- and γ-activity'.[79]

The designers of the first Soviet nuclear test prepared a number of means to collect samples for the determination of explosion parameters. First, two specially shielded battle tanks were driven to ground zero approximately 10 minutes after the test. One of the tanks conducted radioactivity measurements and collected samples of soil that had been fritted (i.e. fused to form glass).[80]

Second, a more systematic sample-collection effort was conducted the day after the test, on 30 August 1949. A three-person team equipped with personal protection gear, a metal scoop, tarpaulin sacks and a normal car

[75] Dubasov, Yu. V., [Radiochemical studies of consequences of nuclear explosions, serious radiological accidents and radioactive waste disposal], *Proceedings of the V. G. Khlopin Radium Institute*, vol. 10 (2003), p. 33 (in Russian).

[76] Mikhailov, ed. (note 61), p. 195.

[77] Mikhailov, V. N. (ed.), [Nuclear tests in the USSR], vol. 2 (RFYaTs-VNIIEF: Sarov, 1997) (in Russian).

[78] Mikhailov, ed. (note 77); and Kruglov, A. K., [How the nuclear industry was created in the USSR] (TsNIIAtominform: Moscow, 1995), p. 157 (in Russian).

[79] Dokuchaev, Ya. P., [Concerning the test of the first nuclear bomb], *Ohrana Prirody Yuzhnogo Urala*, 2008, p. 74.

[80] Kruglov (note 78), p. 158; and Holloway, D., *Stalin and the Bomb: The Soviet Union and Atomic Energy, 1939–1956* (Yale University Press: New Haven, CT, 1994), pp. 214, 217–18.

collected samples at ground zero and at various radial distances from it. The team spent about 5–7 minutes there collecting samples and reported that they found at ground zero an area of molten and resolidified sand with a radius of approximately 250 metres. The same procedure was repeated by another team the next day.[81] Collected samples contained coagulated drops of metal from the bomb tower that had been vaporized and then condensed with inclusions of radioactive debris (this substance is colloquially known as kharitonki or kharitonchiki, after Yuli B. Khariton, the chief designer of the Soviet bomb). The RIAN group headed by Tolmachev used those to determine the device's efficiency and explosion yield.[82]

Third, in 1947 the leader of the Soviet nuclear weapon project, Igor V. Kurchatov, had requested that the Aerosols Laboratory of the Karpov Institute of Physical Chemistry in Moscow develop a method to collect radioactive aerosols from the radioactive cloud that would form after the nuclear weapon test. The Aerosols Laboratory, headed by Igor V. Petryanov-Sokolov, developed a specialized synthetic fibrous filter material for this purpose.[83] This work was probably completed by 1948 because in that year the Central Aerohydrodynamic Institute (Tsentralniy Aerogidrodinamicheskiy Institut, TsAGI) in Zhukovsky, Moscow Oblast, had begun to develop 'aerogondolas' (i.e. enclosed compartments suspended from, in this case, an aeroplane) for the YaK-9v aircraft, which would contain the filter material manufactured by the Karpov Institute.[84] For the purposes of sample collection from the radioactive cloud, the YaK-9v, a training fighter aircraft, was modified to be remotely controlled and unmanned. Five such aircraft were manufactured by August 1949. All suffered from landing difficulties and sustained damage in trial runs. Given the complicated weather conditions on the day of the test and the less-than-ideal performance of the YaK-9v aircraft, Kurchatov cancelled their use for sample collection.[85]

The RIAN methodology was not designed for remote detection of nuclear explosions. It was supposed to provide one of a number of measurement techniques to characterize the first Soviet nuclear test. How-

[81] Kruglov (note 78), p. 159.

[82] Vasil'ev, A. P., *Rozhdennaya Atomnym Vekom* [Created by the nuclear age], vol. 1 (Self published, Moscow, 2002), p. 208; and Mikhailov, ed. (note 61).

[83] Ogorodnikov, B. I., 'The development of technical means for aerosol method of detection and control of nuclear explosions', Proceedings of the Second International Symposium on the History of Atomic Projects (HISAP'99), International Institute for Applied Systems Analysis (IIASA), Laxenburg, 4–8 Oct. 1999, p. 1. The proceedings of HISAP'99 were prepared but never published. Referenced texts are provided to SIPRI by the Kurchatov Institute, Moscow. Information on the symposium is available at <http://webarchive.iiasa.ac.at/Admin/INF/PR/PR-99-10-08.html>.

[84] Dyachenko, A. A. (ed.), *Opalennye v bor'be pri sozdanii yadernogo schita Rodiny* [Scorched in the struggle to create the nuclear shield of the motherland] (Poligraf-Service: Moscow, 2008), p. 348.

[85] Dyachenko, ed. (note 84), p. 348; and Zhuchikhin, V. I., *Pervaya atomnaya* [First nuclear bomb] (IzdAT: Moscow, 1993), p. 93.

ever, analysis of the measurement results conducted shortly after the test suggested the possibility of remote detection of nuclear explosions. First, the efficiency and yield of the device were determined, as described above. Second, analysis of radioactive debris from the first test demonstrated the possibility that the type of fissile material used in a weapon could be determined.[86] Third, radioactivity measurements conducted for health and safety purposes were able to detect radioactivity from the test fallout as far as 720–750 km from the test site.[87]

The aerial collection of fallout was successfully tried by the Soviet Union again in 1951. By the autumn of 1951 Li-2 military transport aircraft were equipped with filtering aerogondolas similar to those developed for the YaK-9v.[88] The yield of the second Soviet nuclear weapon test, conducted on 24 September 1951, was established radiochemically by determining the fraction of fissioned plutonium. The results were verified by an independent method of measuring the volume and temperature of the nuclear fireball.[89] It is not clear if an aircraft was used in the debris collection on this occasion.

Unlike the two previous tests, in the third Soviet test, on 18 October 1951, the nuclear weapon was dropped from a Tu-4 aircraft and detonated at a height of 400 metres. The Li-2 collected debris on its filters from various points of the 'stem' of the mushroom cloud 27 minutes after the explosion. Additional measurements were taken by a battle tank at the hypocentre 80 minutes after the explosion. Radiochemical analysis determined that the yield was 41.2 kilotons. This result was again verified by independent methods that involved measuring the size and temperature of the fireball and the shock wave parameters.[90]

Debris analysis after the first three domestic tests confirmed the viability of the Soviet radiochemical methods in general and served as the basis for further development of more sophisticated procedures. By 1953 the early methodology developed by Tolmachev at RIAN had been adapted for the analysis of thermonuclear tests by Vyacheslav N. Ushatskii.[91] In addition to RIAN, in 1952 the Laboratory of Measurement Equipment (Laboratoriya Izmeritel'nykh Priborov Akademii Nauk, LIPAN, later known as the Kurchatov Institute) began independent 'radiochemistry and radiometry analysis of the fallout'.[92]

[86] Vasil'ev, ed. (note 82), p. 184.
[87] Vasil'ev, ed. (note 82), p. 6; and Mikhailov, ed. (note 61), p. 196.
[88] Dyachenko, ed. (note 84), p. 350.
[89] Dyachenko, ed. (note 84), p. 350.
[90] Ogorodnikov (note 83), p. 2; and Dyachenko, ed. (note 84), p. 350.
[91] Vasil'ev, ed. (note 82), p. 7.
[92] Vasil'ev, ed. (note 82), p. 7.

Another group of methods, developed by Ovsei I. Leipunskii and his colleagues, exploited so-called 'neutron indicators'.[93] If a stable isotope is placed in a predefined position within the nuclear explosive device, it will be 'hit' by a neutron in the explosion, and transformed by a process of neutron activation into a new, unstable isotope (i.e. activation product). Activation products can later be detected in debris, and their relative amount allows calculation of neutron flux at that predefined position, as well as the neutron yield of the whole device. By choosing stable isotopes that are only activated by neutrons of a certain energy, information can be obtained about the energy spectrum of neutrons inside the nuclear explosive device.[94] Some activation products are only produced by neutrons originating from fusion reactions and can serve as indicators of thermonuclear yield (see chapters 6 and 8). According to the official history of the Soviet nuclear tests, 'high-threshold neutron indicators were used for registration of thermonuclear neutrons in the first thermonuclear explosion on 12 August 1953'.[95] The nuclear weapon tested on that day in the Soviet Union is known to be of a *sloika* (layer cake) design, which uses lithium deuteride as thermonuclear fuel and where several layers of uranium-238 are interleaved with sheets of thermonuclear fuel.[96] Another method of radiochemical analysis of weapon debris, reportedly developed for characterization of the same test, was based on registration of ^{7}Be, which is produced in the reaction of deuterium with lithium isotopes.[97]

Analysis of debris from the US tests

The Soviet Union received initial confirmation that long-range detection of nuclear tests was possible in 1949. The first clue was provided by a statement released on 23 September 1949 by US President Harry S. Truman that 'within recent weeks an atomic explosion occurred within the U.S.S.R.'.[98] A few weeks after that, a Soviet spy, Harold 'Kim' Philby, obtained access to detailed information on the methods by which the UK and the USA had been able to detect the explosion.[99] It is not clear, however, how widely that information was distributed within the Soviet Union.

Successful radiochemical analysis of debris from the first three Soviet weapon tests gave further impetus to the Soviet programme of long-range

[93] Mikhailov, ed. (note 77).

[94] Mikhailov, ed. (note 77); and De Geer, L.-E., *CTBT Relevant Radionuclides*, Comprehensive Nuclear-Test-Ban Treaty Organization (CTBTO) Provisional Technical Secretariat (PTS) Technical Report PTS/IDC-1999/02 (CTBTO: Vienna, Apr. 1999), p. 19. See also chapter 6 in this volume.

[95] Mikhailov, ed. (note 77) (author's translation).

[96] Holloway (note 80), pp. 306–307.

[97] Holloway (note 80), p. 304.

[98] Ziegler and Jacobson (note 2), p. 211.

[99] Ziegler and Jacobson (note 2), pp. 214–15. See also chapter 2 in this volume.

detection of US nuclear explosions.[100] The first Soviet studies of the dissemination of radioactive aerosols in the atmosphere were initiated by Igor Kurchatov at LIPAN in 1951.[101] (Parallel studies were conducted by RIAN.) The work at LIPAN comprised two parts. First, Isaak K. Kikoin led the development of methods and equipment for collection, radioactivity measurement and analysis of radioactive fallout for the purpose of determining the date of an explosion.[102] This work included verification of the empirical Way–Wigner equation on the rate of decay of fission products, published in 1948.[103] Second, Boris V. Kurchatov (Igor Kurchatov's brother) led the activities related to the study of 'radiochemical and radiometric' debris analysis for the purpose of determining the design parameters of foreign nuclear and thermonuclear weapons.[104]

On 23 August 1952 two nuclear weapon designers, Andrei D. Sakharov and David A. Frank-Kamenetskii, submitted the first detailed proposal in the Soviet Union to use flights by regular aircraft equipped with filters along predefined routes as a means of long-range detection of foreign nuclear explosions. Their proposal included calculations of debris dilution in the atmosphere; discussion of ^{237}U as an indicator of fusion of deuterium and tritium (a D-T reaction), due to (n,2n) reaction on 14-MeV neutrons; analysis of the feasibility of direct detection of tritium in the atmosphere; and practical recommendations concerning measurements of gamma activity, fission products, tritium and ^{237}U in filters.[105]

Systematic daily measurements of atmospheric fallout began in 1953 in the Leningrad Oblast (probably using stationary passive debris collectors). Samples were first categorized on the spot by measuring their total beta activity, and then the 'informative' ones were sent to RIAN for detailed radiochemical characterization.[106] At that time the date of the explosion was determined by the total rate of decay of fission products, calculated on the basis of a modified Way–Wigner equation.[107]

Further radiochemical studies aimed to determine the nuclear explosive material (whether HEU or plutonium), weapon type (whether nuclear or

[100] Vasil'ev, ed. (note 82), p. 8.

[101] Vasil'ev, A. P., 'An initial stage of creation in the USSR of the system of detection of nuclear explosions', HISAP'99 (note 83), p. 1.

[102] Dubasov, Yu. V., Rimskii-Korsakov, A. A. and Ushatskii, V. N., 'Development in RIAN of the aerosol method for control and registration of foreign nuclear tests', HISAP'99 (note 83), p. 1.

[103] Way, K. and Wigner, E. P., 'The rate of decay of fission products', *Physical Review*, vol. 73, no.11 (1 June 1948) pp. 1318–30.

[104] Vasil'ev (note 101), p. 1; and Lobikov, E. A. et al., 'Development in the USSR of the physical methods of long-distance detection of nuclear explosions', HISAP'99 (note 83), p. 1.

[105] Vasil'ev, A. P., *Rozhdennaya Atomnym Vekom* [Created by the nuclear age], vol. 3 (Self published, Moscow, 2002), pp. 218–26.

[106] Dubasov, Rimskii-Korsakov and Ushatskii (note 102), pp. 1, 3.

[107] Dubasov, Rimskii-Korsakov and Ushatskii (note 102), p. 1.

thermonuclear) and the presence of ^{238}U tamper, as well as to calculate the fraction of the yield contributed by nuclear and thermonuclear reactions.[108] The methodology was developed under the direction of Boris Kurchatov.[109] It was based on determination of fission product yield curves and measurements of ^{239}Np and ^{237}U.[110] The fission product yield curves for uranium and plutonium differed in shape depending on the energy of the neutrons that caused fission. This difference can be quantified as a ratio of quantities of radionuclides located at the trough (such as ^{111}Ag, ^{89}Sr and ^{103}Ru) to the quantities of radionuclides located at the peak of the curve (such as ^{140}Ba). For example, according to the data available in the Soviet Union at the time, if ^{235}U is fissioned by a thermal neutron, the ratio of ^{111}Ag to ^{140}Ba will be about 0.01, and in the case of 'thermonuclear neutrons' that ratio will increase to 0.17.[111]

By 1953 the Soviet Union had accumulated enough information and experience to confirm the feasibility of the long-range detection of foreign nuclear tests and analysis of their features. This allowed it to launch a large-scale effort to create a network of ground-based and airborne fallout collectors. By 1954 the Soviet Union had deployed a network of 120 stationary passive debris collectors ('gauze trays'). It also launched regular sample-collection flights (at altitudes of 3–7 km) between Leningrad and Odessa and from bases in China, equipped with TsAGI filtering aerogondolas, similar to those used for debris collection after the first Soviet tests.[112] Characterization of samples was conducted at RIAN and LIPAN by radiochemists. The results were transferred to nuclear weapon designers at Design Bureau no. 11 (Konstruktorskoe Byuro no. 11, KB-11) in Arzamas-16, Gorky Oblast, for final interpretation.[113] In order to control and coordinate multiple monitoring efforts, which included debris collection and analysis, on 4 March 1954 the Soviet Minister of Defence, Nikolai A. Bulganin, ordered that a special service for monitoring of foreign nuclear weapon tests (Special Monitoring Service) be created within the Armed Forces' Main Intelligence Directorate (Glavnoe Razvedyvatel'noe Upravlenie, GRU).[114]

All these assets were deployed in time to detect the US Operation Castle test series at the Marshall Islands in the Pacific, carried out between 1 March and 14 May 1954.[115] The first foreign nuclear test detected in the

[108] Dubasov, Rimskii-Korsakov and Ushatskii (note 102), pp. 2–3.
[109] Lobikov et al. (note 104), p. 2.
[110] See note 56; and e.g. Saey (note 7), p. 43.
[111] Lobikov et al. (note 104), p.2; and Dubasov, Rimskii-Korsakov and Ushatskii (note 102), p. 2.
[112] Vasil'ev, ed. (note 82), p. 8; and Vasil'ev (note 101), p. 2.
[113] Dubasov, Rimskii-Korsakov and Ushatskii (note 102), p. 3.
[114] Vasil'ev, ed. (note 82), p. 9.
[115] Hansen (note 28), p. 100.

Soviet Union by analysis of radioactive debris was the Castle Bravo test shot.[116] Due to the damage caused by this test (e.g. to Japanese fishermen by the fallout), its precise date was widely known from press reports.[117] This allowed Evgeniy A. Lobikov at LIPAN to calibrate an existing model (based on the Way–Wigner law) for calculation of the explosion time from the decay rate of the fission products. The calibration resulted in improvement of accuracy to ±1 day. (Before calibration the Soviet method had an accuracy of ±5 days.)[118]

All six Operation Castle tests were reportedly detected and dated using this method. Three of them—Bravo, Romeo and Yankee—were correctly identified as thermonuclear by detection of uranium-237.[119] The failure to detect thermonuclear yield in the case of the Koon test was because it was a 'fizzle'. The Union and Nectar tests were successful and were, indeed, thermonuclear, but they were not identified as such in the Soviet Union.[120] Summarizing the results of analysis of radioactive fallout collected by their institute in the Leningrad area in 1954–55 (see table 7.1), and comparing it with published histories of nuclear testing, the RIAN experts concluded that 'not all explosions conducted at the time were identified individually, but all test series were detected, except for the British tests in Australia in November 1954. This fact convincingly demonstrates that any explosion in the atmosphere can be reliably detected and dated even by relatively outdated means of fallout analysis—regardless of the distance, but in the same hemisphere'.[121]

By 1955–57 the Soviet Union had created an early version of a system for long-range detection of nuclear weapon tests, which was coordinated by the Special Monitoring Service. It was based on four physical principles: registration of very low frequency radio waves (around 12 kilohertz), seismic waves, radioactive aerosols and infrasound. The radionuclide component of this system comprised stationary debris collectors (designed to determine that a nuclear test had occurred and its date) and aircraft-mounted filter equipment (designed to ascertain the specific features of the exploded devices).[122] These events can be seen as the end of the first,

[116] The only known earlier instance of debris collection reportedly involved collection of snow in the Leningrad area in Nov. 1952. It was assumed that the snow would contain debris from the Ivy Mike test, conducted by the USA on 31 Oct. 1952, but samples were reportedly discarded by accident before characterization. Sakharov, A., *Memoirs* (Hutchinson: London, 1990), p. 158. Ivy Mike, however, was still the first foreign nuclear test detected by the Soviet Union using technical means—detection of seismic waves. Vasil'ev, ed. (note 82), p. 7.

[117] See chapter 9 in this volume.

[118] Lobikov et al. (note 104), p. 2.

[119] Lobikov et al. (note 104), p. 3.

[120] Hansen (note 28), p. 100; and Lobikov et al. (note 104), p. 3.

[121] Dubasov, Rimskii-Korsakov and Ushatskii (note 102), p. 3.

[122] Vasil'ev, ed. (note 82), pp. 10–11.

Table 7.1. Results of Soviet radiochemical analysis of explosion products, 1953–56

The tests are non-Soviet (US) unless otherwise indicated.

Test date	Yield	Isotope content ratio to Ba-140 U-237	Np-239	Ag-111	Sr-89	Ru-103	Weapon type (contemporary Soviet conclusion)
1953							
12 Aug.[a]	400 kt	4.6	–	0.06	0.73	–	Thermonuclear with U-238 tamper
1954							
28 Feb.	15 Mt	0.94 ±0.2	–	0.073 ±0.010	0.58 ±0.06	1.15	Thermonuclear with U-238 tamper
26 Mar.	11 Mt	1.65	–	0.045	0.59	1.0	Thermonuclear with U-238 tamper
4 May	13.5 Mt	1.9 ±0.2	–	0.044 ±0.004	0.70 ±0.05	1.1	Thermonuclear with U-238 tamper
1955							
22 Feb.	2 kt	–	–	$\leq 4\times10^{-2}$	0.8	–	Nuclear with HEU
7 Mar.	43 kt	0.078 ±0.070	–	0.026 ±0.005	0.55 ±0.03	1.2 ±0.1	Nuclear with HEU
12 Mar.	4 kt	0.078	–	0.0145	0.63	0.65	Nuclear with HEU
22 Mar.	8 kt	0.15 ±0.01	–	0.082 ±0.010	0.60 ±0.07	2.5	Nuclear with Pu+(d,t) boosting
29 Mar.	14 kt	0.036 ±0.001	–	0.030 ±0.002	0.95 ±0.03	0.9	Nuclear with Pu and HEU
6 Apr.	3 kt	0.072	–	0.04	0.90	1.8	Nuclear with Pu and HEU
15 Apr.	22 kt	0.052	–	0.01	0.91	0.53	Nuclear with Pu and HEU
5 May	29 kt	0.057 ±0.008	–	0.020 ±0.005	0.74 ±0.07	0.9 ±0.1	Nuclear with Pu and HEU
1956							
27 May	3.5 Mt	2.0	–	0.06	0.6	–	Thermonuclear with U-238 tamper
16 June	1.7 Mt[b]	2.28 ±0.4	0.60 ±0.06	0.048 ±0.002	1.21 ±0.14	–	Thermonuclear with U-238 tamper
10 July	4.5 Mt	4.1 ±1.4	0.38	0.11 ±0.03	0.93 ±0.12	–	Thermonuclear with U-238 tamper
20 July	5.0 Mt	2.3 ±0.5	1.12 ±0.17	0.128 ±0.002	0.57 ±0.12	–	Thermonuclear with U-238 tamper

HEU = highly enriched uranium; kt = kiloton; Mt = megaton.

[a] This was a Soviet test.

[b] The yield of this test was 1.7 kt. This figure of 1.7 Mt may be an error in the source.

Source: From an internal Kurchatov Institute report, issued in 1958, as presented in Lobikov, E. A. et al., 'Development in the USSR of the physical methods of long-distance detection of nuclear explosions', Proceedings of the Second International Symposium on the History of Atomic Projects (HISAP'99), International Institute for Applied Systems Analysis (IIASA), Laxenburg, 4–8 Oct. 1999, p. 3.

exploratory, stage of development of the Soviet long-range detection programme. The further studies and organizational changes undertaken by the Soviet Union after 1957 aimed to transform this system into a routinely functioning control mechanism.[123]

Notably, the available open sources seem to suggest that the analysis of radioactive isotopes of noble gases did not receive as much attention in the Soviet Union in the 1950s as it did in the USA. The publications that are available began to discuss analysis of gaseous products of nuclear explosions (including noble gases and iodine isotopes) only in connection with the need to detect underground tests after the 1963 Partial Test-Ban Treaty (PTBT).[124]

[123] For detailed information on further development of the Soviet long-range detection system up to 2000 see Vasil'ev, ed. (note 82), in particular pp. 11–41.

[124] Vasil'ev, ed. (note 82), pp. 188, 31; Dyachenko (note 84), p. 355; and Treaty Banning Nuclear Weapon Tests in the Atmosphere, in Outer Space and Under Water (Partial Test-Ban Treaty, PTBT), signed by 3 original parties 5 Aug. 1963, opened for signature by other states 8 Aug. 1963, entered into force 10 Oct. 1963, *United Nations Treaty Series*, vol. 480 (1963).

8. The origins of nuclear forensic analysis II: analysis of nuclear weapon debris in Sweden

LARS-ERIK DE GEER

In 1943, five years after nuclear fission was experimentally demonstrated at the Germany's Kaiser Wilhelm Institute for Chemistry, Sweden obtained intelligence information from Berlin that Germany was developing 'atom disintegration weapons' that could effectively destroy armoured vehicles and kill people close to a detonation by shock and thermal waves and also those at a greater distance via 'oxygen extinction'.[1] This did not cause an uproar in Stockholm. The director of the Nobel Institute for Physics, when asked to comment on it, responded to the Swedish Government 'that it is not possible to make an atomic bomb'.[2] It was only when the United States actually demonstrated nuclear bombs to the world on 6 and 9 August 1945 that the authorities in Sweden and in other countries started to prioritize consideration of the option of nuclear armaments. The newly created National Defence Research Establishment (Försvarets forskningsanstalt, FOA) was tasked not only with studying the effects of nuclear weapons and how to defend against them, but also with investigating the possibility of a Swedish nuclear bomb.

The FOA research programme peaked in the late 1950s. It started to be scaled back from the early 1960s, as public opinion changed and as Sweden started to be an active player in nuclear disarmament negotiations, especially as an original member of the Geneva-based Eighteen Nation Committee on Disarmament (ENCD). This committee concentrated on a nuclear test ban and nuclear non-proliferation and Sweden came to play a special role in its capacity as the only non-aligned county in the ENCD with a deep knowledge of nuclear weapons.

[1] The designer of the 1938 experiment, Lise Meitner, had fled Germany for Sweden in Dec. 1938, and together with her nephew, Otto Frisch, had devised the concept of nuclear fission while on a Christmas Day walk near Kungälv on the west coast of Sweden. Meitner, L. and Frisch, O. R., 'Disintegration of uranium by neutrons: a new type of nuclear reaction', *Nature*, vol. 143, no. 3615 (11 Feb. 1939), p. 239–40. Enrico Fermi had actually, already in 1934, conducted a similar experiment but he failed to draw the right conclusion. He even rejected it 3 months later when the German chemist Ida Noddack suggested that splitting the uranium atoms could have caused his somewhat unorthodox results. This was a pivotal point in time: if Dr Noddack had been taken seriously, then nuclear weapons could have been developed 4 years earlier than they actually were, which would have had a dramatic impact on the course of World War II.

[2] *FOA och kärnvapen: dokumentation från seminarium 16 november 1993* [FOA and nuclear weapons: documentation from the seminar of 16 November 1993], Försvarets Forskningsanstalt (FOA) VET om försvarsforskning no. 8 (FOA Veteranförening: Stockholm, 1995).

By signing the 1968 Non-Proliferation Treaty (NPT) Sweden formally abandoned its nuclear weapon programme and the focus of FOA's research moved towards nuclear defence and the means to protect civilians and troops during a nuclear attack.[3] However, as weapon physics is sometimes needed to understand the effects of nuclear weapons and to design appropriate protection, the demarcation line between the two can be blurred. This has sometimes led to accusations in the press, both in Sweden and abroad, of a secret nuclear weapon programme; but such suspicions seem now, rightly, to have died out.[4]

Nuclear test ban verification techniques came to be a new line of research with the goal of demonstrating the feasibility of verification of a comprehensive nuclear test ban and to further develop such means. The focus was on seismic techniques and on methods to collect and characterize radioactive debris from clandestine explosions. This chapter is mainly about the latter and describes how FOA developed techniques for remote sensing of nuclear fission and fusion events (section I) and how these have been applied to detect atmospheric nuclear tests and leaks from nuclear testing underground (section II), with a detailed look at the Chinese nuclear test series (appendix 8A). It then describes the development of the Swedish radionuclide verification system (section III) and Swedish detection of radioactive leakages from underground nuclear explosions (section IV). Section V focuses on the application of nuclear forensics in Sweden to non-nuclear weapon events, and section VI on the implications of Swedish experience and expertise for the 1996 Comprehensive Nuclear-Test-Ban Treaty (CTBT).[5]

I. Remote sensing of nuclear fission and activation radionuclides

Nuclear test explosions and other nuclear events produce radioactive residues that can be dispersed in the atmosphere and other environments and in that way publicize things that were not necessarily meant to be publicly known. The first occasions on which radionuclides were searched for with this intention was in the autumn of 1944 when the USA looked for possible plutonium-producing reactors in Germany. The focus at the time was on

[3] Treaty on the Non-Proliferation of Nuclear Weapons (Non-Proliferation Treaty, NPT), opened for signature 1 July 1968, entered into force 5 Mar. 1970, IAEA Information Circular INFCIRC/140, 22 Apr. 1970, <http://www.iaea.org/Publications/Documents/Treaties/npt.html>. Sweden signed on 19 Aug. 1968 and ratified on 9 Jan. 1970.

[4] Larsson, C., 'Historien om en svensk atombomb 1945–1972' [The story of a Swedish atomic bomb 1945–1972], *Ny Teknik*, no. 17 (25 Apr. 1985); and Coll, S., 'Neutral Sweden quietly keeps nuclear option open', *Washington Post*, 25 Nov. 1994.

[5] Comprehensive Nuclear-Test-Ban Treaty (CTBT), opened for signature 24 Sep. 1996, not yet in force, <http://treaties.un.org/Pages/CTCTreaties.aspx?id=26>.

the noble gas nuclide xenon-133, but nothing was found. Five years later, a US reconnaissance flight from Japan to Alaska picked up fresh fission products from the first Soviet nuclear test, on 29 August 1949.[6]

In the aftermath of that detection, nuclear testing gathered speed. An average of 16 nuclear tests, all above ground, were conducted worldwide per year between 1951 and 1954 and in each of the next three years the test rate increased by about 50 per cent. In 1958, the number of tests more than doubled to 101. By the end of that year, when the three existing nuclear powers agreed a nuclear-testing moratorium, a total of 276 non-underground nuclear weapon tests had taken place since the first test, Trinity, on 16 July 1945.[7]

The majority of these explosions took place in the northern hemisphere, from Christmas Island (at 2° N), at that time a British territory near the equator, to Novaya Zemlya (at 74° N), a Soviet island north of the Arctic Circle. Only the United Kingdom tested its atomic weapons south of the equator in the period 1951–58: it conducted 10 fission bomb tests in Australia, and three thermonuclear tests at Malden Island (at 4° S), then a British territory in the Pacific Ocean. The amount of fission products injected into the northern hemisphere by the end of 1958 has been estimated to correspond to a total fission yield of about 47.3 Mt. Of this, '2.3 Mt' went into the troposphere (the lowest atmospheric layer, up to an altitude of 10–15 km) and '45 Mt' into the stratosphere (the next layer, extending to about 50 km).[8] The differentiation between the two layers is important as debris remains in the troposphere for about a month due to fallout (both dry fallout and, in particular, wet fallout in rain and snow), while debris can remain for several years in the stratosphere (where the temperature increases with the altitude, which makes it much more stable than the troposphere). The hemispheric differentiation is also important as little pollution crosses the equator. These facts set the scene for the atmospheric radionuclide surveillance system that was built up in Sweden in the 1950s.

The effects of globally dispersed nuclear weapon debris were first noted in Sweden in mid-1951 by Professor Rolf Sievert as a background nuisance in his work with a whole-body counter.[9] In subsequent years large hydro-

[6] See chapter 7 in this volume.

[7] This total does not include the 2 combat explosions in Japan in 1945. Fedchenko, V., 'Nuclear explosions, 1945–2013', *SIPRI Yearbook 2014: Armaments, Disarmament and International Security* (Oxford University Press: Oxford, 2014), table 6.16. 'Non-underground tests' are almost all atmospheric tests but include a few crater, surface, underwater and high-altitude tests that some evaluations count separately.

[8] United Nations Scientific Committee on the Effects of Atomic Radiation (UNSCEAR), *Sources and Effects of Ionizing Radiation*, 2000 Report to the General Assembly, with scientific annexes, E.00.IX.3 (United Nations: New York, 2000).

[9] Sievert was a famous pioneer of health physics whose name has been commemorated as a unit of equivalent, effective and committed doses, as well as dose commitments.

gen bombs were tested and, when the fallout increased, Sievert wrote a letter to the prime minister, Tage Erlander, in April 1954 requesting that something should be done about it (he was certainly also influenced by the Castle Bravo fallout in early March 1954 and the international protests that followed).[10] At that time FOA had been following the wet fallout for a year by measuring the beta activity in the residues of evaporated rainwater and analysing the accumulating deposition of strontium-90, the fission product that at the time was considered the most detrimental to the health of humans due to its high-energy beta radiation, long half-life and tendency to accumulate in the skeleton.

By 1955 the system had developed to such an extent that debris was collected on glass fibre filters and the radiation in those filters was measured by NaI:Tl scintillation crystals.[11] That meant that the radiation could be sorted according to its energy in different bins that together make up a spectrum. Bins filled by more events than others signify, by their corresponding energies and number of events, individual radionuclides and their concentrations in the sampled air mass. In the mid-1950s the electronics for this data sorting were not readily available and FOA therefore developed its own so-called Hutchinson–Scarrot analyser. Colleagues from that time describe it as including 20 metres of nickel wires that were mounted around the walls of the room to function as delay lines.

A spectrum produced in this way could only resolve a limited amount of data. In a spectrum from a recent nuclear test, FOA's experts could typically see ^{131}I, ^{132}Te, ^{132}I, ^{103}Ru, ^{95}Zr, ^{95}Nb, and ^{140}La. The analyses were done by fitting so-called standard spectra of individual nuclides to the sample spectrum. With that technology some quite badly resolved peaks, such as that for ^{95}Zr and ^{95}Nb, could also be analysed. As ^{95}Zr decays with a half-life of 64 days, into ^{95}Nb (with a half-life of 35-day) that is not itself produced in the fission process, this pair makes a perfect 'clock' that can date and connect a single sample to an individual test for several months. However, in the 1950s and the early 1960s, when so many atmospheric tests were carried out close in time, it could be difficult to ensure that a sample was from just one test. Often the solution was to localize a single so-called hot particle that must have come from a single explosion in order to do the dating.

A national network of air-filtering stations was established in Sweden (figure 8.1). At these sites generally 1000 cubic metres per hour (m^3/h) samplers were operated with glass fibre filters that were changed two to three times a week. For many years a station at Grindsjön, near Stockholm, sampled 5400 m^3/h through five filters that were changed twice a week.

[10] On the Castle Bravo test see chapter 9 in this volume.

[11] On thallium-doped sodium iodide scintillators see chapter 4 in this volume.

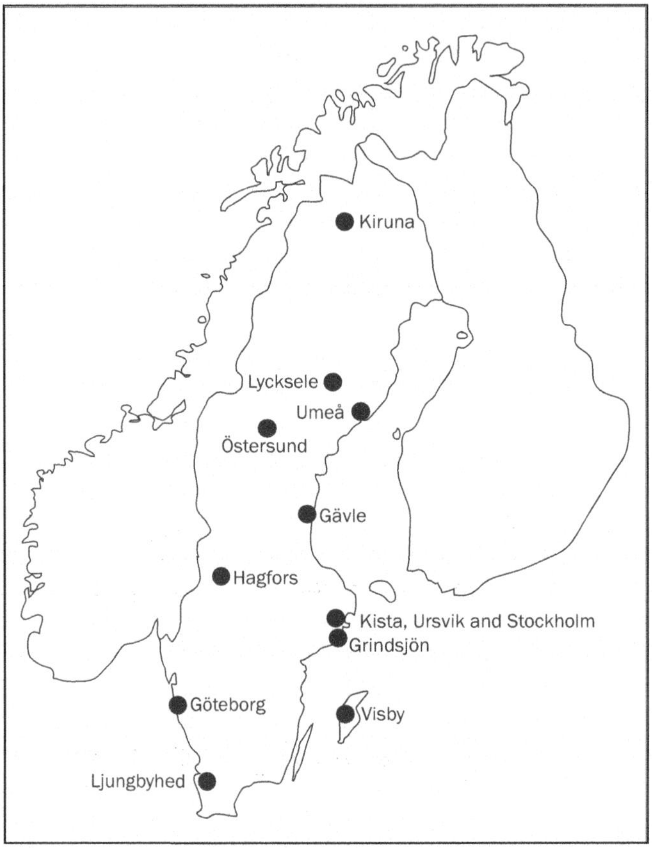

Figure 8.1. Ground level sampling stations that have been active in Sweden's National Defence Research Establishment (Försvarets forskningsanstalt, FOA) network

Note: The stations that are active at the time of writing (2015) are the ones at Gävle, Kiruna, Kista, Ljungbyhed, Umeå and Visby. Kista also hosts aerosol and noble gas stations belonging to the CTBTO's global network.

Not all stations have been active all the time, and those close to Stockholm have been relocated owing to reorganizations of FOA. Stations at Gävle and Visby were established after the accident at the Chernobyl reactor in Ukraine, Soviet Union, in 1986—Gävle to follow re-suspension (i.e. radionuclide transport from fallout back to the air) in a heavy fallout region, and Visby on the Baltic island of Gotland because the Chernobyl cloud first crossed Gotland. Stations at Gothenburg, Lycksele and Östersund were closed in the 1970s for financial reasons and the lower rate of atmospheric nuclear testing. The Grindsjön station closed in 2003 when a new station with a 1600 m^3/h capacity was built at Kista.

In order to make more detailed studies of individual bomb residues, samplers were developed that were mounted under the wings of Swedish military aircraft. This system enabled the collection of quite hot samples, sometimes corresponding to almost 100 picograms (pg) of split uranium or plutonium.[12] The aircraft used were the Saab J29 Tunnan and the Saab 32 Lansen. Even though the last Lansen was built 50 years ago, one is still ready to make sampling flights today. However the Lansen has only been on active duty twice since the last atmospheric test (in Lop Nur, China, on 16 October 1980): in 1984 to sample for possible residues from a Soviet satellite reactor that was close to crashing in Sweden, and in 1986 during passage of the Chernobyl cloud.

In the 1950s FOA focused on following the Soviet and US testing programmes. The peaks in atmospheric radionuclide concentrations detected by FOA were shown to be consistent with records from a seismograph operated by Professor Markus Båth at Uppsala University and with other information from around the world. Because the thermonuclear bomb was developed and tested in the Soviet Union, the UK and the USA, this meant that not only fission products were disseminated around the world, but also activation products created by the enormous neutron fluxes in such explosions.

A radionuclide such as ^{237}U, with its 6.75 day half-life, could dominate a spectrum with four times higher activity than the most prominent fission products. ^{237}U is created by (n,2n)-reactions induced in ^{238}U by high-energy neutrons (up to 14.1 MeV) originating in the thermonuclear explosion. ^{238}U, in the form of natural or depleted uranium, is used as the outer shell of fission pits to tamp explosion and increase the time of criticality and, thus, the burn-up and explosive yield. In a thermonuclear device the same material is often used in a similar way by encompassing the thermonuclear fuel where it still tamps but, more importantly, it acts as a pusher to squeeze the lithium deuteride with extreme pressure driven by the X-rays from the primary (the first stage of a thermonuclear weapon).[13] Even though the ingenious invention of the hydrogen bomb (H-bomb) mechanism by Edward Teller and Stanislav Ulam (known as the Teller–Ulam design) in 1951 was not fully understood in Sweden until the late 1970s, it became obvious quite early that a significant part (*c.* 50 per cent) of the explosive yield of H-bombs derived from fission of ^{238}U induced by high energy neutrons. These bombs were therefore dubbed fission–fusion–fission (3F) bombs. More than 90 per cent of all fission debris that contaminated the atmosphere was due to this final step.

[12] While 100 pg of e.g. plutonium may seem like a small amount, fully split it corresponds to *c.* 10 000–40 000 Bq each of the more prominent short-lived fission products, such as Mo-99, I-131, Te-132 and Ba-140.

[13] See figure 6.2 in chapter 6 in this volume.

A phenomenological marker of a full-scale thermonuclear test became what was called the (n,γ) : (n,2n) ratio, which was defined as the atom ratio of ^{239}Np ($T_{½}$ = 2.356 days) to ^{237}U ($T_{½}$ = 6.75 days), both nuclides that could easily be analysed by gamma spectroscopy in a sample. The former is born as ^{239}U by neutron capture at lower energies, which with a half-life of just 23.45 minutes rapidly decays to ^{239}Np. Experience showed that high-yield H-bombs produce an (n,γ) : (n,2n) ratio at explosion time that was close to 1, while in debris from a classical atom bomb (A-bomb) it was 30 or more.

As early as the 1950s the Swedish airborne-radionuclide surveillance system picked up signals from radiological accidents that were only indirectly connected to weapon testing. The tritium needed for the British thermonuclear bomb pushed the limits of one of the plutonium production reactors at Windscale in Cumberland on the Irish Sea. The main cause was the so-called Wigner effect, in which energy is accumulated by neutron-produced dislocations in the graphite moderator. Such energy can then be spontaneously released and cause overheating of the fuel. This problem was known and was usually dealt with, but the change to tritium production caused overheating that went unnoticed, and the reactor caught fire on 10 October 1957. That fire released an estimated 740 terabecquerels (TBq) of iodine-131 and 22 TBq of caesium-137.[14] The winds were blowing from the west and ^{131}I and ^{132}Te were detected in Stockholm after two days, at levels of about 40 mBq/m^3.

Two interconnected topics stand out from the research carried out by FOA on the collected nuclear weapon residues in the 1950s: the concept of fractionation and its dependence on the type and environment of the explosion, and micro-particle studies carried out by means of reversible autoradiography.

Fractionation

It was soon noted that a sample of debris, even from a single nuclear detonation, often did not contain the mixture of radionuclides that would be expected from the amounts known to be produced in the fission process. Instead, the mixture was a complicated function of a number of factors, especially environmental ones, such as whether the explosion was a surface burst, airdrop or took place on a barge, but also on the total explosive yield. Only very large tests in which the fireball had no contact with the ground and the cloud rose well into the stratosphere created debris with an essentially unchanged composition. Fractionation was the term

[14] To put it in perspective, this is no more than half a per mille of what was released at Chernobyl nearly 30 years later. Cooper, J. R., Randle, K. and Sokhi, R. S., *Radioactive Releases in the Environment: Impact and Assessment* (Wiley: Chichester, 2003).

used for the phenomenon, and it is conveniently described by a fractionation factor,

$$f_{\text{A-R}} = \frac{[N_{\text{A}} / N_{\text{R}}]_{\text{experimental}}}{[N_{\text{A}} / N_{\text{R}}]_{\text{fission}}},$$

where the numerator is the atom ratio in the sample (calculated back to explosion time) of two nuclides of mass A and R, respectively, and the denominator is the corresponding fission yield ratio.[15] R is a reference chain that in Sweden used to be mass 95, with ^{95}Zr and ^{95}Nb as the measurands. The USA, which had the resources to collect more fresh samples, often used mass 99 with its more short-lived measurand ^{99}Mo ($T_{1/2}$ = 66 hours). Given access to young samples, molybdenum is a much better choice as it minimizes the problem of mixed debris from several explosions in the same sample. In Sweden, fractionation studies therefore became more fruitful after the 1963 Partial Test-Ban Treaty (PTBT) entered into force in 1963.[16] After this, only one country, China, carried out atmospheric tests in the northern hemisphere, and it did so at the comparably low rate of one, two or, occasionally, three per year (see section II).

General fractionation behaviour could, however, be discerned already in the 1950s and the mechanisms behind it became clear.[17] Debris that reached Sweden within a few weeks from a rather small airdrop test where the fireball had not touched the ground (i.e. the altitude at which the explosion had taken place in metres was higher than 55 times the cube root of the explosion yield in kilotons) was generally enhanced in refractory elements.[18] This was because larger particles (0.1–10 μm) descended during transport from a distant test and were the first to hit the samplers. At the same time, these larger particles were the ones that had started to condense first during the cooling of the cloud and therefore contained more of the refractive elements, as compared to the volatile elements that had to wait longer to form an aerosol. Such a sample was said to be 'normally' fractionated.

Debris that arrived later had travelled for a longer time, the larger particles had deposited on the ground, and the samples exhibited what was dubbed 'opposite' or 'reversed' fractionation with a strong enhancement of

[15] Edvarson, K., Löw, K. and Sisefsky, J., 'Fractionation phenomena in nuclear weapons debris', *Nature*, vol. 184, no. 4701 (1959), pp. 1771–74.

[16] Treaty Banning Nuclear Weapon Tests in the Atmosphere, in Outer Space and Under Water (Partial Test-Ban Treaty, PTBT), signed by 3 original parties 5 Aug. 1963, opened for signature by other states 8 Aug. 1963, entered into force 10 Oct. 1963, *United Nations Treaty Series*, vol. 480 (1963).

[17] Edvarson, Löw and Sisefsky (note 15).

[18] Refractory elements here refer to elements or element oxides with boiling points (i.e. condensation points) above about 1500°C and often very much above that. The reference elements zirconium and molybdenum are 2 of the most refractive elements, which condense at 4377°C and 4612°C, respectively. That is, of course, one reason why they serve so well as references.

the volatile elements instead. As the phenomena resulted from the same process, the two sample types were sometimes called mirror particulates. In the case of an explosion near the ground, such as on a tower, a lot of ground material would be drawn into the cloud, which would preferentially scavenge refractive elements and deposit them close to ground zero. Typical radionuclides in the refractive group, apart from molybdenum and zirconium, are isotopes of barium, cerium, neodymium and yttrium. Radionuclides typical for the volatile group (not counting krypton and xenon, which remain gases down to below −100°C) are iodine, caesium, tellurium and often ruthenium (actually as an oxide).

Other elements are intermediate as their condensation points, or that of their oxides, are closer to the condensation points of the oxides of matrix materials such as aluminium, iron and silicon, which make up the bulk of the particles. Whether the element or the oxide should be considered depends on the temperature at which the oxide decomposes. An example is barium peroxide, which has a low melting point but cannot be formed above about 500°C, unlike barium dioxide, which has melting and boiling points around 2000°C. Ruthenium, a very clear example of where the oxide is decisive, condenses at 3900°C but behaves more like iodine in oxidized debris due to its tetroxide, which condenses at 40°C. Some variations in the fractionation pattern can also appear due to the cooling time and physics of the initial cloud. In the free atmosphere the cloud cools down to particle-formation temperatures within a few tens of a second. That is, for example, the time frame in which a substantive part of the mass 140 resides as xenon and caesium but not yet barium. Thus, the detailed cooling history can have a decisive impact on fractionation.

Studies of hot particles

Hot particles are particles in the debris that carry substantive amounts of radionuclides and are large enough to be studied in an optical microscope, that is, with a diameter above about 0.2 μm. In studying the details of fall-out and nuclear weapon remnants for various reasons, detailed analyses of hot particles are of great interest. Such particles could be identified on an exposed filter with a small Geiger–Muller counter or by autoradiography, where a beta radiation-sensitive film is exposed for a time on top of the filter. The resulting black spots will then guide the scientist. However, a more sophisticated method was developed at FOA.[19] After initial standard autoradiography, the pieces of the filter that had caused spots were punched out with a 4-millimetre hollow tool. The filter cores were then

[19] Sisefsky, J., 'Method for photographic identification of microscopic radioactive particles', *British Journal of Applied Physics*, vol. 10 (Dec. 1959), pp. 526–29.

stirred in a celluloid solution and the resulting paste, diluted with acetone if necessary, was spread across a celluloid-coated glass plate. A beta-sensitive nuclear emulsion was then poured over the plate and exposed for a few days, the number depending on the activity of the particles. (A thousand or so disintegrations within each particle are needed during exposure.) The plate was then reversal developed and in the microscope the hot particles appeared in the centre of circular optical holes in the plate.

With this method it was possible to study how the particle activity varied with the diameter of the particles. An illustrative example that involved a lot of work was the analysis of a quite strong sample, about 100 000 Bq in total, collected at an altitude of 12 km over central Sweden on 17 October 1958.[20] By radionuclide analyses and seismic records it was shown that it was totally dominated by debris from two thermonuclear tests on the morning of 30 September at Novaya Zemlya, 2300 km north north-west from the sampling track. (The total yield is now known to have been 1.2 + 0.8 = 2.0 Mt and they were airdrops at altitudes of 1.5 and 2.3 km.[21]) Almost 1000 particles were located and individually analysed. The particles were colourless to reddish, translucent, and more or less spherical. The reddishness could have indicated steel (iron) in the casing and mechanisms of the bomb. When the activity of all particles was plotted against their diameter it was seen that the activity was proportional to the volume, which very well fits the picture given above of refractive species being incorporated into the particles during the full condensation process. Sometimes when surface detonations that had sucked massive amounts of ground material into the cloud were analysed, the particles were irregular, as they had not gone through the full melting–solidification cycle. That also meant that the radionuclides deposited on the surface of the particles and the activity became more proportional to the area of the particles.

II. Detecting atmospheric nuclear tests

Atmospheric nuclear testing, 1958–63

On 31 October 1958 a moratorium on nuclear testing that had been agreed by the nuclear weapons states at that time—the Soviet Union, the UK and the USA—went into effect. Nonetheless, the Soviet Union conducted two small tests on 1 and 3 November, but thereafter all test sites went silent. The moratorium was partly a propaganda competition between the Soviet Union and the USA to satisfy worldwide opinion but, of course, it also fitted the development programmes in both countries, which had just finished

[20] Sisefsky, J., 'Debris from tests of nuclear weapons', *Science*, vol. 133, no. 3455 (1961), pp. 735–40.
[21] Mikhailov, V. N. (ed.), *Catalog of Worldwide Nuclear Testing* (Begell Atom, LLC: New York, 1999).

major test series and needed time to analyse the results. The UK was the only participant that was not really ready for a moratorium and it had to accelerate its thermonuclear programme during the autumn of 1958.

On 13 February 1960 France entered the nuclear club with a 100-metre tower test of a 40–80 kt fission device (Gerboise Bleue) at Reggane, Algeria.[22] Debris from the test, including hot particles, was detected in both precipitation and in air in south-western Sweden.[23] The cloud entered Swedish airspace from the west with a delay of more than two weeks during which the cloud had circumvented the North Pole and even passed the North American continent. Three more atmospheric tests with yields of an order of magnitude smaller then followed at the same site within 15 months.

On 1 September 1961 the Soviet Union broke the moratorium and started a period of unprecedented nuclear testing in the atmosphere that did not cease until late December 1962. These tests included the so-called Tsar Bomba, which exploded with a yield of 50 Mt above the Soviet northern test site at Novaya Zemlya, and 30 other thermonuclear explosions at the same site ranging from 1 to 24.2 Mt. In 1961–62 the world experienced 136 Soviet, 39 US and 1 French non-underground nuclear explosions.[24] These injected 2.5, 4.4 and 0.0004 Mt of fission products, respectively, into the troposphere and 68, 14 and 0 Mt of fission products into the stratosphere.

It was a time of mental terror (underlined also by the Cuban crisis in October 1962) and the FOA surveillance system was therefore focused on the health threats posed by the heavy testing at Novaya Zemlya, just 1200 km outside Sweden. Fallout in food, such as radioiodine in milk and ^{137}Cs in meat from lichen-grazing reindeer, was closely observed. In the winter of 1965, when the transport from the stratospheric reservoir filled up by the 1961–62 tests had peaked, the caesium concentration in reindeer meat was 10 times higher than the first limit for human consumption of 300 Bq/kg imposed in Sweden after the Chernobyl accident (it was later changed to 1500 Bq/kg).

Two effects, however, reduced the risks of very high regional fallout from Soviet tests at Novaya Zemlya. First, the many high-yield tests lifted a greater part of their debris into the stratosphere, where it decayed substantially and was distributed around the hemisphere before it entered the troposphere and was deposited on the ground during the following years (mainly in spring). Second, the Tsar Bomba reportedly had one or more

[22] International Atomic Energy Agency (IAEA), *Radiological Conditions at the former French Nuclear Test Sites in Algeria: Preliminary Assessment and Recommendations*, IAEA Radiological Assessment Reports Series no. 6 (IAEA: Vienna, 2005).

[23] Lindblom, G., *Advection over Sweden of Radioactive Dust from the first french nuclear test explosion*, Försvarets Forskningsanstalt (FOA) 4 Report no. C 4155-4127(Rev) (FOA: Stockholm, 1961).

[24] Fedchenko (note 7).

tamper–pushers made of lead instead of uranium to limit its fission yield.[25] It was one of the 'cleanest' tests reported with only 3 per cent of the 50 Mt yield deriving from fission.

In response to the mounting popular protests around the world, in 1963 the Soviet Union and the USA signed the PTBT, which from 10 October that year banned all nuclear explosions 'in the atmosphere; beyond its limits, including outer space; or under water, including territorial waters or high seas; or in any other environment if such explosion causes radioactive debris to be present outside the territorial limits of the State under whose jurisdiction or control such explosion is conducted'.[26] In Sweden the paradigm shifted and the goal of radionuclide surveillance became to carry out voluntary verification of the PTBT and put pressure on the nuclear powers to extend the treaty so that it would become comprehensive. That would be vital to curb the arms race, something that is clear today, when it is known that more than 75 per cent of all nuclear detonations took place underground after the PTBT entered into force.

Chinese nuclear tests in the atmosphere, 1964–80

In addition to voluntary verification of the PTBT, the Swedish long-range-detection programme received additional focus when China conducted its first nuclear test, on 16 October 1964—a tower-mounted HEU-based fission device with a 20 kt yield. It was the first in a series of 22 atmospheric tests carried out over a period of 16 years near Lop Nur in the Taklimakan Desert of China's Xinjiang province.

Radioactive debris was found and analysed in Sweden from all of these tests after it had travelled 20 000 km from Lop Nur downwind across the Pacific, North America and the Atlantic fairly regularly for about 10 days. The detail in which these tests could be studied so far away is amazing and was due to several factors. First, the jet streams around the tropopause acted like a bus, carrying the debris at speeds up to 100 km per hour. Second, germanium gamma detectors became available, which immensely increased the sensitivity of the analysis and allowed detection of mixed radionuclide samples with up to hundreds of highly resolved peaks in the spectra. Third, the low frequency of testing limited the complexity of the samples that, previously, had contained mixed debris from several explosions. Fourth, the reversal autoradiography method was fully developed and ready to be used. Fifth, a seven-station national sampling network was operating, including the extremely high-volume (5400 m^3/h) sampler at

[25] The number of tamper–pushers is uncertain since there are reports that the Tsar Bomba might have had 2 thermonuclear stages, possibly even with the second stage split into several parts.

[26] Treaty Banning Nuclear Weapon Tests in the Atmosphere, in Outer Space and Under Water (note 16), Article I.

Grindsjön. Seventh, the Swedish Air Force was on standby to chase radioactive clouds up to altitudes of 14 km anywhere in a country whose length is equal to nearly half of Europe in the north–south direction. Finally, a great technological advantage was a pure research project, Online Separation of Isotopes at a Reactor in Studsvik (OSIRIS), in which FOA participated in the late 1960s and early 1970s, that provided priceless experience with the new detector types and even more so with the emerging minicomputers and online measurement techniques. Software was written in assembler language and an interactive multichannel analysis programme was created in 8K of 16-bit memory that in many respects was superior to current commercial and much larger systems. Sweden thus became effective at catching Chinese debris, which led to the detection of all 22 atmospheric tests at levels at least 1000 times above FOA's detection limit.

Details of these Chinese tests as detected in Sweden are given in appendix 8A. The series illustrates the road a country has to travel to acquire advanced nuclear arms. The Swedish monitoring revealed that China's tests included 'simple' fission bombs, boosters, perhaps a layer cake, failures, partial or possibly voluntarily terminated thermonuclear tests, and a number of fully fledged two-stage thermonuclear explosions. A few tests also seem to have been trials of bombs produced for China's arsenal. After the first 26 tests, which included 4 underground tests, 18 more tests were carried out underground in shafts and tunnels up until 29 July 1996, less than two months before China signed the CTBT on the day it opened for signature.

French nuclear tests in the atmosphere, 1966–74

After its first test on 13 February 1960, France conducted a further three atmospheric tests during 1960 and 1961, near Reggane, Algeria. Thirteen underground tests then followed between 1961 and 1966 in tunnels dug into the Taourirt Tan Afella granite massif near the In Ekker fort, 575 km east south-east of Reggane. The total yield of these underground tests was 270 kt. All but the first two were carried out after Algeria gained independence from France on 5 July 1962. This was permitted by a clause in the Évian Accords that had laid the framework for Algeria's independence.[27]

By 1966 France had moved its nuclear test activities to French Polynesia and the atolls of Mururoa and Fangataufa, where 46 atmospheric tests with a total yield of 10.1 Mt were carried out in annual campaigns (excluding 1969) up to and including 1974. Very little radioactive debris from these

[27] Government Declarations of 19 Mar. 1962 on Algeria, *Journal Officiel de la République Française*, 20 Mar. 1962, Chapter III (in French).

tests was expected to appear in the northern hemisphere as in general no winds pass the equator. For Sweden to be able to study the French tests in the Pacific it was therefore necessary to sample in the southern hemisphere. A FOA employee knew the director of the Swedish shipping company Hillerström, and the idea was put forward of installing a sampler on one of the company's ships that sailed in the Pacific and of the crew changing the filters and posting them to Stockholm when in port. The company agreed and a ship was selected that regularly sailed around Australia, a merchant vessel called M/S *Milos* based in Hong Kong. Thus, from 1971 to 1974, 40 × 40-cm glass fibre or Microsorbane filters were exposed for (typically) two to five days and sent to FOA for analysis (24 in 1971, 19 in 1972, 23 in 1973, 13 in 1974 and 3 in 1975). The sampling rate was 400 m^3/h, which meant that the samples corresponded to 20 000–50 000 m^3 of air.

During 1971–74 France carried out 20 weapon-related atmospheric nuclear test explosions at its Pacific test site (see table 8.1).[28] All but two of these were test devices hanging from balloons above Mururoa at 220–480 metres, altitudes that accommodate yields of 110–1100 kt with no contact between the fireball and the ground or sea. The actual yield interval was 0.05–955 kt and all exploded high enough not to draw any ground or sea material into the cloud.

Most of the filter samples were from between 25 June and 30 October 1971, when the *Milos* sailed south of the equator in the easterly trade winds, and contained fresh fission products. As expected, nothing was seen north of the equator when the ship sailed between ports in Hong Kong, the Philippines and Taiwan. Two filters, in particular, which had been exposed on either side of Australia, showed quite high concentrations (several mBq/m^3) of common fission products. The first of these, collected on 28 June to 1 July between eastern Papua New Guinea and Brisbane, Australia, could accurately be dated by the ^{95}Zr : ^{95}Nb ratio and was shown to derive from the French test on 5 June.

The fractionation profile was slightly opposite (volatiles enhanced) even though 40 hot particles in the range 1–2 μm were isolated. The second sample with a high concentration was collected a month later, between Perth, Australia, and Java, Indonesia, and was dated back to 12 June ±1.3 days, which was consistent with the Encelade test of nearly 0.5 Mt. Due to the long time (2 months) between the explosion and the measurements, very few short-lived nuclides could be seen. It was, however, noted that ^{131}I and ^{103}Ru were depleted, which was consistent with the fact that 11 hot particles in the range 1–3 μm were found on the filter. At the end of August a

[28] An additional 4 above ground safety-related test explosions with a near-zero yield were also conducted but they are not relevant to the current discussion.

Table 8.1. French nuclear tests in the atmosphere, 1971–74

Date[a]	Time[a]	Test designation	Tested device[b]	Height (m)	Yield (kt)
Fifth campaign, 1971					
5 June	10:15	Dione	AN51	275	34
12 June	10:15	Encelade	MR41	450	440
4 July	12:30	Japet	TN60	230	9
8 Aug.	09:30	Phoebe	TN60	230	4
14 Aug.	10:00	Rhea	TN60	480	955
Sixth campaign, 1972					
25 June	10:00	Umbriel	TN60	230	0.5
30 June	09:30	Titania	TN60	220	4
27 July	09:40	Oberon	TN60	220	6
Seventh campaign, 1973					
21 July	09:00	Euterpe	TN60	220	11
28 July	14:00	Melpomene	(TN60)	270	0.05
18 Aug.	09:15	Pallas	(TN60)	270	4
24 Aug.	09:00	Parthenope	(TN60)	220	0.2
28 Aug.	09:30	Tamara	AN52		6
Eighth campaign, 1974					
16 June	08:30	Capricorne	(TN70)	220	4
7 July	14:15	Gemeaux	(TN70)	312	150
17 July	08:00	Centaure	(TN80)	270	4
25 July	08:30	Maquis	AN52	250	8
14 Aug.	15:30	Scorpion	..	312	96
24 Aug.	14:45	Taureau	..	270	14
14 Sep.	14:30	Versau	(TN60)	433	332
Total yield					**2 100**

Notes: All but 2 tests were carried out hanging from balloons at 220–480 metres above Mururoa Atoll. 'Tamara' was dropped from an aircraft about 26 km west of Mururoa and 'Maquis' was dropped from an aircraft about 17 km west south-west of that atoll. Four safety tests with no, or no more than 1 tonne, yield are not included.

[a] The time and date are local (= UTC–9 hours).

[b] AN51 = the tactical 10 or 25 kt nuclear warhead used on the Pluton short-range missile; AN52 = a free fall bomb version of the same device; MR41 = the 500-kt boosted fission charge for the Redoutable class ballistic missile submarines; TN60 = 1000-kt sea-launched and air-to-surface thermonuclear missile; TN70 = 150-kt sea-launched and air-to-surface thermonuclear missile; TN80 = 300-kt sea-launched and air-to-surface thermonuclear missile. Brackets indicate uncertain entries.

Sources: Bouchez J. and Lecomte, R., *The Atolls of Mururoa and Fangataufa (French Polynesia)*, vol. 2, *Nuclear Testing: Mechanical, Lumino-thermal and Electromagnetic Effects* (Direction des centres d'experimentations nucleaires/Commissariat à l'énergie atomique: Paris, 1996); and IAEA, *The Radiological Situation at the Atolls of Mururoa and Fangataufa*, Main Report by an International Advisory Committee, IAEA-STI/PUB/1028 (IAEA: Vienna, 1998), p. 27.

sample was collected that could be quite accurately dated back to the Rhea thermonuclear test two weeks earlier. That sample contained a large irregular and yellowish particle that exhibited a clear ^{237}U signal, which confirmed the thermonuclear character of the test. In general, the particles from the French tests in the Pacific were colourless to light yellow.

During the 1972 campaign only four samples showed fresh fission products: two collected on 16–24 July along the west coast of Australia, heading north, and two on 17–22 August along the west coast of Irian Jaya (New Guinea), Indonesia, heading south. Sampling in 1973 was even less productive, with traces of fission products in only 3 of 23 samples. Meteorological conditions, of course, played an important role, but also the fact that in 1973 the tests were of quite low yields. The next and final year for French atmospheric testing involved higher-yield explosions. Eight of 10 samples taken from mid-July to mid-November contained fresh fission products. Four samples—believed to be due to the Gemaux, Scorpion and Versau tests—showed that high-energy detector elements had been used, yttrium in Gemaux and iridium in the other two. Their presence was indicated by the detection of the reaction products ^{88}Y and ^{192}Ir (see the China-16 test in appendix 8A).

In June 1975 France started to test underground under the rims and lagoons of the Mururoa and Fangataufa atolls. In total, 137 such tests were done with a total yield of 3.2 Mt. There were also 10 safety trials, 7 with zero yield and 3 with a combined yield of about 0.5 kt.[29]

III. Development of the Swedish radionuclide verification system after 1963

The 1963 Partial Test-Ban Treaty fairly effectively stopped the injection of radionuclides into the atmosphere and thereby cut the most significant route for doses from nuclear testing to the general population worldwide. France conducted a further 46 atmospheric tests before it stopped in September 1974, while China conducted 22 before it stopped in October 1980. These tests added about 10 per cent to the doses that had earlier been committed by the Soviet Union, the UK and the USA.[30] The PTBT did not halt nuclear weapon developments, and nearly four times as many nuclear tests took place underground after 1963 than in the atmosphere before the treaty.[31]

The Eighteen Nation Committee on Disarmament, which had held its first meeting in Geneva on 14 March 1962, had been a prelude to the PTBT.

[29] International Atomic Energy Agency (IAEA), *The Radiological Situation at the Atolls of Mururoa and Fangataufa*, Technical Report, vol. 3, *Inventory of Radionuclides Underground at the Atolls: Report by an International Advisory Committee (Working Group 3)*, IAEA-MFTR-3 (IAEA: Vienna, 1998).

[30] For a definition of 'dose commitment' see the Glossary in this volume.

[31] Fedchenko (note 7).

The Swedish delegation, led by the future Nobel Peace Prize laureate Alva Myrdal, represented one of eight non-aligned states in the ENCD that submitted a joint proposal on verification of a comprehensive nuclear test ban on 16 April 1962. That proposal included a substantial number of seismic control posts, an international commission to analyse and report on relevant findings, and the ideas of challenge inspections and inspection by invitation. For a while the prospects for a comprehensive test ban looked quite positive, but when the USA sought up to seven inspections per year and the Soviet Union fewer than four the two countries entered into bilateral talks in the summer of 1963, which very quickly resulted in nothing more than the PTBT. That was a disappointment to many countries, and Sweden decided to continue to push for a comprehensive ban. A seismic observatory was built at Hagfors in western Sweden and its atmospheric radionuclide surveillance system was geared to detect violations of the PTBT (which had no verification regime) and to demonstrate how such a system could be an effective part of a future CTBT verification system.

The particulate system that collected aerosols on filters was improved by increasing the sampling rate, mostly to 1000 m^3/h and at one station to 5400 m^3/h. At this time semiconductor detectors with germanium crystals cooled to liquid nitrogen temperatures had been invented and were being developed, which substantially increased the capability to see even faint details in the samples. The concentrations that could be observed with this system became as low as 0.1 $\mu Bq/m^3$, which for the most important fission products from the detection point of view corresponds to around 0.1 $atoms/m^3$ (equivalent to 10 atoms in a normal living room).

Through the years this sensitive particulate surveillance system has detected many smaller radionuclide emissions, such as ^{131}I from hospitals, sometimes via sewage treatment plants, ^{123m}Te from a Swedish incinerator (in 1975), ^{75}Se from hospitals or agricultural tracer experiments (in 1976), ^{140}La from decontamination exercises at Bourges south of Paris, France (in 1990), primarily ^{106}Ru from an accident at the reprocessing plant at Tomsk, Russia (in 1993), ^{82}Br used for calibration of gas flow meters in the gas network of Stockholm (in 2002), and ^{192}Ir from Studsvik, Sweden, at the time a production site for radioactive sources (in 2004).[32] A widely noted series of

[32] Kolb, W. and Weiss, W., 'Occurrence of lanthanum-140 in ground-level air', *Journal of Environmental Radioactivity*, vol. 13, no. 1 (1991), pp. 79–83; Söderström, C. et al., *History of the Sampling Station at Grindsjön with Quarterly Report on Measurements of Radionuclides in Ground Level Air in Sweden: Fourth Quarter 2003*, Totalförsvarets forskningsinstitut (FOI) Report no. FOI-R-1262-SE (FOI: Stockholm, 2004); Söderström, C. et al., *Quarterly Report on Measurements of Radionuclides in Ground Level Air in Sweden: Fourth Quarter 2002*, FOI Report no. FOI-R-0801-SE (FOI: Stockholm, Feb. 2003); and Söderström, C. et al., *Detection of ^{192}Ir in Ground Level Air in Ursvik 20 September 2004 with Quarterly Report on Measurements of Radionuclides in Ground Level Air in Sweden: Fourth Quarter 2004*, FOI Report no. FOI-R-1691-SE (FOI: Stockholm, Sep. 2005).

detections, primarily of ^{239}Np and ^{99}Mo, was made in 1976.[33] The air masses on these occasions came from the Soviet Union, and the international press speculated whether the observed radionuclides possibly derived from a nuclear explosion-driven generator built to power a charged-particle beam accelerator.

On one occasion in 1975 tritium, which was believed to have come from a Soviet nuclear test at Novaya Zemlya, was also detected in Hagfors. Atmospheric tritium surveillance started at FOA in 1975 and continued for a little more than a year.[34] Atmospheric hydrogen was catalytically combusted to water and trapped in a zeolite column with fine dispersed palladium powder in its cavities. This water containing HTO (i.e. tritiated water, ^{3}HOH) was then reduced back to hydrogen and measured in a low-level gas counter.

In 1978, when it was understood that China would soon stop atmospheric testing, FOA started a programme to develop a sensitive system for analysis of radioactive isotopes of xenon in the atmosphere. Xenon is a noble gas, which means that it is difficult and very expensive to fully contain it underground after a nuclear test. For this reason and because xenon isotopes are among the most plentiful radionuclides produced in fission, they are good indicators and verification agents of a clandestine nuclear test under a comprehensive test ban. Krypton is also a noble gas with isotopes that are abundantly produced in fission but none of these has suitable half-life and radiation. Xenon has a quartet—two isotopes, ^{133}Xe and ^{135}Xe, and two isomeric states, ^{131m}Xe and ^{133m}Xe—with half-lives between 9 hours and 12 days and gamma, X-ray, beta and conversion electron-radiation that can be used for sensitive analysis.

A xenon system analysing just ^{133}Xe operated in Stockholm in 1980–82, a period often dominated by large xenon emissions from a Swedish reactor with unusually large fuel leaks.[35] After that the system was mothballed for financial reasons until after the Chernobyl accident, when the government's sensitivity regarding radionuclides had naturally grown. The system went online again in October 1990 just in time to detect ^{133}Xe from the last Soviet Union test, which was carried out underground at Novaya Zemlya on 24 October. The system had been improved and rebuilt to work at room temperature instead of in the conventional way of having its activated charcoal columns cooled by liquid nitrogen or a low-temperature freezer. A version was also built to be part of an International Surveillance of Atmos-

[33] De Geer, L.-E., 'Airborne short-lived radionuclides of unknown origin in Sweden in 1976', *Science*, vol. 198, no. 4320 (1977), pp. 925–27.

[34] Bernström, B., *Tritium in Atmospheric Hydrogen Gas at a Swedish Sampling Station at Hagfors*, Försvarets Forskningsanstalt (FOA) Report no. C 40062-T2 (FOA: Stockholm, 1977).

[35] Bernström, B. and De Geer, L.-E., *Mätning av små mängder xenon-133 i luft* [Measuring small amounts of xenon-133 in air], Försvarets Forskningsanstalt (FOA) Report no. C 20515-A1 (FOA: Stockholm, 1983).

pheric Radionuclides (ISAR) station, a container-carried control post developed by FOA and fielded in Ussuriysk near Vladivostok, Russia, in 1991 to serve as a demonstration station for a CTBT.

Since the CTBT was opened for signature in 1996 the development of the Swedish xenon analyser has continued and its throughput has been increased. Instead of a low-energy germanium detector, it has been equipped with a beta–gamma coincidence detector, a small plastic cylinder that acts both as a container of sample gas and a detector of beta and conversion electrons inside a sodium iodide detector that registers X- and gamma rays. The detection limit for three of the xenon isotopes is about 0.2 mBq/m^3 and somewhat higher for ^{135}Xe, about 0.9 mBq/m^3 due to its shorter half-life (9.14 hours). Recently, work has been done to improve the system significantly by essentially eliminating the so-called memory effect whereby a small percentage of the xenon gas diffuses into the walls of the beta cell and acts as a background for the next sample.[36] A significant development was finding a commercial partner that could produce the system, now called the Swedish Automatic Unit for Noble Gas Acquisition (SAUNA), to be fielded by the Comprehensive Nuclear-Test-Ban Treaty Organization (CTBTO) in its global network.[37] As of November 2014 the CTBT verification system had 15 SAUNA systems installed worldwide, together with 12 French online air sampling and analysis systems for quantifying xenon (Systéme de Prélèvements et d'Analyse en Ligne d'Air pour quantifier le Xénon, SPALAX) and 3 Russian Analyser of Radioactive Isotopes of Xenon (ARIX) systems.

IV. Swedish detection of radioactive leakages from underground nuclear explosions

Although all underground nuclear explosions leaking radionuclides that can be detected outside the borders of the testing country are forbidden for the 126 states parties of the PTBT, such detections have not been uncommon. Countries downwind of test sites, such as Canada and Japan, have detected such emissions and other countries, such as the Soviet Union and the USA, have also detected leaks by flying sampling missions downwind of foreign test sites to search for residues from underground explosions. As of 2014, FOA has been involved in detecting leaks from 10 under-

[36] Bläckberg, L. et al., 'Investigations of surface coatings to reduce memory effect in plastic scintillator detectors used for radioxenon detection', *Nuclear Instruments and Methods in Physics Research, Section A: Accelerators, Spectrometers, Detectors and Associated Equipment*, vol. 656, no. 1 (Nov. 2011), pp. 84–91.

[37] Ringbom, A. et al., 'SAUNA—a system for automatic sampling, processing and analysis of radioactive xenon', *Nuclear Instruments and Methods in Physics Research, Section A: Accelerators, Spectrometers, Detectors and Associated Equipment*, vol. 508, no. 3 (Aug. 2003), pp. 542–53; and Gammadata Sauna Systems AB, Uppsala, Sweden.

ground explosions: 4 in the Soviet Union, 4 in the USA and 2 in North Korea (see table 8.2). In 5 of these cases the leaking agent was xenon, but there is little doubt that xenon was also a major component in the other 5 explosions, although no xenon samplers were deployed at the time.

The first eight leaks were detected by the national Swedish network: six by the particulate system, one by a tritium station and one by a xenon network. Three of these were so-called peaceful nuclear explosions (PNEs). The Schooner test at the USA's Nevada test site on 8 December 1968 was a test of a device aimed at excavating large holes in the ground to study, for example, how a harbour in Alaska or canals, such as a new one across Panama, could be constructed economically by one or a series of nuclear detonations.[38] To minimize radiation doses to the population, the Schooner device had a tungsten pusher that created minimal residual radiation. Even so, Sweden and other countries in Europe detected the tungsten isotope ^{181}W that was activated in the pusher of this test.

The Soviet Union conducted two PNEs with the same purpose, on 18 December 1966 and on 23 March 1971.[39] Typical for these was that the fission product debris was particularly enhanced in ^{140}Ba. The reason was that the charges were buried at shallow depths and, after the explosions, mass chain 140 was rapidly injected as short-lived ^{140}Xe ($T_{1/2}$ = 13.60 seconds). Being a noble gas, it escaped re-deposition during ejecta fallback—in stark contrast to the destiny of most other fission products. In such scenarios, detecting ^{140}Ba in filters as a granddaughter of ^{140}Xe can sometimes be seen as 'the poor man's noble gas system'.

Accidental releases occurred on 18 December 1970, 26 February 1987 and 2 August 1987, when cracks unexpectedly opened up escape routes for primarily noble gases and volatile radionuclides (isotopes of iodine, caesium and tellurium) with boiling points less than around 1000°C. When these accidents occurred, on-site personnel were evacuated to avoid radiation injuries. The accident on 2 August 1987 at Novaya Zemlya, for example, was similar to an earlier accident at the same site, on 14 October 1969, when testing personnel were not evacuated and were therefore irradiated with doses in the range of 1 sievert.[40] Meteorological conditions meant that the 1969 test was not detected in Sweden. The leak on 21 October 1975, when tritium was detected at Hagfors, is mentioned above. The final three

[38] Persson, G., *Radioactive Tungsten in the Atmosphere following Project Schooner*, Försvarets Forskningsanstalt (FOA) 4 Report no. C 4460-28 (FOA: Stockholm, 1971).

[39] Persson, G., 'Fractionation phenomena in activity from an underground nuclear explosion', *Health Physics*, vol. 16, no. 4 (1968), pp. 515–23; and Eriksen, B., *Undersökningar av luftburet radioaktivt material härrörande från en underjordisk kärnladdningsexplosion i USSR den 23 mars 1971* [Investigations of airborne radioactive material originating from an underground nuclear device explosion in the USSR on March 23 1971], Försvarets Forskningsanstalt (FOA) 4 Report no. C 4502-A1 (FOA: Stockholm, 1972).

[40] Bjurman, B. et al., 'The detection of radioactive material from a venting underground nuclear explosion', *Journal of Environmental Radioactivity*, vol. 11, no. 1 (1990), pp. 1–14.

Table 8.2. Underground nuclear explosions with leaked radionuclides detected and/or analysed in Sweden[a]

Date, local time (difference from UTC)	Yield (kt)	National designation	Depth (m)	Radionuclides detected/ Comments
1. Semipalatinsk, Soviet Union				
18 Dec. 1966 10:58 (UTC+6)	20–150	#262 Shaft 101	427	Ba-140 (*c.* 40 mBq/m^3), Cs-137, Sr-89, Sr-90; PNE; all progeny of short-lived xenon and krypton isotopes
2. Nevada, United States				
8 Dec. 1968 08:00 (UTC−8)	30	#577 Schooner	107	W-181 (*c.* 20 μBq/m^3); (n,2n)-activation of a tungsten pusher in a low-radiation PNE device
3. Nevada, United States				
18 Dec. 1970 07:30 (UTC−8)	10	#666 Baneberry	278	I-131 (*c.* 1 mBq/m^3); accidental containment failure
4. North of Perm, Soviet Union				
23 Mar. 1971 12:00 (UTC+5)	3 × 15	#335 Taiga	127	Ba-140 (*c.* 2 mBq/m^3), I-131; PNE; excavation test for construction of a Pechora-Kama canal
5. Novaya Zemlya, Soviet Union				
21 Oct. 1975 15:00 (UTC+3)	1200	#432 adit A-12	..	Tritium gas (*c.* 200 mBq/m^3); a 5-explosion salvo
6. Semipalatinsk, Soviet Union				
26 Feb. 1987 10:58 (UTC+6)	10	#669 adit 130	..	I-131 (*c.* 50 μBq/m^3)[b]
7. Novaya Zemlya, Soviet Union				
2 Aug. 1987 06:00 (UTC+4)	150	#682 adit A-37A	..	I-131 (*c.* 2 mBq/m^3), I-133, Te-132; a 5-explosion salvo
8. Novaya Zemlya, Soviet Union				
24 Oct. 1990 17:58 (UTC+3)	70	#715 adit A13-N	..	Xe-133 (*c.* 25 mBq/m^3); an 8-explosion salvo
9. Mount Mantap, North Korea				
9 Oct. 2006 10:35 (UTC+9)	<1 kt	#1 adit	..	Xe-133 (*c.* 7 mBq/m^3), Xe-133m; sampled in South Korea
10. Mount Mantap, North Korea				
12 Feb. 2013 11:58 (UTC+9)	10 kt	..	..	Xe-133 (*c.* 3 mBq/m^3), Xe-131m (*c.* 0.5 mBq/m^3); sampled by the CTBTO in Japan and Russia

PNE = peaceful nuclear explosion.

[a] Of these 10 events, 8 were detected in Sweden, 1 detected with Swedish involvement abroad, and 1 detected by the Comprehensive Nuclear-Test-Ban Treaty Organization (CTBTO) and analysed with the involvement of the Swedish Defence Research Agency (FOI).

[b] The Institute for Atmospheric Radioactivity (IAR) in Freiburg, Germany, measured 1600 mBq/m^3 of Xe-133. Those findings in Germany helped convince the Swedish Government to finance the re-establishment of radioxenon surveillance.

Sources: Persson, G., 'Fractionation phenomena in activity from an underground nuclear explosion', *Health Physics*, vol. 16, no. 4 (1968), pp. 515–23; Persson, G., *Radioactive Tungsten in the Atmosphere following Project Schooner*, Försvarets Forskningsanstalt (FOA) 4 Report no. C 4460-28 (FOA: Stockholm, 1971); Eriksen, B., *Undersökningar av luftburet radioaktivt material härrörande från en underjordisk kärnladdningsexplosion i USSR den 23 mars 1971* [Investigations of airborne radioactive material originating from an underground nuclear device explosion in the USSR on March 23 1971], Försvarets Forskningsanstalt (FOA) 4 Report no. C 4502-A1 (FOA: Stockholm, 1972); Bergström, B., *Tritium in Atmospheric Hydrogen Gas at a Swedish Sampling Station at Hagfors*, Försvarets Forskningsanstalt (FOA) Report no. C 40062-T2 (FOA: Stockholm, 1977); Bjurman, B. et al., 'The detection of radioactive material from a venting underground nuclear explosion', *Journal of Environmental Radioactivity*, vol. 11, no. 1 (1990), pp. 1–14; Ringbom, A. et al., 'Measurements of radioxenon in ground level air in South Korea following the claimed nuclear test in North Korea on October 9, 2006', *Journal of Radioanalytical and Nuclear Chemistry*, vol. 282, no. 3 (2009), pp. 773–79; and Ringbom, A. et al., 'Radioxenon detections in the CTBT international monitoring system likely related to the announced nuclear test in North Korea on February 12, 2013', *Journal of Environmental Radioactivity*, vol. 128 (Feb. 2014), pp. 47–63.

cases all dealt with leaks of xenon isotopes that were also detected by xenon stations. The first, in 1990, was detected by the Swedish national xenon system and the second, in 2006, by a Swedish SAUNA system that was rapidly moved to the north-eastern corner of South Korea in the week following the first North Korean nuclear test.[41] The third case was detected by the CTBTO network in Takasaki, Japan, and Ussurijsk, Russia, 7.5 to 8.5 weeks after the test in February 2013, probably in connection with re-entry into the testing chamber.[42] Swedish experts participated significantly in the analysis of this event.

There is an additional case, which is not yet acknowledged by all international experts, related to a series of xenon and xenon progeny isotopes that were detected in Japan, Russia and South Korea in the period 13–22 May 2010 and interpreted as coming from a very low yield (around 3 tonnes) nuclear test at the North Korean Mount Mantap test site just after midnight UTC between 11 and 12 May 2010. The dynamics of the leak indicate decoupling, which based on previous experience means that the real yield could have been significantly higher. In turn this implies that such a test could have been a step in making the weapons physically smaller in order to fit North Korean missiles, something that North Korea actually claimed to have succeeded in doing after its test in February 2013.[43]

41 Ringbom, A. et al., 'Measurements of radioxenon in ground level air in South Korea following the claimed nuclear test in North Korea on October 9, 2006', *Journal of Radioanalytical and Nuclear Chemistry*, vol. 282, no. 3 (2009), pp. 773–79. See also chapter 9 in this volume.

42 Ringbom, A. et al., 'Radioxenon detections in the CTBT international monitoring system likely related to the announced nuclear test in North Korea on February 12, 2013', *Journal of Environmental Radioactivity*, vol. 128 (Feb. 2014), pp. 47–63.

43 De Geer, L.-E., 'Radionuclide evidence for low-yield nuclear testing in North Korea in April/May 2010', *Science & Global Security*, vol. 20, no. 1 (2012), pp. 1–29; De Geer, L.-E., 'Reinforced evidence of a low-yield nuclear test in North Korea on 11 May 2010', *Journal of Radioanalytical and Nuclear Chemistry*, vol. 298, no. 3 (2013), pp. 2075–83; Schaff, D. P., Kim, W.-Y. and Richards, P. G.,

V. Nuclear forensics applied in Sweden to non-nuclear explosion events[44]

Operating a sampling network and a laboratory for airborne radionuclide surveillance, especially in an institutional environment that to varying degrees focused on deep understanding of nuclear weapons and defence and protection against their effects, proved to be fruitful, even when unanticipated problems arose. Fulfilling a high-tech routine, where occasionally detections and events require special analyses, leads to development of a type of forensic thinking and the acquisition of hardware that also create preparedness for unexpected situations. The following rather amazing chain of events illustrates that context.

First, in October 1981 at the peak of the cold war a Soviet Whiskey-class submarine, designated U-137 by Sweden (and known in the media as Whiskey on the Rocks) went aground inside Swedish military waters.[45] The decision was taken to scan the hull with a handheld instrument to detect possible radiation from a nuclear charge that, in fact, was not believed to be present. However, when the scan was done by the present author, the detection of an eightfold increase in gamma radiation near the top of the port torpedoes heated up the situation considerably and, during a following night, a more advanced gamma detector was taken to the site from the air surveillance laboratory. The measurement proved the presence of uranium-238 and, via secondary effects, neutrons—a combined signal that could hardly be due to anything other than a plutonium bomb. It could also be shown that for a simple fission charge the uranium tamper would effectively shield low-energy gamma rays from either ^{235}U or ^{239}Pu. In the aftermath of the incident, the success of the mission led to the FOA laboratory getting the opportunity to acquire one of the first commercially available portable germanium-detector systems. This system was then, among other things, calibrated and used to measure densities of

'Seismological constraints on proposed low-yield nuclear testing in particular regions and time periods in the past, with comments on "Radionuclide evidence for low-yield nuclear testing in North Korea in April/May 2010" by Lars-Erik De Geer', *Science & Global Security*, vol. 20, no. 2-3 (2012), pp. 155–71; Miao Zhang, M. and Wen, L., 'Seismological evidence for a low-yield nuclear test on 12 May 2010 in North Korea', *Seismological Research Letters*, vol. 86, no. 1 (Jan./Feb. 2015); Wotawa, G., 'Meteorological analysis of the detection of xenon and barium/lanthanum isotopes in May 2010 in Eastern Asia', *Journal of Radioanalytical and Nuclear Chemistry*, vol. 296, no. 1 (2013), pp. 339–47; and Wright, C. M., 'Low-yield nuclear testing by North Korea in May 2010: assessing the evidence with atmospheric transport models and xenon activity calculations', *Science & Global Security*, vol. 21, no. 1 (2013), pp. 3–52.

[44] This section draws on the personal experience of the present author.

[45] Leitenberg, M., 'The case of the stranded sub', *Bulletin of the Atomic Scientists*, vol. 38, no. 3, pp. 10-13; and Sundström, O., 'U137 och U238—en fastställd kombination' [U137 and U238—a proven combination], *Foa tidningen*, vol. 19, no. 4 (1981), pp. 6–9; and De Geer, L.-E., *Nonintrusive detection of nuclear weapons on ships*, Försvarets Forskningsanstalt (FOA) Report no. C20817–4.1 (FOA: Stockholm, 1999).

radionuclides deposited on the ground, which is referred to as field-gamma spectroscopy.

Second, on 7 February 1983 a nuclear reactor from a Soviet radar reconnaissance satellite, Cosmos 1402, disintegrated over the south Atlantic.[46] Warnings were issued in late December 1982 and there was a clear possibility that Sweden would be hit (as Canada had been hit in 1978 by a similar reactor).[47] At FOA, search operations to look for heavily contaminated pieces of debris were soon set up and tested in early January in cooperation with a private company, Swedish Geological AB (SGAB). The company usually flew a 16-litre sodium iodide crystal detector in a small fixed-wing aircraft to conduct geological surveys. In this case, the detector was installed in a helicopter and smaller instruments were tested in a slow-flying military aeroplane, with both flying at various altitudes above fields and forests where controlled ^{137}Cs sources had been placed. No radioactive debris hit Sweden, but a great deal of experience was gained during that month in flying search operations across large areas and on the sensitivities of different detectors flown at different altitudes.

Third, in the first half of the 1980s small concentrations of especially ^{95}Nb, but also ^{95}Zr, were detected at Swedish sampling stations. To help localize the emission point, meteorological backward trajectory calculations were often requested from the Swedish Meteorological and Hydrological Institute (SMHI). These are simple services while the 'weather' data remains in active memory but are a bit more complex if data has to be read back from secondary memory. A routine was therefore initiated where these analyses were automatically done every day and for all sampling stations in the network. Most trajectories for the mass-95 detections pointed to Ignalina in Lithuania, Soviet Union, where the largest reactors in the world at the time were situated. When that information was published, the Soviet news agency referred to the FOA group as agents of the US Central Intelligence Agency (CIA).

On 26 April 1986 the Chernobyl accident occurred. For the FOA radionuclide surveillance group a period of very hectic work started but, due to the three events described above, FOA was especially well-equipped to address the situation. By early the next morning, the radioactive cloud had entered Swedish airspace over the Baltic islands of Gotland and Öland. A network of ionization chambers operated by the Radiation Protection Institute (Statens strålskyddsinstitut, SSI) failed to report the increased radiation levels due to communication problems. Instead, the cloud was first

[46] Broad, W. J., 'Satellite's fuel core falls "harmlessly"', *New York Times*, 8 Feb. 1983.

[47] Jasani, B., 'The military use of outer space', *SIPRI Yearbook 1979: World Armaments and Disarmament* (Taylor & Francis: London, 1979), pp. 270, 274-78; Lay, F., 'Nuclear technology in outer space', *Bulletin of the Atomic Scientists*, vol. 35, no. 7, pp. 27-31; and Weiss, G. W., 'The life and death of Cosmos 954', *Studies in Intelligence*, spring 1978.

noted by a Swedish nuclear power plant the next day, whose scientists initially thought something had happened at their own site.[48] FOA was alerted, and a filter was immediately taken down from the Stockholm sampler; within 10 minutes it was clear from the trajectories, which had already automatically been delivered by computer from SMHI, that the emission point was to be found along a track crossing the western Soviet reactor belt. A strong ^{134}Cs signal showed that a reactor accident had taken place and the $^{133}I : ^{131}I$ activity ratio could have been used to date the accident and thereby pinpoint Chernobyl. That was not done that day, however, as the small FOA laboratory was showered with telephone calls from the media and from organizations such as SSI and the Swedish Nuclear Power Inspectorate (Statens kärnkraftinspektion, SKI). Furthermore, in the afternoon there was an alarm from Gotland about excessive radiation. FOA's portable germanium detector system was then flown by military helicopter to Gotland, where it could be demonstrated that the alarm had been quite exaggerated. Flying across the Baltic, the helicopter crossed the rain cloud which later that night caused high deposition (e.g. up to *c.* 200 kBq/m^2 of ^{137}Cs) in the region around Gävle. In the days and months that followed, a large amount of fallout mapping was done by SGAB in which the experience from three years earlier was useful for planning operations and interpreting data. SGAB's detectors were calibrated against field gamma measurements done by the FOA group with the portable system, still the only one in Sweden, during several helicopter tours around the country.

VI. Implications for the Comprehensive Nuclear-Test-Ban Treaty

When the cold war ended it was finally possible to conclude a comprehensive test-ban treaty, and half a year after the CTBT was opened for signature, a provisional technical secretariat was inaugurated in Vienna, Austria, in March 1997. At the time of writing, nearly two decades later, the CTBT has been signed by 183 states, including 163 ratifications, of a total of 196 possible signatories. Under the terms of the treaty, entry into force requires ratification by 44 named states, but 8 of these have yet to do so: China, Egypt, India, Iran, Israel, North Korea, Pakistan and the USA.

For many years Sweden worked together with the international community to negotiate a CTBT and to demonstrate how such a treaty could be reliably verified. The Group of Scientific Experts (GSE)—a Swedish initiative that was created in 1976 under the umbrella of the Geneva-based Com-

[48] In fact, the SSI was just about to decide to close down the quite aged ionization chamber network.

mittee on Disarmament (CD, the successor to the ENCD, from 1979 called the Conference on Disarmament)—carried out the tedious work on the verification regime even during the years when the political prospects for achieving a treaty were dim.[49] Thus, when the political barriers went down in the early 1990s, the GSE had managed to develop verification science to a degree that was pivotal to the success of the negotiations. Much of the work of the GSE focused on seismic verification technology and several exercises were carried out to test international data centres that could collect data and conduct global analyses.

In the early 1980s interest turned to monitoring radionuclides, and in 1982 Sweden tabled a proposal in the CD, based on scientific arguments and studies, to deal with the radionuclide component of a CTBT verification regime. A global network of 50 to 100 radionuclide control posts, capable of carrying out particulate and xenon sampling and analysis, was proposed—together with one radionuclide laboratory on each continent. The final outcome of the Geneva negotiations was very similar to the Swedish proposal: 80 particulate radionuclide stations were to be established, with half able to conduct xenon analysis. Some states believed in the efficacy of xenon-analysis technology and some did not, and the final agreement was basically a compromise. Subsequent developments, such as the tests conducted by North Korea that leaked xenon, make it difficult to deny the wisdom of equipping not half, but all, of the 80 radionuclide stations with xenon-analysis capabilities. Actually the Protocol to the Treaty says that 'At its first regular annual session, the Conference [of the states parties] shall consider and decide on a plan for implementing noble gas monitoring capability throughout the network'.[50]

The CTBT verification system is a unique global utility that is shared by 93 per cent of the states of the world. Its network, the International Monitoring System (IMS), comprises 50 primary and 120 auxiliary seismic stations, 11 hydroacoustic, 60 infrasound and 80 radionuclide stations as well as 16 radionuclide laboratories, all evenly spread around the earth. These report, many of them continuously, via satellite links to the International Data Centre (IDC) in Vienna. It is designed and optimized to detect clandestine nuclear tests anywhere on the globe, but it can clearly also be useful for many other purposes, as it was in the wake of the nuclear reactor disaster at Fukushima, Japan, in 2011. Fukushima and the 2004 Indian Ocean tsunami were events that really opened the eyes of the CTBT states signatories to the fact that the CTBTO can provide useful data for disaster warn-

[49] Dahlman, O., Mykkeltveit, S. and Haak, H., *Nuclear Test Ban: Converting Political Visions to Reality* (Springer: Dordrecht, 2009); and Dahlman, O. et al., *Detect and Deter: Can 'Countries Verify the Nuclear Test Ban?* (Springer: Dordrecht, 2011).

[50] Comprehensive Nuclear-Test-Ban Treaty (CTBT), opened for signature 24 Sep. 1996, not yet in force, <https://treaties.un.org/pages/CTCTreaties.aspx?id=26&subid=A&lang=en>, Protocol, Part I.

ing and mitigation. Today tsunami warning centres in 11 countries and international and national institutions that are responsible for aviation and maritime safety receive relevant data from the CTBTO seismic and hydroacoustic networks; likewise they receive data from the infrasound stations that warn of erupting volcanoes and the breaking up of ice shelves. Pure science also benefits from CTBTO data, such as when a meteor broke up over the Ural Mountains in February 2013 and caused the largest infrasound wave ever recorded by the IMS. On 12 August 2000 the hydracoustic subsystem had also picked up signals from the explosion in the Russian submarine *Kursk*. In summary, the CTBT verification system has proved to be a valuable asset for humanity even before the treaty has entered into force.

Appendix 8A. Detection by Sweden of Chinese nuclear tests in the atmosphere

LARS-ERIK DE GEER

China conducted a total of 26 tests between 1964 and 1980. These were numbered from China-1 up to China-26 by the Swedish National Defence Research Establishment (Försvarets forskningsanstalt, FOA).[1] All but four of the test—China-9, -17, -20 and -24—were atmospheric (for details of the tests see table 8A.1). The 22 atmospheric tests presented a unique opportunity to follow the development of a full-scale nuclear weapon programme. In the following sections each test is summarized as seen from Sweden at the time.

China-1, 16 October 1964

This test was the first one where individual particles could be studied in Sweden from a single near-surface nuclear explosion. The first debris appeared in ground level samples after 11 days and showed a clear opposite (or reversed) fractionation (i.e. it was enhanced in volatile nuclides).[2] Reversal autoradiography was used to study 43 individual particles with diameters between 1 and 5 μm. A large number of them were perfectly spherical and they ranged in colour from colourless to yellow-red-brownish. The specific activity was substantially (100 times) lower than the laboratory had experienced before from mainly large Soviet and US tests. This was interpreted to be the result of a near-surface explosion that sucked a lot of ground material, estimated at 1000 tonnes, into the fireball.

China-2, 14 May 1965

China-2 was probably a weaponized version of the device tested in 1964. Debris was first observed around the tropopause on 24 May and continued to appear in high-altitude samples for nearly a month.[3] Fractionation

[1] Some compilations also include an aborted atmospheric test on 13 Sep. 1979, making this total 27.

[2] Sisefsky, J., 'Debris particles resulting from the Chinese nuclear bomb test', *Nature*, 12 June 1965, pp. 1140–41.

[3] Persson, G., 'Fractionation phenomena in debris from the Chinese nuclear explosion in May 1965', *Nature*, 19 Mar. 1966, pp. 1193–95; and Sisefsky, J., 'Debris particles from the second Chinese nuclear bomb', *Nature*, 11 June 1966, pp. 1143–44.

Table 8A.1. Chinese nuclear tests, 1964–80

FOA test designation	Date	Time[a]	Yield	Description of device
China-1	16 Oct. 1964	13:00	22 kt	On a 102-m steel tower, fission device, based on implosion of U-235
China-2	14 May 1965	08:00	35 kt	Dropped from a Hong-6 bomber, exploding at about 500 m, fission device based on implosion of U-235
China-3	9 May 1966	14:00	200–300 kt	Dropped from a Hong-6 bomber, fission of U-235, some thermonuclear material (Li-6)
China-4	27 Oct. 1966	07:10	12 kt	Delivered by a Dongfeng 2 ballistic missile from the Shuangchengzi missile test range, *c.* 850 km east of the test site, fission of U-235, airburst at 569 m
China-5	28 Dec. 1966	10:00	122 kt	On a 102-m. steel tower, first test of a partial yield two-stage H-bomb, U-235, U-238 and Li-6 deuteride, too massive to be a deliverable weapon
China-6	17 June 1967	06:19	3.3 Mt	Airdrop from a Hong-6A bomber, parachute retarded airburst at 2960 m., first full-scale 3F (fission-fusion-fission) H-bomb, U-235, U-238 and Li-6 deuteride
China-7	24 Dec. 1967	10:00	10–20 kt	Airdrop from a Hong-6 bomber, U-235, U-238 and Li-6 deuteride, apparently a failed thermonuclear test
China-8	27 Dec.1968	13:30	3 Mt	Airdrop from a Hong-5 bomber, new thermonuclear device with first use of plutonium (China's first reactor went critical in Oct. 1966), Pu-239, U-235, U-238 and Li-6 deuteride
China-9	23 Sep. 1969	22:15	19.2 kt	In tunnel, first Chinese underground test
China-10	29 Sep. 1969	14:40	3 Mt	Airdrop from a Hong-6 bomber, thermonuclear
China-11	14 Oct. 1970	13:30	3 Mt	Airdrop from a Hong-6 bomber, thermonuclear
China-12	18 Nov. 1971	12:00	10–15 kt	Partially buried, fission Pu-239 and U-235
China-13	7 Jan. 1972	13:00	5–10 kt	Airdrop from a Qian-5 attack jet, fission, Pu-239
China-14	18 Mar. 1972	12:00	150–200 kt	Airdrop from a Hong-6 bomber, thermonuclear, apparent failure
China-15	27 June 1973	10:00	2.5 Mt	Airdrop from a Hong-6 bomber, burst at high altitude, thermonuclear
China-16	17 June 1974	12:00	0.2–1 Mt,	Air burst, thermonuclear (the yield was probably close to 1 Mt)
China-17	27 June 1975	07:00	15 kt	Test in tunnel, fission
China-18	23 Jan. 1976	12:00	2–20 kt	Near-surface, fission

FOA test designation	Date	Time[a]	Yield	Description of device
China-19	26 Sep. 1976	12:00	(20–)200 kt	Airburst, fission, possible failure of a fusion device
China-20	17 Oct. 1976	11:00	10 kt	Tunnel, fission
China-21	17 Nov.1976	12:00	4 Mt	Airdrop from a Hong-6 bomber, test of new thermonuclear design
China-22	17 Sep. 1977	13:00	20 kt	Airburst, fission
China-23	15 Mar. 1978	11:00	6–20 kt	Surface burst, fission
China-24	15 Mar. 1978	16:00	5 kt	Shaft, fission
China-25	14 Dec. 1978	..	10 kt	Surface burst, fission
China-X	13 Sep. 1979	..	0 kt	Surface burst, parachute did not deploy, full failure, not counted at the time
China-26	16 Oct. 1980	10:40	700 kt	Airburst, last atmospheric nuclear test by any actor

FOA = Swedish National Defence Research Establishment, kt = kiloton, Mt = megaton.

[a] The times given are local (UTC + 8 hours).

Sources: The information is from US Government agencies at the time and from a tabulation based on remarks by Chinese scientists during 2 visits in 1990 to nuclear weapon establishments in China that was compiled by Danny Stillman, then Division Leader of the International Technology Division at Los Alamos National Laboratory. Reed, T. S., 'The Chinese nuclear tests, 1964–1996', *Physics Today*, vol. 69, no. 9 (Sep. 2008), pp. 47–53. The time-of-the-day data is taken from Norris, R. S., Burrows, A. S. and Fieldhouse, R. W., *Nuclear Weapons Databook: British, French and Chinese Nuclear Weapons*, vol. 5 (Westview Press: Boulder, CO, 1994).

profiles for these samples showed a nice continuous change from 'normal' fractionation at early times to a highly 'reversed' fractionation towards the end of the period—consistent with a picture of larger particles early on that settle with time, leaving the smaller ones behind. About 130 particles from the early samples were studied individually. The colours of the particles were remarkably uniform, from colourless via yellow to somewhat reddish yellow. All were transparent and largely spherical, with a tendency to elongation. The specific activity was much higher than after the first Chinese test, more like it had been during the large test series in the Soviet Union and the USA. These specific activities suggested the presence of about 10 tonnes of bulk material in the device.

China-3, 9 May 1966

Debris was first discovered eight days after the China-3 test in samples taken at high altitudes.[4] Three weeks later a second circumnavigation of

[4] Persson, G., Sisefsky, J. and Lindblom, G., *Detektion av kinesiskt kärnladdningsstoft över Sverige maj-juni 1966* [Detection of Chinese nuclear debris in Sweden May–June 1966], Försvarets Forskningsanstalt (FOA) 4 Report no. C 4260-23 (FOA: Stockholm, 1966); and Sisefsky, J., *Study of*

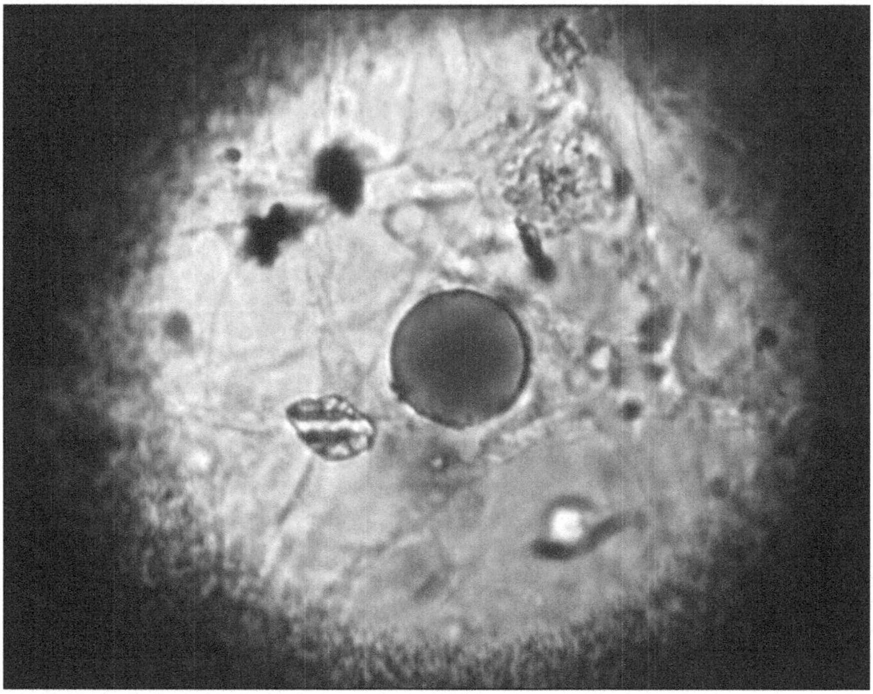

Figure 8A.1. Reversal autoradiograph photo of an 11 μm diameter hot particle produced in the China-5 test and collected in Sweden

Sources: Sisefsky, J., Försvarets Forskningsanstalt (FOA).

the globe by the cloud was also apparent at high altitudes. A clear ^{237}U signal showed that the test involved thermonuclear reactions although it was not a fully fledged H-bomb. Dr Danny Stillman, a former leader of the International Technology Division at the USA's Los Alamos National Laboratory, related that it was a boosted HEU-based device and probably the first test of a Chinese thermonuclear primary, the first stage of a thermonuclear weapon.[5] The use of ^{6}Li was confirmed. However, it did not fit a 'conventional' booster based on deuterium–tritium (DT) gas. The test was reminiscent of the early Soviet design of a so-called *sloika* (layer cake) bomb that was tried in the fourth Soviet nuclear test in 1953, delivering a yield of 400 kt, quite similar to China-3.[6] The particle properties of China-3 were studied and found to be quite similar to China-2, but with slightly higher specific activities.

Debris Particles from the Third Chinese Nuclear Test, FOA 4 Report no. C 4271-23 (FOA: Stockholm, 1966).

[5] Reed, T. S., 'The Chinese nuclear tests, 1964–1996', *Physics Today*, vol. 69, no. 9 (Sep. 2008), p. 52.

[6] See chapter 7 in this volume.

China-4, 27 October 1966

Debris from China-4 was first discovered in Sweden 13 days after the test was carried out.[7] The particles resembled those from China-2, mostly yellow or yellowish, some with a brownish tint, transparent and without any optical activity. The specific activity was also found to be about the same as in the China-2 debris, which now became a standard for explosions without ground contact in the lower troposphere.

China-5, 28 December 1966

The China-5 test was detected after only 8 days.[8] The samples were typically quite weak although a great number of particles could be identified. Compared to earlier tests, the particle spectrum was shifted against larger sizes in both high-altitude and ground-level air samples. At ground level, sizes larger than 6 μm were common and even particles up to 11 μm were found (see figure 8A.1). The specific activities were extremely low, which corroborated the information that it was a tower-mounted experiment. The observation of quite large particles at such a long distance indicated a low-density matrix. Some tests of chemical resistance were made on the particles, and they were found to easily withstand chemical attack. All this suggested that the particles consisted mainly of silicon dioxide (density about 2.3 g/cm^3) from the ground. Most of the particles were perfect spheres, with no signs of wrinkles or similar surface irregularities. In some cases, smaller particles in the 0.2–0.5 μm range were attached to the surface of larger ones. The colour varied from completely colourless and transparent via yellow, orange and brown to nearly black. A small fraction was of another type: cherry red, transparent, not so perfectly spherical and with 100 times higher specific activity. It is reasonable to assume that these particles condensed in a part of the cloud dominated by iron residues from the steel tower and other parts of the set-up.

China-6, 17 June 1967

No activity from this test arrived within reach of the Swedish surveillance system until a single 2.3 μm-diameter hot particle was caught at ground level on 12 August, nearly two months after the event.[9] More particles of the same size range appeared in the following days and months in samples

[7] Sisefsky, J., *Studies of Debris Particles from the Fourth and Fifth Chinese Nuclear Tests,* Försvarets Forskningsanstalt (FOA) 4 Report no. C 4327-28 (FOA: Stockholm, 1967).

[8] Sisefsky (note 7).

[9] Persson, G. and Sisefsky, J., 'Debris from the sixth Chinese nuclear test', *Nature,* 12 July 1969, pp. 173–75.

taken both at high altitudes and at ground level, but often only one by one. Eventually, smaller particles were collected and the number of them increased, but still not more than a few tens per 100 kg of air.

The late detection and the fact that the debris was dominated by hot particles were consistent with a high-yield test with the cloud rising well up in the stratosphere. If produced in a test, hot particles will then descend gravitationally, led by the largest ones. A simple calculation shows that it would take a 2 μm-diameter spherical particle with a density of 5 g/cm^3 (iron oxide) around two months to descend from 18 km down to ground level. Microbarograph registrations in Sweden clearly showed that the device exploded in the troposphere, which is consistent with the formation of hot particles. The specific activity of the particles was higher than in any earlier Chinese tests and also higher than mostly found in Soviet and US thermonuclear tests. The appearance of the particles resembled that of earlier Soviet thermonuclear tests with structureless and spherical colourless ones and more irregular darker ones, sometimes with wrinkles and other surface marks.

China-7, 24 December 1967

For the first time, China did not announce that a test had occurred, which was immediately interpreted as an indication of some kind of failure. Debris reached Sweden nine days post-shot in a fairly concentrated cloud passing at an altitude of 10 km.[10] Many hundreds of hot particles were identified and found to be rather uniformly red, often strongly elliptical or almond shaped. Also crystalline particles with more or less quadratic or rhombic cross sections were common. The particle activities were fairly proportional to volume with specific activities similar to the earlier low-yield airdrops at Lop Nur. The (n,γ) : (n,2n) ratio was around 40, which clearly showed that it was not a functioning fission–fusion–fission (3F) bomb.[11] The larger hot particles found in the range of 0.4–4.5 μm showed normal fractionation and the smaller ones, accordingly, an opposite fractionation.

[10] Sisefsky, J. and Persson, G., *Investigations on Nuclear Debris from the 7th Chinese Test*, Försvarets Forskningsanstalt (FOA) 4 Report no. C 4386-28 (FOA: Stockholm, 1969); and Sisefsky, J. and Persson, G., 'Fractionation properties of nuclear debris from the Chinese test of 24 December 1967', *Health Physics*, vol. 18, no. 4 (1970), pp. 347–56.

[11] See chapter 8, section I, for a description of this ratio's significance.

China-8, 27 December 1968

The (n,γ) : (n,2n) ratio in this test was found to be less than 2.2, which firmly confirmed its thermonuclear character.[12] As after China-6, the first signs of debris were found quite late, 24 days post-shot. China-8 was the first in a series of Chinese thermonuclear tests that did not exhibit any fractionation effects, at least not in bulk samples. This certainly was the result of detonations at reasonably high altitudes where the air is thin, the fireball expands more, and the plasma and gas of debris will occupy such a large volume that they seriously hamper the 'classic' particle condensation process. Rather, the atoms will then, in due course, attach to the natural aerosol with the same efficiency for all except the noble gases and tritium. No more than 29 particles could be identified and they were in the size range 0.2–2.3 μm and nearly colourless spheres with record-high specific activities. From them and an assumption of aluminium oxide as the matrix of the particles, it was estimated that the device contained roughly 300 kg of aluminium.

China-9, 23 September 1969

China-9 was the first Chinese underground test.

China-10, 29 September 1969

China-10 was very similar to China-8, with a 'thermonuclear' (n,γ) : (n,2n) ratio of 1.4.[13] Just 16 hot particles were found, all below 0.5 μm and looking the same and having the same specific activities as the ones from China-8. One difference to China-8 was that its debris was found sooner, just 8 days after the test, at an altitude of 14 km.

China-11, 14 October 1970

China-11 was also very similar to China-8, with the first debris detected quite late, a (n,γ) : (n,2n) ratio of 1.4 and with very few hot particles found. Bulk samples were unfractionated. Because of this, a special study was made on unfractionated samples from China-8, -10 and -11.[14] With the help of chemical separation, the radiation from nuclides produced near sym-

[12] Persson, G. and Sisefsky, J., 'Radioactive particles from the eighth Chinese nuclear test', *Health Physics*, vol. 21, no. 3 (1971), pp. 421–28.

[13] Sisefsky, J. and Persson, G., 'Debris over Sweden from the Chinese nuclear weapon test in September 1969', *Health Physics*, vol. 21, no. 3 (1971), p. 463.

[14] Bernström, B., *Analyses of Fresh Radioactive Debris from the Chinese Nuclear Tests of December 27th, 1968, September 29th, 1969, and October 14th 1970*, Försvarets Forskningsanstalt (FOA) 4 Report no. C 4549-A1 (FOA: Stockholm, 1973).

metric fission were analysed. They are generally low but also very sensitive to the neutron energy. One example was ^{111}Ag, where a ^{95}Zr to ^{111}Ag ratio between 15 and 30 was observed in all three tests. Experiments were made with 14, 7, 4 MeV and fission spectrum neutrons irradiating ^{238}U, which showed the same ratio to be 7, 25, 51 and 82. An average (between the fission threshold at about 1 and 14 MeV) neutron energy through the tamper–pusher uranium thus seemed to be about 7 MeV in all three explosions.

China-12, 18 November 1971

This test turned out to be a very interesting case as, despite the low explosive yield, it showed a $(n,\gamma) : (n,2n)$ ratio of 1.8 ± 0.3, which clearly indicated a thermonuclear device.[15] Perhaps it was a test of a uranium-tamped booster that was not intended to be an H-bomb primary. Debris was first detected in ground level air two weeks after the test. Not a single hot particle was found and all samples exhibited a quite strong enhancement of volatile components. The explanation is probably that the test threw large amounts of ground materials into the cloud and the falling soil scavenged most of the larger particles and left more of the mirror particulate to be globally disseminated.

China-13, 7 January 1972

The China-13 test was first detected 10 days post-shot on a flight at 10 km. This test exhibited a quite conventional picture: 82 hot particles in the size range of 0.6–4 μm with colours from yellow, red via reddish to orange brown.[16] Three almost colourless particles showed the highest fractionation, which had also been found in earlier tests. Assuming that those were composed of mainly aluminium oxide and the coloured ones mainly of iron oxide, this is in accordance with theories of fractionated condensation.

China-14, 18 March 1972

This test produced an unusually high number of hot particles that were able to remain airborne until reaching Sweden.[17] However, unlike earlier cases with many hot particles, China-14 debris contained particles with

[15] De Geer, L.-E., Forslund, K. and Sisefsky, J., *Debris in Sweden from the Low-yield Nuclear Tests Performed by the People's Republic of China on November 18, 1971 and January 7, 1972*, Försvarets Forskningsanstalt (FOA) Report no. C40068-T2(A1) (FOA: Stockholm, 1977).

[16] De Geer, Forslund and Sisefsky (note 15).

[17] Sisefsky, J. and Arntsing, R., *Particle Properties of Debris Appearing in Sweden from the Chinese Nuclear Test of March 18, 1972*, Försvarets Forskningsanstalt (FOA) Report no. C 40113-T2(A1) (FOA: Stockholm, 1980).

quite high specific activity, indicating that the material was not derived from soil. A total of 630 particles of size 0.7–3.5 μm were studied and classified as brownish or opaque, orange, yellow or colourless and where the specific activity slightly increased in that order. In particular, the concentration of mass chain 95 was found to vary substantially with colour—colourless particles having the highest content. The (n,γ) : (n,2n) ratio was 6.5 ± 1, which is indicative of a fission device with some contribution from thermonuclear reactions, a booster or a not fully functioning two-stage weapon (see China-3).

China-15, 27 June 1973

With (n,γ) : (n,2n) = 1.4, very few hot particles, and those found very small, below 0.7 μm, this seemed to be another test in the China-8, -10, -11 series.[18] In 1972 FOA started to routinely use the newly developed Ge(Li)-detector technology and with the much-improved resolution it became possible to observe more radionuclides in the samples. One was ^{54}Mn, that is an (n,p)-reaction product from high-energy neutrons hitting ^{54}Fe in steel. The shape of the cross section curve is not too different from the fission cross section curve of ^{238}U so with no fractionation and a fission–fusion ratio of 2 (US estimate), the data suggested that about 7 grams of ^{54}Mn had been produced.

China-16, 17 June 1974

The China-16 test was detected 17 days after the explosion in a high-altitude sample. Again, with (n,γ) : (n,2n) = 1.4 and with no hot particles, it appeared similar to the previous test. A new feature was the detection of ^{88}Y, which—with a half-life of 106.6 days and being injected into the stratosphere—stayed detectable at ground level for more than a year. This was the first high-energy neutron fluence detector used in a nuclear weapon test that was noted in Sweden. If a certain amount of natural yttrium (^{89}Y) is added to a particular location of interest within the device, ^{88}Y as well as ^{87}Y ($T_{1/2}$ = 79.8 hours) would form in single and double (n,2n)-reactions in the extremely high fluence of high-energy neutrons during the explosion. The fluence at the point of interest could then be calculated from the ratio of the two isotopes and their (n,2n) cross sections.

18 Arntsing, R., De Geer, L.-E. and Vintersved, I., *Radioactivity from Nuclear Explosions in Ground Level Air at Three Swedish Sampling Stations: Ge(Li) Measurements up to Midyear 1975*, Försvarets Forskningsanstalt (FOA) Report no. C 40038-T2(A1) (FOA: Stockholm, 1976). Also published in US Energy Research and Development Administration (USERDA), Health and Safety Laboratory (HASL), Quarterly report (1 Sep. 1976 through 1 Dec. 1976), HASL-315, UC-11, 1 Jan. 1977, <http://www.osti.gov/scitech/servlets/purl/7126236>.

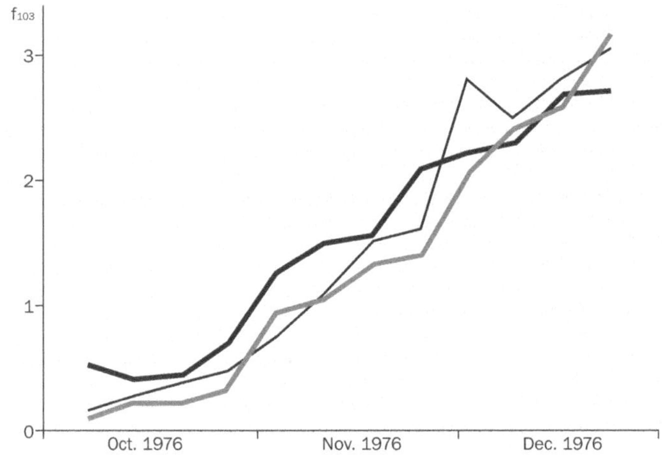

Figure 8A.2. Fractionation factor of Ruthenium-103 after the China-19 explosion

The measurements in ground-level air were made at three stations, Kiruna, Grindsjön and Ljungbyhed, Sweden, during 3 months after the test, on 26 Sep. 1976.

Source: De Geer, L.-E. et al., *Particulate Radioactivity, Mainly from Nuclear Explosions, in Air and Precipitation in Sweden Mid-year 1975 to Mid-year 1977,* FOA report C 40089-T2(A1) (Försvarets Forskningsanstalt: Sundbyberg, Nov. 1978), <http://www.iaea.org/inis/collection/NCLCollectionStore/_Public/11/543/11543720.pdf>. Also published in *Environmental Quarterly,* Report EML-349 (Environmental Measurements Laboratory: New York, 1979).

Detailed fluence values can, however, only be usefully read by the testing party for two reasons. First, they are the only ones who know where in the device the stable yttrium, ^{89}Y, was initially inserted. Second, they are reasonably the only ones that can take samples strong enough to detect the (n,2n)[2] product ^{87}Y. If the problem is simplified such that both (n,2n) cross sections are the same, σ m^2, the neutron energy is constant and well above the (n,2n)-threshold and burnout is not considered, the number of ^{88}Y nuclides formed during t seconds in a flux density (i.e. fluence rate) of φ s^{-1}m^{-2} neutrons is $N_{Y\text{-}88} = N_{Y\text{-}89} \times \varphi \sigma t$. A simple integration then gives the number of ^{87}Y nuclides as $N_{Y\text{-}87} = N_{Y\text{-}89} \times (\varphi \sigma t)^2 / 2$, which finally solves the fluence as $\varphi t = (2 / \sigma)(N_{Y\text{-}87} / N_{Y\text{-}88})$. Note that very high fluences are required to obtain a measurable quotient. With $\sigma = 1$ barn and an atom quotient of, for example, 0.5, the fluence needs to be of the order of 10^{28} high-energy neutrons per m^2 (about 2 moles/cm^2), something that on earth can only be reached in a fusion charge.

China-17, 27 June 1975

This was the second underground test.

China-18, 23 January 1976

Debris from this test started to appear at ground level around two weeks after the shot.[19] Nothing was found at high altitudes. The samples clearly displayed an opposite fractionation pattern, and not a single hot particle was found. This was the same as after China-12 with one important difference: this time the (n,γ) : (n,2n) ratio was greater than 30 and the device was probably a pure fission one.

China-19, 26 September 1976

The 19th Chinese test explosion caused the heaviest fallout of short-lived radionuclides in Sweden since the large Soviet test series in 1962. In the southern part of the country hundreds of becquerels per m^2 of ^{131}I and ^{140}Ba were deposited with the precipitation.[20] Debris first occurred at ground level and at high altitudes 8–10 days after the test and was then the major contributor to anthropogenic radionuclides in the air until the end of the year. The (n,γ) : (n,2n) ratio was 31 ± 3, signifying a fission device, but at the same time steel activation products, such as ^{54}Mn and ^{58}Co, bore witness to high-energy neutrons. The device could have been a pure booster or a booster primary, together with a terminated main thermonuclear stage. In the latter case, it could have indicated preparation for future underground tests where the full yield would have to be limited.

Nearly 300 hot particles in the range 0.7–7.8 μm were examined, a great majority of them red. Most were spherical, although some were elliptical. Fractionation effects were typical for the particulate found. Changes with time during the last quarter of 1976 of f_{103}, the fractionation factor of ^{103}Ru (which via its oxide is extremely volatile), were clearly measured at three ground level stations (see figure 8A.2). Hot spherical particles in October showed a low f_{103} that then, when the mirror particulate fraction gradually took over, increased by 0.3 units per week up to the end of the year.

[19] De Geer, L.-E. et al., *Particulate Radioactivity, Mainly from Nuclear Explosions, in Air and Precipitation in Sweden Mid-year 1975 to Mid-year 1977*, FOA report C 40089-T2(A1) (Försvarets Forskningsanstalt: Sundbyberg, Nov. 1978), <http://www.iaea.org/inis/collection/NCLCollectionStore/_Public/11/543/11543720.pdf>. Also published in *Environmental Quarterly*, Report EML-349 (Environmental Measurements Laboratory: New York, 1979).

[20] De Geer et al. (note 19).

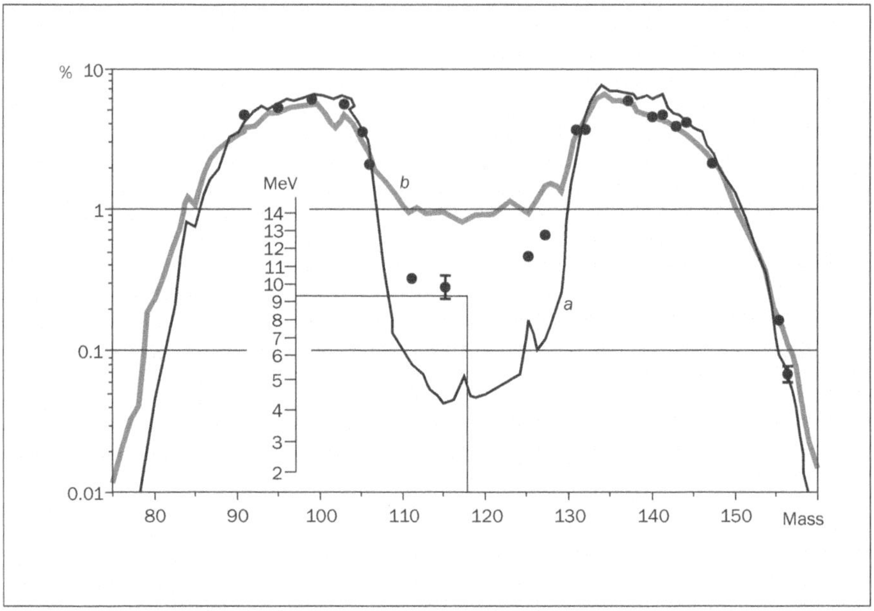

Figure 8A.3. Cumulative mass yields as determined from the 26 November 1976 sample

These yields (in percentages) were normalized to a mass-95 yield of 5.07%. Most error bars are small enough to be hidden behind the data points. For comparison the mass–yield curves for fission neutron (*a*) and 14 megaelectronvolts (MeV) neutron (*b*) fission of uranium-238 are given. The inset scale gives the yield for symmetric fission as a function of incident neutron energy. According to this scale, the average neutron energy causing fission in China-21 was about 9–10 MeV, a little higher than that found in China-11. This can be interpreted as 30% fissions induced by 14 MeV neutrons and 70% induced by fission spectra neutrons.

Source: De Geer, L.-E. et al., *Particulate Radioactivity, Mainly from Nuclear Explosions, in Air and Precipitation in Sweden Mid-year 1975 to Mid-year 1977*, FOA report C 40089-T2(A1) (Försvarets Forskningsanstalt: Sundbyberg, Nov. 1978), <http://www.iaea.org/inis/collection/NCLCollectionStore/_Public/11/543/11543720.pdf>. Also published in *Environmental Quarterly*, Report EML-349 (Environmental Measurements Laboratory: New York, 1979).

China-20, 17 October 1976

This was the third underground test.

China-21, 17 November 1976

This was the largest Chinese thermonuclear test ever (with (n,γ) : (n,2n) = 1.34 ± 0.04), and the cloud entered Swedish airspace 8 days after the detonation.[21] A sample collected at 14 km on 26 November was unusually strong,

[21] De Geer et al. (note 19).

corresponding to more than 10^{11} fissions, and it was carefully studied by gamma and alpha spectroscopy for more than a year. At ground level, however, very little was noted until the spring of 1977. No hot particles were found and the debris was essentially unfractionated. In the very rich sample, 25 fission products (not counting any daughter nuclides of an already identified nuclide) and 12 activation products were identified and quantified. Due to the non-fractionation, the mass-yield curve could be analysed (see figure 8A.3). Cobalt and manganese isotopes were detected in ratios consistent with an exposure of steel by a neutron fluence of thermonuclear origin. As in China-16, yttrium was used as a high-energy neutron fluence detector.

The measurements of alpha radiation revealed the presence of ^{238}Pu, $^{239+240}$Pu and ^{241}Am. The ^{239}Pu formed in the tamper–pusher via one neutron absorption and subsequent rapid decays could be resolved via the ^{239}Np activity measured when the sample was fresh. These relations of important transuranic elements were food for thought about the Teller–Ulam H-bomb mechanism that FOA had not fully grasped at the time. Additionally, since experts—such as Carson Mark, a former director of the Theoretical Division at Los Alamos, and Herbert York, the first director of the Lawrence Livermore Laboratory in the 1970s—claimed that it was possible to deduce elements of the Teller–Ulam design from the debris, China-21 appeared to provide a good chance to try to do that.

The first thoughts (based on the wrong assumption of a uranium trigger) interestingly enough led to an idea similar to the one applied in the first Soviet thermonuclear device, the *sloika* (layer cake).[22] Later, when the principles of thermonuclear processes leaked out from civilian fusion programmes, and when some weapon test reports were accidentally declassified in the USA, the enigma started to be clarified. Two ideas were then born about what Mark and York might have had in mind, this time with the assumption of a plutonium primary.[23]

The first idea was based on an analysis of China-21, which showed that there was more residual plutonium-239 from the trigger than would be expected if the plutonium charge were in close contact with the thermonuclear burn region. The second idea was simpler and based on the measured very low ^{238}Pu : ^{237}U mass ratio of 0.0002. Both are (n,2n)-products of major fuels with fairly similarly shaped cross-section curves that differ by just a factor of five quantitatively. With a reasonable plutonium mass of 5 kg in the trigger, that would give an estimated ^{238}U tamper–pusher mass of $5 / (5 \times 0.0002) = 5000$ kg, corresponding to 250 litres. This appeared to be too much and a new indication was provided that the trig-

[22] See chapter 7 in this volume.

[23] De Geer, L.-E., 'The radioactive signature of the hydrogen bomb', *Science & Global Security*, vol. 2, no. 4 (1991), pp. 351–63.

ger plutonium had experienced far fewer fusion neutrons than the tamper-pusher. More recent literature gives a factor of about 2.5, instead of 5, between the cross sections and this strengthens the argument as the estimate of the tamper-pusher mass doubles to 10 tonnes (0.5 m^3).

Thus, both ideas suggest that the trigger had been placed separately and probably shielded from the thermonuclear zone. Managing to get away from the close match-fuel paradigm was the key to the Teller–Ulam breakthrough and it initiated thinking that fairly soon resulted in understanding of radiation implosion. That process, however, benefited from knowledge of the extreme compression of the thermonuclear fuel when exploded. That compression can be estimated quite simply with the neutron fluence and the fusion yield as parameters.

When the neutron fluence through the pusher of the first thermonuclear explosion, code-named Mike, was published as greater than 2 moles/cm^2 in 1962, the compression could have been estimated to be of the order of 100, something that greatly overshadows the compressions reached in a fission bomb. It would then have been obvious that something very different was needed. Actually, this could have been understood as early as 1955, when the discovery of the transuranium elements einsteinium (atomic number 99) and fermium (atomic number 100) in Mike debris was published. These elements could only have been produced at very high fluences by multiple neutron capture in ^{238}U and subsequent decay.

It is interesting to note how scientists acted as potential proliferators when they managed to declassify their findings by referring to ongoing accelerator experiments at the Nobel Institute in Stockholm, which was on the verge of producing some lighter isotopes of these elements.[24] Claiming first discovery of a new element gives high prestige and also the right to name the element, so it was deemed necessary to publish. It has been reported that the US group later acted out of slight guilt and in a chivalrous way when it agreed on the name nobelium for element 102, although it had been discovered by them.

China-22, 17 September 1977

Signs of fresh debris were first seen at all ground-level stations after 11–13 days. Many fairly long-lived radionuclides only marginally increased above the background still present from China-21 the year before. The samples were first fairly unfractionated with a tendency to be enhanced in volatiles in late October, which indicated that the explosion had occurred

[24] Hoffman, D. C., Ghiorso, A. and Seaborg, G. T., *The Transuranium People: The Inside Story* (Imperial College Press: London, 2000).

at a fairly high altitude. The fission character was corroborated by a (n,γ) : (n,2n) ratio of 27 ± 3.

China-23, 15 March 1978

Debris from this test was first detected at all stations after 10–11 days. The debris-collection aeroplane first crossed the cloud on 28 March, at an altitude of 8 km. The debris was fractionated to a large extent with volatile mass chains enhanced up to a factor of 13. Heavy close-in fallout must have occurred, which would have depleted the debris available for long-range transport in refractive mass chains and this is a sign of a near-ground explosion. The ration (n,γ) : (n,2n) was 30 ± 5, which is consistent with a fission device.

China-24, 15 March 1978

This was the fourth underground test.

China-25, 14 December 1978

Fairly weak samples of fresh fission products first appeared on filters exposed at ground level on 18–25 December. The samples were heavily enhanced in volatile nuclides and as such were very similar to the samples collected after the preceding atmospheric test in March.

China-X, 13 September 1979

China-X was a full failure that FOA chose not to count. The parachute failed to deploy.

China-26, 16 October 1980

This was the last atmospheric test in China and occurred exactly 16 years after the first such test. Debris was first detected after 10 days at an altitude of 14 km. The next day the strongest sample, corresponding to 10^{10} fissions, was collected at the same altitude. The samples showed a mix of fission and activation products that was typical of a relatively high-yield thermonuclear explosion. A (n,γ) : (n,2n) ratio of 1.56 ± 0.03 was observed. The samples were practically unfractionated and a mass-yield analysis was made, as for China-21 (see figure 8A.4). With better yield data available to compare to the measured data, it was concluded that about 57 per cent of all fission events were induced by fission spectrum neutrons on uranium-238, 30 per cent were induced by high-energy neutrons on uranium-238,

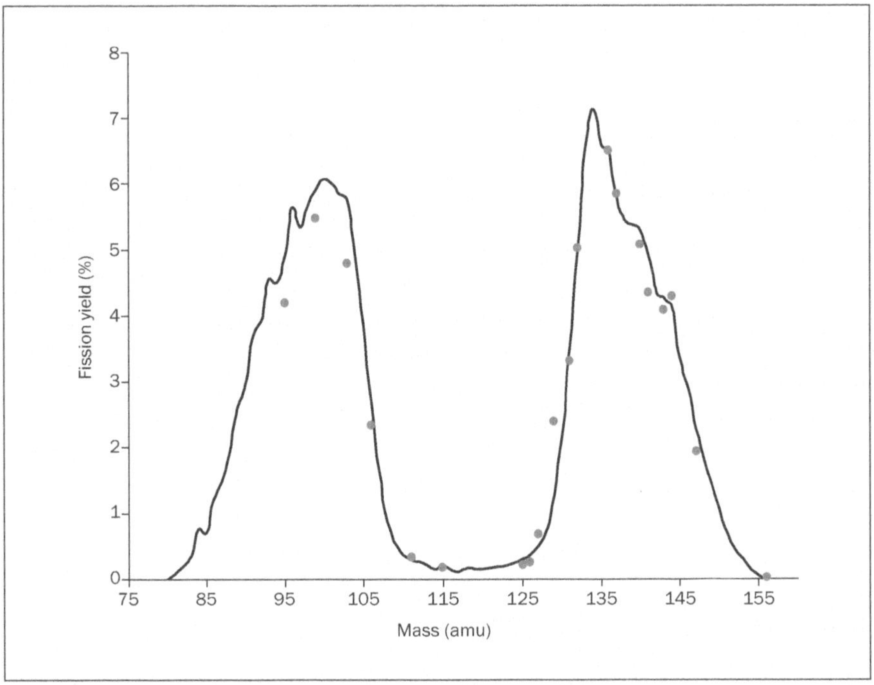

Figure 8A.4. The experimental chain yield data from China-26

The experimental data are compared to a fitted composite yield curve of 57 per cent for U-238 created by fast neutron-induced fission, 30 per cent for U-238 created by high-energy neutron-induced fission and 13 per cent for Pu-239 created by fast neutron-induced fission. The mass 95, 99 and 103 yields fall a bit below the fitted curve, which could be a sign of a slight reversed fractionation, as their main elements are zirconium, niobium, molybdenum, technetium and ruthenium, all with very high condensation temperatures (in the range of 4000 to 5000 °C)

Source: De Geer, L.-E., Försvarets Forskningsanstalt (FOA), 1981.

and 13 per cent were induced by fission spectrum neutrons on plutonium-239. Interestingly, this agreed well with the fission-neutron fission to high-energy neutron fission ratio that had been found in China-21. A final detail of the China-26 test was the detection of niobium-92m, which is a (n,2n)-product of stable niobium. It was first believed to be a fluence detector but was later reinterpreted as a result of using niobium as a stabilizer in uranium. The China-26 spectrum was used as a base for a proficiency test by the Comprehensive Nuclear-Test-Ban Treaty Organization (CTBTO) laboratories in 2003.[25]

[25] Karhu, P. et al., 'Proficiency test for gamma spectroscopic analysis with a simulated fission product reference spectrum', *Applied Radiation and Isotopes*, vol. 64, nos 10–11 (2006), pp. 1334–39.

After China-26

China carried out a further 18 tests underground, in shafts and tunnels, until its last test, on 29 July 1996. Six of the first underground tests after China-26 dealt with development of a neutron bomb and one (China-34, probably 10 kt on 26 May 1990) is believed by some to have been a test by or for Pakistan at Lop Nur of a device derived from the one tested at 12 kt in China-4.[26] The other 11 tests were of warheads for sea- and land-based ballistic missiles, including safety features, such as insensitive high explosives in the primaries.

[26] Reed, T. C. and Stillman, D. B., *The Nuclear Express: A Political History of the Bomb and Its Proliferation* (Zenith Press: Minneapolis, MN, 2009).

9. Applications of nuclear forensic analysis

VITALY FEDCHENKO AND ROBERT KELLEY

The final chapter of this book focuses on individual cases where nuclear material analysis has been applied in order to address international or national security concerns. In many of these examples, in particular the more historical ones, the term 'nuclear forensics' is either anachronistic or may have seemed inappropriate to the analysts directly involved. However, the techniques of nuclear material analysis applied all fall under the definition of nuclear forensics that is followed here (see chapter 1). The examples here therefore serve to demonstrate that the techniques of nuclear forensic analysis are used widely in practice in a variety of different frameworks and illustrate the many contemporary and future applications of this broad discipline.

In order to systematically illustrate that point, cases in this chapter are grouped according to the international legal framework that they can be used to help verify. Section I describes individual cases of the use of nuclear forensic techniques in verification of the 1968 Non-Proliferation Treaty (NPT) in Iraq and Iran, illustrating the importance of quality control and the interpretation of measurement results. Section II discusses techniques that could be useful for verification of a future fissile material cut-off treaty (FMCT)—a proposed treaty that would ban production of fissile materials—if and when it is negotiated. Section III describes instances of nuclear forensic analysis following a nuclear explosion, including in the framework of the 1996 Comprehensive Nuclear-Test-Ban Treaty (CTBT). Section IV describes a nuclear forensic investigation of three cases of trafficking of nuclear materials.

I. Verification of the Non-Proliferation Treaty

Among other things, the 1968 Non-Proliferation Treaty prohibits any non-nuclear weapon state—defined as any state that had not exploded a nuclear weapon prior to 1 January 1967—from manufacturing or acquiring nuclear weapons.[1] The treaty requires each non-nuclear weapon state party to conclude a comprehensive safeguards agreement (CSA) with the International Atomic Energy Agency (IAEA), which give the IAEA a right to inspect

[1] Treaty on the Non-Proliferation of Nuclear Weapons (Non-Proliferation Treaty, NPT), opened for signature 1 July 1968, entered into force 5 Mar. 1970, IAEA Information Circular INFCIRC/140, 22 Apr. 1970, <http://www.iaea.org/Publications/Documents/Treaties/npt.html>.

nuclear facilities in the state. Comprehensive safeguards are based on a combination of nuclear material accountancy and containment and surveillance techniques, with a view to preventing diversion of nuclear energy from peaceful uses to nuclear weapons. In implementation of CSAs, the IAEA draws on many techniques of nuclear forensic science.

Uranium particles on hostages' clothing in Iraq

The 1990–91 Gulf War started on 2 August 1990 when Iraq, led by President Saddam Hussein, invaded its oil-rich neighbour Kuwait.[2] Iraq soon found that international opinion was strongly against this action and faced a large build-up of Western forces. In an attempt to protect its key military facilities from aerial bombardment, it placed Western hostages in them. On 17 August 1990 the Iraqi Government announced that it will 'hold the nationals from any countries threatening Iraq until the threats ceased'.[3]

The 661 hostages were distributed among many militarily critical facilities.[4] Iraq's action resulted in number of technically savvy engineers from the oil industry and other high-technology projects suddenly being billeted at critical facilities—these included facilities for uranium enrichment in support of Iraq's then undiscovered nuclear weapon programme. When these hostages were released in December 1990, their home countries quickly debriefed them, including on questions such as where they had been held and what processes were under way there.[5] Significantly, clothing and possessions they had with them while being held at secret sites were collected for analysis.[6] The United States had also been able to monitor the movement of some hostages during their period of detention and had identified about 55 of the 'most important' (but not necessarily nuclear-related) facilities throughout Iraq.[7]

Some of the hostages were held at the huge Al Tuwaitha Nuclear Research Centre outside Baghdad. Al Tuwaitha was declared as a nuclear facility to the IAEA under the 1973 CSA between Iraq and the IAEA.[8]

[2] Posen, B. R., 'Military mobilization in the Persian Gulf conflict', *SIPRI Yearbook 1991: World Armaments and Disarmament* (Oxford University Press: Oxford, 1991).

[3] Wilson, J., *The Politics of Truth* (Carroll & Graf: New York, 2004), pp. 133–34.

[4] *The Independent*, 9 Oct. 1990, cited in Hiro, D., *Desert Shield to Desert Storm: The Second Gulf War* (iUniverse: Bloomington, IN, 2003), pp. 157, 221.

[5] For another historical instance of detainees providing valuable information on nuclear weapon activities of the state that detained them see Maddrell, P., *Spying on Science: Western Intelligence in Divided Germany 1945–1961* (Oxford University Press: Oxford, 2006), pp. 205–21.

[6] Albright, D. and Hibbs, M., 'Iraq's nuclear hide-and-seek', *Bulletin of the Atomic Scientists*, vol. 47, no. 7 (Sep. 1991), pp. 15–16.

[7] Wilson (note 3).

[8] Agreement between the Republic of Iraq and the International Atomic Energy Agency for the Application of Safeguards in Connection with the Treaty on the Non-Proliferation of Nuclear Weapons, signed and entered into force 29 Feb. 1972, IAEA Information Circular INFCIRC/172, 22 Feb. 1973.

Before the Iraqi invasion of Kuwait, the IAEA routinely carried out safeguards inspections at a few buildings on this immense site.[9] But there were about 100 other buildings that the IAEA could not visit and Western intelligence agencies were keen to know what was going on there.

Hostages were also held at the Al Tarmiya site north of Baghdad, which was not declared to the IAEA under Iraq's CSA. This site was an enigma to Western intelligence agencies. It was known as a 'concrete Rorschach test' to intelligence analysts, because missile analysts thought it was for missiles, chemical analysts thought it was for chemical weapons and nuclear analysts thought it was a gas centrifuge enrichment plant for enriching uranium to weapon-grade levels.

Nuclear forensics was able to contribute to understanding of the Iraqi nuclear programme by analysing particles obtained on at least two occasions. First, the analyses of samples of clothing from hostages held at the Iraqi nuclear facilities between August and December 1990 provided some surprising results. Particles enriched in uranium-235 and simultaneously depleted in uranium-234 and uranium-238 were found. Second, the US intelligence community managed to obtain a sample 'from [Al] Tuwaitha before the war', containing uranium particles that were highly depleted in the isotope ^{235}U, taken to be 0.06 per cent weight.[10] The analysis could detect 'essentially none of the other isotopes' among those particles.[11] The isotopic composition of uranium particles found in both cases could serve as a signature, revealing the isotope separation technique that produced it (especially if combined with the particles' chemical composition).

First, for economic reasons, commercial-scale uranium enrichment plants do not normally discharge depleted uranium with tails containing less than approximately 0.2 per cent of ^{235}U.[12] Thus, such a plant can be

[9] These inspections were regulated by the safeguards agreement (note 8), Article 80. In practice there were reportedly 2 inspections per year at 4 facilities at Al Tuwaitha. Before Aug. 1990, 25 inspections had been performed by a total of 13 inspectors. Kokoski, R., SIPRI, *Technology and the Proliferation of Nuclear Weapons* (Oxford University Press: Oxford, 1995), p. 101.

[10] Nichols, J., 'Uncovering the secret program (I): the initial inspections', Transcript from the Institute for Science and International Security conference 'Understanding the Lessons of Nuclear Inspections and Monitoring in Iraq: A Ten Year Review', Washington, DC, 14–15 June 2001, <http://isis-online.org/nichols>. The exact isotopic composition of the particles described is not known from available open sources. However, a later measurement was made on calutron components obtained from the Al Tarmiya site in July 1991 by the 3rd post-Gulf War IAEA inspection in Iraq. One of the equipment components, a graphite collector, was contaminated with uranium depleted to 0.06% weight in U-235. Donohue, D. L. and Zeisler, R., 'Behind the scenes: scientific analysis of samples from nuclear inspections in Iraq', *IAEA Bulletin*, vol. 34, no. 1 (Jan.–Mar. 1992), p. 27.

[11] Nichols (note 10). See also US Congress, Office of Technology Assessment, *Environmental Monitoring for Nuclear Safeguards*, OTA-BP-ISS-168 (Government Printing Office: Washington, DC, Sep. 1995), p. 26.

[12] Neff, T., 'Dynamic relationships between uranium and SWU prices: a new equilibrium', *World Nuclear Association: Annual Symposium 2006*, London, 6–8 Sep. 2006 (World Nuclear Association: London, 2006). Any enrichment facility, no matter which process is used, separates an incoming feed material stream into 2 outgoing streams: a product stream in which the process material is enriched to some degree in the desired isotope and a tails stream (sometimes inappropriately called

ruled out as a likely origin of uranium particles containing around 0.06 per cent of ^{235}U (such as the pre-war Al Tuwaitha sample).[13] Second, enrichment processes can be categorized according to the way they separate various isotopes of the same chemical element. 'Selective' processes (such as laser enrichment and electromagnetic isotope separation, EMIS) are designed in such a way that they enrich only the desired isotope, while simultaneously depleting all others. 'Mass difference' processes (such as gaseous diffusion, gas centrifuges and aerodynamic separation methods) do not single out any particular isotope, but rather make broader distinctions between isotopes based on their relative mass. Thus, mass difference schemes can concentrate isotopes that are either all heavier or all lighter than the targeted isotope.[14] In other words, in the case of uranium some enrichment technologies (EMIS and lasers) distinguish between ^{234}U and ^{235}U, while others pass them through together.[15]

Particles picked up from the hostages' clothing were enriched in ^{235}U but simultaneously depleted in heavier (^{238}U) and lighter (^{234}U) isotopes. This clearly suggests that they were a product of a selective enrichment process. The particles said to have been obtained from Al Tuwaitha before the war had isotopic composition consistent with the tailings of a selective enrichment process, because the content of the isotopes ^{234}U and ^{235}U was too low for any mass difference process.[16]

Essentially, all techniques other than laser enrichment and EMIS were ruled out by isotopic composition analysis. Although there is no clear indication of this in published sources, the final determination of the enrichment technique that had been used could probably have been made by determining the chemical and elemental composition of uranium particles. EMIS uses uranium in the form of uranium tetrachloride, whereas laser enrichment processes use either uranium metal (atomic vapour laser isotope separation, AVLIS) or uranium hexafluoride (molecular laser isotope separation, MLIS).[17]

The results of these samples were not widely disseminated in the intelligence community in the USA, and were not known to the IAEA for some time. As a result, it fell to satellite imagery to give the first clues about what uranium isotope separation scheme was really under development in Iraq. Satellite imagery showed large discs, each about 4 metres in diameter, being removed from bombed Iraqi nuclear facilities. John Googin, a scien-

the waste stream), which is somewhat depleted in this isotope. Krass, A. S. et al., SIPRI, *Uranium Enrichment and Nuclear Weapons Proliferation* (Taylor & Francis: London, 1983), p.9.

[13] See note 10.

[14] For detailed descriptions of these isotope separation methods see Krass et al. (note 12), pp. 121–91.

[15] US Congress, Office of Technology Assessment (note 11), p. 12.

[16] Nichols (note 10).

[17] Krass et al. (note 12), pp. 188–89.

tist from the USA's Oak Ridge National Laboratory, whose experience dated back to the Manhattan Project, explained that these were magnet discs for EMIS.[18]

Both indications of the existence of the EMIS enrichment programme in Iraq—provided by nuclear forensics and by satellite imagery analysis—were disputed both at the IAEA and within the US intelligence community.[19] The idea that the Iraqi Government may have decided to use EMIS for uranium enrichment had been put forward by Nerses Krikorian, a group leader of the International Technology Division in the USA's Los Alamos National Laboratory, a number of years before the Gulf War.[20] His idea was not accepted at the time, because EMIS was considered to be too energy-intensive and too inefficient in terms of material losses, and therefore obsolete. However, computers and particle accelerator technologies developed by the 1990s made the use of the electromagnetic enrichment method attractive for weapon programme purposes.[21] It turned out that the chief Iraqi scientist, Jafar Dhia Jafar, had done PhD work on high-energy physics at the University of Birmingham, United Kingdom, had experimented with the proton synchrotron at the UK's Rutherford High Energy Laboratory and had worked with large magnets at CERN (Organisation Européenne pour la Recherche Nucléaire, European Organization for Nuclear Research)—all of which contributed to his ability to implement the EMIS technique using technology upgraded to modern standards.[22] Subsequent field inspections in Iraq revealed dozens of discs and the results of the analysis of the particles on the hostages' clothes were validated.

It should be noted that the EMIS process that produced the discovered particles was still in the development stage and was plagued with problems at the time of the Gulf War.[23] It was not easy for Iraq to expand it to an industrial scale and it was not on track at the time to produce usable quantities of weapon-grade uranium. It might well have been abandoned in favour of centrifuges if the war had not intervened.

[18] Thorne, L., 'IAEA nuclear inspections in Iraq', *IAEA Bulletin*, vol. 34, no. 1 (Jan.–Mar. 1992), pp. 19–20; and Perricos, D., 'Uncovering the secret program—initial inspections', Transcript from the Institute for Science and International Security conference 'Understanding the Lessons of Nuclear Inspections and Monitoring in Iraq: A Ten Year Review', Washington, DC, 14–15 June 2001, <http://isis-online.org/perricos>.

[19] Richelson, J. T., *Spying on the Bomb: American Nuclear Intelligence from Nazi Germany to Iran and North Korea* (W. W. Norton & Co.: New York, 2006), p. 451.

[20] For a short biography of Krikorian see 'Side-by-side as equals: an unprecedented collaboration between the Russian and American nuclear weapons laboratories to reduce the nuclear danger', *Los Alamos Science*, no. 24 (1996), pp. 42–43.

[21] Thorne (note 18), p. 21.

[22] Stone, R., 'Profile: Jafar Dhia Jafar', *Science*, 30 Sep. 2005, pp. 2158–59.

[23] IAEA, General Conference, 'The implementation of United Nations Security Council resolutions relating to Iraq', Report by the Director General, GC(40)/13, 12 Aug. 1996, Annex, para. 6.

Cross-contamination issues in Iraq

Among the conditions imposed on Iraq after its defeat in the Gulf War was an intrusive inspection regime by the IAEA.[24] Prior to this, the IAEA had only done bulk sampling with methods such as high-resolution gamma spectrometry (HRGS) and X-ray fluorescence analysis (XRF), and never sensitive particle analysis.[25] In support of the new inspection regime the US Government offered to expand the IAEA's particle analysis capabilities. For the first time, the IAEA was offered an opportunity to use fission-track thermal ionization mass spectrometry (FT-TIMS), a sensitive technique for analysis of microparticles.[26] The US Air Force Technical Application Center (AFTAC) carried out all such particle analyses on behalf of the IAEA. Samples were collected on swipe papers or cloths by the IAEA and sent for analysis to AFTAC's Advanced Nuclear Applications laboratory at General Electric's Vallecitos Nuclear Center in California, USA.[27]

The first samples taken in Iraq during the first inspection, in May 1991, produced surprising results. Particles of 93 per cent enriched uranium were found at the Al Tuwaitha site. In later cases, samples taken at sites that were still under construction and had never had nuclear materials, such as the planned enrichment plant at Ash Sharqat, were found to contain HEU. The process of analysing the samples and producing results took weeks or months because of the slowness of the fission track process and the delicate manipulation of particles found. Hence, the particles were found in the summer of 1991, at the same time as Iraq's massive clandestine calutron-based (EMIS) enrichment programme was being unmasked.[28] Iraq claimed that it had not been successful at enrichment. In particular, the highest enrichments achieved were declared by Iraq 'to have been 17% for gram quantities and 45% for milligram quantities'.[29]

The particle analysis results contradicted Iraqi declarations and were used to suggest that Iraq was lying about its progress. Subsequent analysis showed that Iraq had been truthful. The use of swipe sampling and particle

[24] UN Security Council Resolution 687, 3 Apr. 1991.

[25] Zendel, M. et al., 'Nuclear safeguards verification measurement techniques', eds A. Vértes et al., *Handbook of Nuclear Chemistry*, 2nd edn, vol. 6 (Springer: Heidelberg, 2011), p. 2986.

[26] See chapter 3 in this volume; and US Congress, Office of Technology Assessment (note 11), p. 26.

[27] US Congress, Office of Technology Assessment (note 11), p. 26.

[28] Most information on the Iraqi EMIS programme was revealed during the 2nd, 3rd and 4th IAEA inspections in Iraq under UN Security Council Resolution 687, i.e. from June to Aug. 1991. IAEA, Consolidated report on the first two IAEA inspections under Security Council Resolution 687 (1991) of Iraqi nuclear capabilities, 12 July 1991, annex to UN document S/22788, 15 July 1991, pp. 11–12; IAEA, Report on the third IAEA on-site inspection in Iraq under Security Council Resolution 687 (1991), 7–18 July 1991, 25 July 1991, annex to UN document S/22837, 25 July 1991, pp. 6–9; and IAEA, Report on the fourth IAEA on-site inspection in Iraq under Security Council Resolution 687 (1991), 27 July–10 August 1991, 27 Aug, 1991, annex to UN document S/22986, pp. 5–9.

[29] IAEA, Report on the fourth IAEA on-site inspection (note 28), para. 9.

analysis in Iraq was an ad hoc measure. The sampling procedures used at the time were not thoroughly tested or quality assured by the IAEA. Many of the samples taken from Iraqi nuclear facilities in 1991 were taken by US inspectors, many of whom worked in the US national laboratories at Livermore, Oak Ridge and Los Alamos. It was later proved that many of the particles found in Iraq came from cross-contamination from these inspectors.[30] This case demonstrated both the extreme sensitivity of environmental sampling and the importance of careful collection procedures.

Environmental sampling as a safeguards measure

The Iraqi experience showed that the IAEA safeguards regime had to be better equipped to be able to provide assurance of correctness and especially completeness of a state's nuclear programme declarations. In April 1993 the IAEA Director General's Standing Advisory Group on Safeguards Implementation (SAGSI) put forward a number of recommendations for improving the safeguards system. On the basis of these recommendations, the IAEA launched a development programme of improvement of safeguards methods and procedures, known as Programme 93+2. Events in Iraq made it certain that environmental sampling (of which the analysis of particles from swipe samples is a major part) was seen as a powerful tool for obtaining unambiguous information about past and current nuclear activities.[31] Programme 93+2 also addressed the cross-contamination risks associated with this technique by developing and documenting sampling, data storage and analysis procedures and protocols and testing them in field trials.[32] The IAEA Board of Governors formally approved environmental sampling as a safeguards measure in 1995, and it began to be used from January 1996.[33]

Collection procedures were greatly improved as a result of these developments. First, new protocols demanded the use of 'sampling kits'—sets of items needed for taking samples, preassembled in a clean laboratory to guarantee the absence of contamination.[34] Second, a sampling team

[30] For a definition of 'cross-contamination' see the Glossary in this volume.

[31] Hooper, R., '"Programme 93+2": IAEA development programme for strengthened and more cost-effective safeguards', Institute of Nuclear Materials Management (INMM), *36th Annual Meeting of the Institute of Nuclear Materials Management*, Palm Desert, CA, 9-12 July 1995 (INMM: Deerfield, IL, 1995).

[32] Kuhn, E., 'Environmental monitoring for safeguards applications', Institute of Nuclear Materials Management (note 31); and Zendel et al. (note 25), p. 2987.

[33] Kuhn, E., Fischer, D. and Ryjinski, M., 'Environmental sampling for IAEA safeguards: a five year review', IAEA-SM-367/10/01, *Symposium on International Safeguards, Verification and Nuclear Material*, Vienna, 29 Oct.–2 Nov. 2001 (IAEA: Vienna, 2001).

[34] The basic IAEA swipe sampling kit, used everywhere except for hot cells and one enrichment facility, consists of 6 cotton swipes, each measuring 10 cm square. Mini-grip bags of 2 different sizes are included to individually bag and double-bag the swipes after sampling. Each sampling kit also

consisting of at least two inspectors must perform the swipe sampling.[35] Great care is taken to ensure that sterile swipe samples are deposited in small sealed plastic bags. Only one inspector—the 'clean inspector'—touches the bags and takes notes. The other—the 'dirty inspector'—touches only the swipes and the inspected surface. Both inspectors take background samples on themselves before entering a nuclear site so that any tiny particles carried into the site can be eliminated from concern.

Once collected, environmental samples are shipped to the Clean Laboratory for Safeguards of the IAEA's Safeguards Analytical Laboratories (SAL) in Seibersdorf, Austria. These samples typically contain six cotton swipes, four of which are archived for reference purposes and two of which are analysed.[36] The samples are assigned code numbers to conceal their origin before being screened at SAL for the presence of radioactive isotopes by HRGS and for the presence of uranium and plutonium by XRF. On the basis of the screening results and according to the IAEA inspectors' requirements, the IAEA identifies methods and chooses laboratories for further sample analysis from its Network of Analytical Laboratories for Environmental Samples.[37]

The Purple Sweater

Other information is also used to document a sample collection, as was illustrated in the run-up to the March 2003 invasion of Iraq. In this case, in which one of the present authors was directly involved, a positive sample was found that could have had significant implications.

The IAEA was aware of a machining factory that was formerly used to cut metal for the Iraqi nuclear programme. No nuclear material had ever been introduced to the site, so no particles of enriched uranium had been discovered by the IAEA in swipe samples. In January–February 2003, however, samples taken at the machining factory were found to contain particles enriched to 2.6 per cent in ^{235}U.

In 1982 Iraq had imported from Italy 1767 kilograms of uranium enriched to 2.6 per cent in ^{235}U in the form of uranium dioxide powder. Since the 1990s the material had been kept under the control of the IAEA, at Location C (a storage complex close to Al Tuwaitha), in the same form as it

contains 2 pairs of clean-room latex gloves, a sample data sheet, pen and labels. Strict procedures are followed during sample kits preparation. Zendel et al. (note 25), pp. 2988–89.

[35] IAEA, *IAEA Safeguards Glossary: 2001 Edition*, International Nuclear Verification Series no. 3 (IAEA: Vienna, 2001), p. 73.

[36] Bevaart, L., Donohue, D. and Fuhr, W., 'Future requirements for the analysis of environmental samples and the evaluation of the results', European Safeguards Research and Development Association (ESARDA), *29th Annual Meeting: Symposium on Safeguards and Nuclear Material Management*, Aix en Provence, 22–24 May 2007 (Office for Official Publications of the European Communities: Luxembourg, 2007).

[37] See chapter 2 in this volume.

was received from Italy.[38] The IAEA had no information of any other site in Iraq containing uranium of this particular enrichment. If the particles with such an enrichment had really been found at the machining factory, it would mean that the material under seal at Location C might have been diverted or used in a way undeclared to the IAEA. At this time, the IAEA's Iraq Action Team was carrying out many inspections in Iraq but finding no evidence of unexpected nuclear activities. A different finding would bolster the case for war, which was being threatened by the USA and some other states. A rapid and careful re-sampling was required; accuracy and care were of the utmost importance in the re-sampling.

Photographs are usually taken during sampling, and it was seen that the purple sweater worn by the inspector had been worn the day before, when he conducted in inspection at Location C near Al Tuwaitha. The sweater was impounded and sampled and proved to have many of the particles in question. This provided strong assurance that cross-contamination was the cause of the uranium discovery at the machining factory. These results illustrated yet again that analysis of fissile particles is extremely sensitive and a great amount of care and attention to procedures is required to avoid cross-contamination, including wearing clean protective clothing where possible. Inspected states are often taken aback by the protective clothing worn by inspectors collecting samples. This clothing is in fact there not to protect inspectors from contamination from the environment, but to protect environmental samples from accidental contamination by inspectors.

Sampling waterways in Iraq

By 1993 the IAEA had discovered and documented almost all the clandestine elements of the Iraqi weapon programme. Its next task was to ensure that no new programme started. However, Saddam Hussein remained in power after the Gulf War and was suspected of still harbouring nuclear weapon ambitions. For example, there were persistent rumours in 1992 that Iraq had built or planned a plutonium production reactor.[39] These reports were eventually thoroughly disproven, but in 1992 it was a nagging doubt.

The fact that the collection of hydrological samples can be an effective technique for detection of nuclear activities, such as reactor operation and nuclear fuel reprocessing, has been well understood since the 1940s.[40]

[38] IAEA, Fourth consolidated report of the Director General of the International Atomic Energy Agency under paragraph 16 of Security Council resolution 1051 (1996), 6 Oct. 1997, annex to UN document S/1997/779, p. 25.

[39] Albright, D., 'A proliferation primer', *Bulletin of the Atomic Scientists*, vol. 49, no. 5 (June 1993), p. 19.

[40] Boni, A. L., 'High sensitivity measurements of ultra-low amounts of radioactivity in the environment', *50 Years of Excellence in Science and Engineering at the Savannah River Site:*

Plutonium separation in particular has been known to release detectable amounts of distinctive radionuclides and chemicals into downstream waterways.[41] According to a study conducted in 1991 at the nuclear fuel reprocessing facility at Sellafield, UK, a small 'emission-controlled' reprocessing facility producing 8 kg of plutonium per year is likely to release annually 12 milligrams of carbon-14 and 2 mg of strontium-90 split between air and water and 125 grams of iodine-129 and 15 grams of technetium-99 in off-site water.[42]

As part of its Programme 93+2, the IAEA asked the Savannah River National Laboratory of the US Department of Energy (DOE) to develop a programme to sample the waterways of Iraq to detect any radioactive effluent that might be introduced by clandestine activities.[43] The programme consisted of four components: (*a*) a geographic study to identify choke points and confluences of rivers to ensure an effective network of sampling locations; (*b*) a system of periodic sample collection, not real-time monitoring; (*c*) water concentration systems to reduce the volume of material to be shipped for analysis; and (*d*) sampling of sludge and vegetation as well as water.

During its 14th post-war inspection, in August and September 1992, the IAEA initiated a survey establishing 'a radionuclide and stable isotope composition baseline in the major watershed regions of Iraq in order to detect changes resulting from aqueous effluents of nuclear related facilities'.[44] This survey had three specific goals: to measure the impact of existing nuclear-related facilities in Iraq on the surface water systems that received their aqueous effluents; to detect possible but as yet unknown nuclear facilities; and to provide a baseline of hydrologic and radiometric conditions from which changes in composition could be easily detectable for interpretation.

Three types of sample were taken: a 100-millilitre water sample; a sediment core; and a filtering column used to concentrate dissolved and particulate matter from a water sample of approximately 300 litres. The analysis of the samples was to include high-sensitivity gamma spec-

Proceedings of the Symposium, Aiken, SC, 17 May 2000 (Westinghouse Savanah River Company: Aiken, SC, 2000); and Alvarez, L. W., *Alvarez: Adventures of a Physicist* (Basic Books: New York, 1989), pp. 119–22.

[41] Paternoster, R. R., *Nuclear Weapon Proliferation: Indicators and Observables*, LA-12430-MS (Los Alamos National Laboratory: Los Alamos, NM, Dec. 1992), pp. 7–9.

[42] US Congress, Office of Technology Assessment (note 11), p. 17.

[43] Boni (note 40), p. 278; and Boni, A. L., 'Environmental sampling in water for verification purposes', IAEA Scientific Forum 'Nuclear Technology in Relation to Water Resources and the Aquatic Environment', Vienna, 22–24 Sep. 1998, <http://www.iaea.org/About/Policy/GC/GC42/SciProg/gc 42-scifor-11.pdf>.

[44] IAEA, Report on the fourteenth IAEA on-site inspection in Iraq under Security Council Resolution 687 (1991), 31 August–7 September 1992, 24 Sep. 1992, annex to UN document S/24593, para. 11; and Kokoski (note 9), p. 135.

trometry, radiochemical separation of the actinides (primarily uranium and plutonium) followed by alpha spectrometry, secondary ion mass spectrometry (SIMS), ultra-low background gas proportional counting for tritium and inductively coupled plasma mass spectrometry (ICP-MS) for stable isotopes.[45]

This 'baseline determination' survey was conducted at a total of 52 sites in Iraq, which were visited during the 14th and 15th IAEA inspections, in late 1992.[46] Seven of those sampling locations were in Kurdish-controlled regions.[47] It became clear that further sampling at those locations would be difficult for personnel security reasons, but since they are at the upper ends of Iraq's waterways this was not considered a problem. In addition, Saddam Hussein had no likelihood of establishing a sensitive nuclear facility in areas beyond his control.

The plans for the long-term monitoring of Iraqi waterways envisaged revisiting about 15 of the 45 accessible sampling sites twice a year. The first of these regular sampling visits was conducted during the 19th IAEA on-site inspection in April–May 1993, and they continued until 1998. Waterway sampling was also used in 2002.[48]

The sensitivity of the measurements was reportedly very high. The system was capable of registering a very small quantity of gamma- or beta-emitting nuclides. For example, it often detected 'the permitted use by Iraq of radioisotopes in medical applications'.[49] Sample analysis was able to distinguish traces of radioactivity introduced by the Chernobyl accident, fall-out from nuclear weapon testing around the world and the treatment of thyroid cancer.[50] This provided strong assurance that a reactor or reprocessing plant was not in operation in Iraq at the time of sampling.

The sampling of waterways was not meant to detect nuclear facilities dealing with natural uranium only. Natural uranium is so ubiquitous in the environment that it is difficult to detect changes in concentrations of uranium unless there is also a man-made change in enrichment or introduction of characteristic isotopes or chemical elements. Overall, the water-sampling programme was judged to be effective at detecting fission products

45 IAEA (note 44), pp.9-10.

46 IAEA, Report on the fifteenth IAEA on-site inspection in Iraq under Security Council Resolution 687 (1991), 8–18 November 1992, 10 Dec. 1992, annex to UN document S/24981, pp. 6–7.

47 Kokoski (note 9), p. 135.

48 IAEA, Report on the nineteenth IAEA on-site inspection in Iraq under Security Council Resolution 687 (1991), 30 April–7 May 1993, 21 June 1993, annex to UN document S/25982, para. 5; IAEA, Update report of the International Atomic Energy Agency to the Security Council pursuant to resolution 1441 (2002), 27 Jan. 2003, annex to UN document S/2003/95, para. 41; and IAEA, Fifteenth consolidated report of the Director General of the International Atomic Energy Agency under paragraph 16 of Security Council resolution 1051 (1996), 11 Apr. 2003, annex to UN document S/2003/422, p. 5.

49 IAEA, Fifteenth consolidated report (note 48), para. 10.

50 Boni (note 43), p. 3.

associated with a reactor or reprocessing. It would have some utility for detecting uranium with an isotopic composition uncommon in nature. It is a model for a similar programme in a cooperative environment and could also be useful in clandestine collections under some circumstances.

Uranium particle analysis to verify Iran's declaration to the IAEA

Media reports published in August 2002 prompted the IAEA to investigate the existence of undeclared uranium-enrichment facilities in Iran.[51] During the visit of a high-level IAEA delegation to Iran in February 2003, Iranian authorities acknowledged the construction of two centrifuge enrichment plants at Natanz, the Pilot Fuel Enrichment Plant (PFEP) and the large Fuel Enrichment Plant (FEP), as well as the existence of a workshop of the Kalaye Electric Company (KEC) in Tehran used for production of centrifuge components. Iran stated that its enrichment programme was indigenous and that, at that time, no enrichment activity involving nuclear material was being conducted at those or other locations.[52]

That was a claim of a considerable significance because Iran—like any other non-nuclear weapon state party to the NPT that has in force a comprehensive safeguards agreement with the IAEA—is required to declare any new nuclear facility before it commences operation and to provide the IAEA with specific information on its design.[53] States do so by completing a design information questionnaire (DIQ). The specific details of the DIQ submission are defined in an annex to the CSA that describes subsidiary arrangements. Such subsidiary arrangements are negotiated by the IAEA separately with each state.

Since 1976 all states have been required to complete a DIQ for any new installation no later than 180 days before the introduction of nuclear material to the facility. In the aftermath of the Gulf War, the IAEA Board of Governors decided to change the subsidiary arrangements in subsequently negotiated CSAs so that the states would have to 'provide design information to the Agency at the time of the decision to construct, or to authorize the construction of, any nuclear facility (i.e. well before construction actually begins) in order to create confidence in the peaceful purpose of the

[51] For a detailed account of the disclosure of Iran's pursuit of sensitive nuclear fuel cycle technologies see Kile, S. N., 'Nuclear arms control and non-proliferation', *SIPRI Yearbook 2004: Armaments, Disarmament and International Security* (Oxford University Press: Oxford, 2004), pp. 604–12.

[52] IAEA, Board of Governors, 'Implementation of the NPT safeguards agreement in the Islamic Republic of Iran', Report by the Director General, GOV/2003/40, 6 June 2003, para. 8.

[53] 'Design information' is defined by the IAEA as 'information concerning nuclear material subject to safeguards . . . and the features of facilities relevant to safeguarding such material'. IAEA (note 35), p. 26.

facility'.[54] However, Iran did not accede to these new rules until 26 February 2003, after the existence of enrichment facilities had been discovered.[55]

Thus, if no nuclear material had been introduced to those facilities before they were discovered, Iran had not committed an act of non-compliance with its CSA. If the material had been introduced, then the failure to declare such a facility would be in contravention of Iran's CSA.[56]

In order to determine whether nuclear material had been introduced into the facilities, the IAEA began to take environmental samples at the Natanz plants in March 2003 and at the KEC workshop in August.[57] The IAEA inspectors noted that there had been 'considerable modification' of the KEC site before they could take samples and that this 'may impact on the accuracy of the environmental sampling and the Agency's ability to verify Iran's declarations'.[58] Despite the interference, samples revealed the presence of uranium particles at both sites that were not consistent with the material in the inventory declared by Iran to the IAEA.

In total, discoveries of particles of natural uranium, low-enriched uranium (LEU) and highly enriched uranium (HEU) up to 70 per cent enrichment (with the majority of the HEU being enriched in the ranges of 32–38 per cent and 50–60 per cent ^{235}U) were reported by the IAEA. This provided conclusive evidence of undeclared activity: either enriched uranium had been imported or enrichment experiments had taken place in Iran. Many of the LEU and HEU particles also had an elevated ^{236}U content, suggesting the use of uranium extracted from spent nuclear fuel. This again indicated either unknown reprocessing activities or an import of enriched material.[59]

When confronted with the evidence, Iran admitted its involvement in both undeclared domestic enrichment experiments and a covert international nuclear trade. In a letter of 21 October 2003 Iran admitted that, contrary to its earlier statements, it had conducted small-scale enrichment experiments between 1999 and 2002. These experiments achieved an

54 IAEA, Board of Governors, 'Strengthening of agency safeguards: the provision and use of design information', GOV/2554/Attachment 2/Rev.2, 1 Apr. 1992, para. 2. See also Hibbs, M., 'Safeguards agreement required early completion of DIQ by Syria', *Nuclear Fuel*, vol. 32, no. 23 (5 Nov. 2007), p. 9; and Schriefer, D., 'The international level', eds R. Avenhaus et al., *Verifying Treaty Compliance: Limiting Weapons of Mass Destruction and Monitoring Kyoto Protocol Provisions* (Springer: Heidelberg, 2006), pp. 437, 452.

55 IAEA, GOV/2003/40 (note 52), 6 June 2003, para. 15.

56 In fact, if Iran had intended to introduce nuclear material into the facilities within 180 days of their discovery, then, technically, Iran would have been in contravention of its CSA. However, there is no way in which such a supposition could be proved.

57 IAEA, Board of Governors, 'Implementation of the NPT Safeguards Agreement in the Islamic Republic of Iran', Report by the Director General, GOV/2003/75, 10 Nov. 2003, annex 1, paras 37 – 46.

58 IAEA, Board of Governors, 'Implementation of the NPT Safeguards Agreement in the Islamic Republic of Iran', Report by the Director General, GOV/2003/63, 26 Aug. 2003, para. 32.

59 IAEA, Board of Governors, 'Implementation of the NPT Safeguards Agreement in the Islamic Republic of Iran', Report by the Director General, GOV/2004/83, 15 Nov. 2004, para. 38.

enrichment level of no more than 1.2 per cent ^{235}U.[60] More importantly, in August 2003 Iran officially admitted that it had imported some centrifuge parts. It suggested that the HEU contamination originated from imported parts and identified Pakistan as a supplier.[61] Pakistan eventually agreed to hand over centrifuge components requested by the IAEA to allow comparison of uranium particles.[62] The IAEA received components on 21 May 2005, took swipe samples and analysed them at its SAL. The results confirmed that most of the contamination was probably of Pakistani origin, as stated by Iran.[63]

The experience in Iran has demonstrated that, although nuclear forensic techniques may be useful for safeguards implementation, they must be complemented with other sources of data, such as open source analysis, satellite imagery and information from IAEA member states. The discovery of enriched uranium in Iran was possible only after the IAEA learned about the Natanz plants from elsewhere.

II. Verification of a fissile material cut-off treaty

In 1995 the Conference on Disarmament (CD) adopted a mandate to negotiate a multilateral FMCT that would ban the production of fissile material for nuclear weapons.[64] Not only the effectiveness but also the acceptability of an FMCT will depend on it being verifiable. The question of whether this is possible has meant that progress in the CD has been slow. A number of techniques already available in the fields of nuclear forensics and IAEA safeguards may be chosen for verification of a future treaty.[65] The following examples describe how nuclear forensic techniques could determine whether plutonium was separated before or after an FMCT enters into force; could measure the amount of plutonium produced over the lifetime of a reactor (and therefore subject to accounting under an FMCT); and could detect prohibited production of fissile material from a distance.

[60] IAEA, GOV/2003/75 (note 57), annex 1, paras 30–35.

[61] Kile, S. N., 'Nuclear arms control and non-proliferation', *SIPRI Yearbook 2005: Armaments, Disarmament and International Security* (Oxford University Press: Oxford, 2005), pp. 558–59.

[62] Bokhari, F., 'Pakistan may hand over nuclear centrifuges', *Financial Times*, 25 Mar. 2005; and 'Centrifuge parts sent to IAEA', *Dawn* (Karachi), 27 May 2005.

[63] IAEA, Board of Governors, 'Implementation of the NPT Safeguards Agreement in the Islamic Republic of Iran', Report by the Director General, GOV/2005/67, 2 Sep. 2005, para. 12.

[64] Conference on Disarmament, Report of Ambassador Gerald E. Shannon of Canada on consultations on the most appropriate arrangement to negotiate a treaty banning the production of fissile material for nuclear weapons or other nuclear explosive devices, CD/1299, 24 Mar. 1995.

[65] For some practical recommendations on implementing an FMCT see Kile, S. N. and Kelley, R. E., *Verifying a Fissile Material Cut-off Treaty: Technical and Organizational Considerations*, SIPRI Policy Paper No. 33 (SIPRI: Stockholm, Aug. 2012); and International Panel on Fissile Materials (IPFM), *Global Fissile Material Report 2008: Scope and Verification of a Fissile Material (Cut-off) Treaty* (IPFM: Princeton, NJ, Sep. 2008).

Plutonium age determination to verify North Korea's initial declaration

The Democratic People's Republic of Korea (DPRK or North Korea) acceded to the NPT in 1985 and, after a significant delay, signed a comprehensive safeguards agreement with the IAEA on 30 January 1992.[66] As required by Article 62 of the CSA, North Korea submitted 'an initial report on all nuclear material subject to safeguards' to the IAEA on 4 May 1992.[67]

The initial report contained a declaration that North Korea had conducted a single experiment in March 1990 at the Radiochemical Laboratory in Yongbyon on separating about 90 grams of plutonium from reportedly damaged spent fuel rods removed from the adjacent 5-megawatt-electric (25-megawatt-thermal) gas-graphite reactor.[68] In May 1992 the IAEA began to conduct ad hoc inspections in North Korea as required by Article 71(a) of the CSA, in order to 'verify the correctness of the information contained in the Initial Report and to assess its completeness'.[69] Six ad hoc inspections were carried out between May 1992 and February 1993.[70] Nuclear forensic analysis was used on two occasions to analyse material collected during those inspections and verify North Korea's claims.

First, North Korea 'presented for verification' material that it described as 'the plutonium product and associated waste solutions resulting from a campaign carried out in 1990, involving the reprocessing of irradiated fuel elements from [North Korea's] 5 [megawatt-electric] Experimental Power Reactor'.[71] Samples of that material were taken to the IAEA's SAL, which divided it into subsamples and sent them on for analysis to 'selected labs' in other countries.[72]

[66] Agreement of 30 January 1992 between the Government of the Democratic People's Republic of Korea and the International Atomic Energy Agency for the Application of Safeguards in Connection with the Treaty on the Non-Proliferation of Nuclear Weapons, signed 30 Jan. 1992, entered into force 10 Apr. 1992, IAEA Information Circular INFCIRC/403, May 1992. See also Lockwood, D. and Wolfsthal, J. B., 'Nuclear weapon developments and proliferation', *SIPRI Yearbook 1993: World Armaments and Disarmament* (Oxford University Press: Oxford, 1993), p. 244.

[67] An 'initial report' is defined by the IAEA as a document required under the CSA containing 'an official statement by the State on all nuclear material subject to safeguards', from which the IAEA 'establishes a unified inventory of all nuclear material . . . for the State'. IAEA (note 35), p. 94.

[68] Smith, R. J., 'N. Korea and the bomb: high-tech hide-and-seek', *Washington Post*, 27 Apr. 1993, p. A1; Kokoski (note 9), p. 223; and Reiss, M., *Bridled Ambition: Why Countries Constrain Their Nuclear Capabilities* (Woodrow Wilson Center Press: Washington, DC, 1995), pp. 242, 294.

[69] IAEA, Report by the Director General of the International Atomic Energy Agency on Behalf of the Board of Governors to all Members of the Agency on the Non-compliance of the Democratic People's Republic of Korea with the Agreement Between the IAEA and the Democratic People's Republic of Korea for the Application of Safeguards in Connection with the Treaty on the Non-Proliferation of Nuclear Weapons (INFCIRC/403) and on the Agency's inability to verify the Non-diversion of Material Required to be Safeguarded', Information Circular INFCIRC/419, 8 Apr. 1993, para. 5.

[70] IAEA, Board of Governors and General Conference, 'Application of safeguards in the Democratic People's Republic of Korea', Report of the Director General, GOV/2011/53–GC(55)/24, 2 Sep. 2011, p. 2.

[71] IAEA, INFCIRC/419 (note 69).

[72] Smith (note 68).

The isotopic ratios of the plutonium isotopes ^{239}Pu, ^{240}Pu and ^{241}Pu were determined for both plutonium in the waste and the separated plutonium.[73] The tests indicated that the isotope ratios did not match, with the $^{239}Pu : ^{240}Pu$ ratio reportedly differing by 0.5–1 per cent.[74] Additionally, the isotopic composition of the material in the samples was not consistent with the declared irradiation history of the fuel.[75] The IAEA interpreted these findings as an indication that the separated plutonium and the plutonium in the waste did not originate from the same fuel rods, and that plutonium might have been separated on other occasions. As Hans Blix, the IAEA Director General, put it, 'We found two gloves, a waste glove and a plutonium glove, and they don't match ... So we concluded there must be two more gloves' that North Korea had not declared.[76]

Second, in order to determine when other separations could have taken place, the IAEA took swipe samples from inside and outside gloveboxes at the end of the reprocessing line at Yongbyon, where freshly separated plutonium is converted from liquid form into oxide compound.[77] The samples were sent to the SAL and then to designated laboratories, including the US AFTAC's McClellan Central Laboratory, to determine their elemental and isotopic composition.[78] This data was then used to calculate the age of the nuclear material.[79] The $^{241}Pu : ^{241}Am$ ratio was reportedly used to determine that the plutonium in the Yongbyon particles was separated in 1989, 1990, 1991 and early 1992.[80]

By revealing these two inconsistencies—that the separated plutonium and the waste had separate origins and that plutonium had been separated on more than one occasion—the nuclear forensic analysis conducted by the IAEA and associated laboratories provided yet more evidence that the North Korean declaration to the IAEA was not entirely correct.

[73] Kokoski (note 9).

[74] Hileman, B., 'North Korea suspected of hiding plutonium', *Chemical & Engineering News*, 11 Apr. 1994, p. 5.

[75] IAEA, INFCIRC/419 (note 69).

[76] Smith (note 68).

[77] Fischer, D., *History of the International Atomic Energy Agency: The First Forty Years* (IAEA: Vienna, 1997), pp. 289, 320; and Albright, D., 'North Korean plutonium production', *Science & Global Security*, vol. 5, no. 1 (Dec. 1994), pp. 66–67, 86–87.

[78] Smith (note 68); and Welch, M., 'AFTAC celebrates 50 years of long range detection', *AFTAC Monitor*, Oct. 1997, pp. 22, 28.

[79] For the definition of 'age' see the Glossary in this volume. See also chapter 2 in this volume.

[80] Hibbs, M., 'Isotopics show three North Korean reprocessing campaigns since 1975', *Nuclear Fuel*, 1 Mar. 1993, p. 9. According to some sources, separation might also have happened in early 1992. Smith (note 68).

Table 9.1. GIRM indicator ratios

Element	Primary measured ratios	Fluence range
Boron	B-10 : B-11	Low
Lithium	Li-6 : Li-7	Low–intermediate
Titanium	Ti-48 : Ti-49	Intermediate–high
Uranium	U-235 : U-238, U-236 : U-238	Low–high
Plutonium	Pu-240 : Pu-239, Pu-241 : Pu-239, Pu-242 : Pu-239	Low–high

Note: 'Fluence range' indicates which levels of fluence that the specified isotope ratios can be used to identify. For example, the ratio B-10 : B-11 is more suitable for use in reactors using low fluence but not intermediate or high fluence.

Source: Gesh, C. J., *A Graphite Isotope Ratio Method Primer: A Method for Estimating Plutonium Production in Graphite Moderated Reactors*', PNNL-14568 (Pacific Northwest National Laboratory: Richland, WA, Feb. 2004), table 2.1.

Determination of the plutonium production and operating history of a reactor

A large proportion of reactors producing military plutonium use graphite as moderator. Most notably, this has been the case with the Russian and US reactors, and with North Korea's plutonium-producing reactor. Once such a reactor is decommissioned, one particular technique of nuclear forensic analysis—the graphite isotope ratio method (GIRM)—can assist 'nuclear archaeology' by determining how much plutonium that reactor produced in its lifetime.[81] GIRM was developed by Russian and US scientists in the context of the Russian–US bilateral disarmament initiatives. The key benefit of GIRM is that it can provide insight into the graphite reactor's cumulative plutonium production even after all of its fuel has been reprocessed or made otherwise unavailable for analysis.[82]

Once put into the reactor, reactor graphite is normally expected to stay there for the lifetime of the reactor. Even highly purified reactor-grade graphite contains impurities at the level of a few parts per million. Neutron irradiation changes the isotopic composition of graphite impurities. After the reactor has been shut down, a sufficient number of graphite samples can be obtained from various points in its core. Mass spectrometry tech-

[81] The term 'nuclear archaeology' was suggested in 1993 to cover all means 'to retrospectively document all nuclear weapons activities, especially the production and disposition of fissile materials'. This activity would necessarily have to use methods of nuclear forensic analysis as it is defined in this volume. Fetter, S., 'Nuclear archaeology: verifying declarations of fissile-material production', *Science & Global Security*, vol. 3, nos. 3–4 (1993), pp. 237–59.

[82] Wood, T. et al., 'Establishing confident accounting for Russian weapons plutonium', *Nonproliferation Review*, vol. 9, no. 2 (summer 2002), pp. 126–37.

niques can then be used to measure the isotopic composition of impurities in the irradiated graphite. That isotopic composition can be compared either to the isotopic composition of impurities in an unirradiated sample of the graphite or, if that is not available, to the literature data. Once changes in impurities have been determined, the total quantity of neutrons needed to cause such change (i.e. the fluence) can be calculated. The total amount of plutonium produced is proportional to the fluence.[83]

Specific isotopic ratios have been determined to be suitable indicators of fluence (see table 9.1). According to most studies, GIRM enables the total plutonium production by a reactor to be estimated to within a few per cent.[84] The use of multiple isotope ratios and a significant number of irradiated graphite samples and the availability of unirradiated graphite reduces the uncertainty associated with a GIRM estimate of plutonium production.

GIRM has been applied on a number of occasions. One proposed application was a detailed description of how to apply GIRM to estimate the total amount of plutonium produced in the graphite reactor at Yongbyon.[85] More recently, a number of attempts have been made to expand GIRM into a generic isotope ratio method (IRM) applicable to other reactor types. IRM was applied to verify the operating history of the Ford Nuclear Reactor at the University of Michigan, USA, a water-cooled research reactor.[86] Similar work was performed at the IRT reactor of the Andronikashvili Institute of Physics, Tbilisi, Georgia.[87] The possibility of using IRM to estimate the cumulative energy production in boiling water reactor (BWR) fuel bundles has also been demonstrated.[88] A number of studies have expanded IRM to heavy-water moderated reactors of the type used for military plutonium production in India (CIRUS) and Pakistan (Khushab-I), as well as the civilian CANDU family of heavy-water

[83] Gesh, C. J., *A Graphite Isotope Ratio Method Primer: A Method for Estimating Plutonium Production in Graphite Moderated Reactors*', PNNL-14568 (Pacific Northwest National Laboratory: Richland, WA, Feb. 2004), p.1. For a definition of 'fluence' see the glossary in this volume.

[84] Heasler, P. G. et al., 'Estimation procedures and error analysis for inferring the total plutonium (Pu) produced by a graphite-moderated reactor', *Reliability Engineering & System Safety*, vol. 91, nos 10–11, (Oct.–Nov. 2006), pp. 1406–13.

[85] Kang, J., 'Nuclear archeology on the 5 MWe graphite reactor at Yongbyon', Institute of Nuclear Materials Management (INMM), *51st Annual Meeting of the Institute of Nuclear Materials Management 2010 (INMM 51)*, vol. 2 (INMM: Deerfield, IL, 2011); and Kang, J., 'Using the graphite isotope ratio method to verify the DPRK's plutonium production declaration', *Science and Global Security*, vol. 19, no. 2 (May–Aug. 2011), pp. 121–29.

[86] Cliff, J. B. et al., 'Isotope ratio method analysis of the Ford Nuclear Reactor', Presentation at the Joint Meeting of the National Organization of Test, Research, and Training Reactors and the International Group on Research Reactors, Gaithersburg, MD, 12–16 Sep. 2007, <http://ncnr.nist.gov/trtr 2005/Proceedings/papers.htm>.

[87] Cliff, J. B. et al., 'Independent verification of research reactor operation', Institute of Nuclear Materials Management (note 85), vol. 4.

[88] Gesh, C. J. et al., 'Application of the isotope ratio method to a boiling water reactor', Institute of Nuclear Materials Management (note 85), vol. 1.

moderated reactors.[89] Finally, it has been proposed that the same idea be applied to verify the operating history of gaseous diffusion enrichment plants.[90]

Remote verification by environmental sampling

A number of techniques useful for the remote verification of nuclear fuel cycle activities have been proposed. In the case of IAEA safeguards, their utility has yet to be fully demonstrated. In the FMCT context, there are promising technologies for detection of undeclared plutonium production and the detection of undeclared uranium-enrichment activities.

Krypton-85 is generated at a high rate during fission of ^{235}U, ^{238}U and ^{239}Pu, which means that it is produced in significant quantities by any reactor. Since krypton is a noble gas, it escapes plutonium-separation facilities in detectable quantities and it would be difficult for a plant operator to filter it out. The half-life of ^{85}Kr is 11 years, which means that there is some background presence of this isotope in the atmosphere. Nonetheless, it has been demonstrated that emissions from individual facilities are distinguishable at a distance of a few tens of kilometres.[91] Although it may be difficult to employ this technique to reasonably assure detection of a previously unknown plutonium-separation facility in the territory of a certain state, it may be possible for an FMCT-implementing body (if and when it is established) to deploy ^{85}Kr detectors around a declared reprocessing plant to which it has been denied access in order to verify that reprocessing is not taking place there.

The remote detection of uranium-enrichment facilities, especially those using centrifuges, is considered to be difficult due to their low emissions into the environment. It has been suggested that attention focus on uranium hexafluoride (UF_6), which is necessary for any enrichment plant oper-

[89] Gasner, A. and Glaser, A., 'Beyond GIRM: nuclear archaeology for heavy-water-moderated plutonium production reactors', Institute of Nuclear Materials Management (note 85), vol. 5; and Broadhead, B. L., 'Nuclear archaeology for CANDU power reactors', Institute of Nuclear Materials Management (INMM), *52nd Annual Meeting of the Institute of Nuclear Materials Management 2011 (INMM 52)*, vol. 5 (INMM: Deerfield, IL, 2012).

[90] Philippe, S. and Glaser, A., 'Nuclear archaeology for gaseous diffusion enrichment plants', Institute of Nuclear Materials Management (INMM), *54th Annual Meeting of the Institute of Nuclear Materials Management 2013 (INMM 54)*, vol. 5 (INMM: Deerfield, IL, 2013).

[91] Saey, P. R. J., 'Ultra-low-level measurements of argon, krypton and radioxenon for treaty verifycation purposes', *ESARDA Bulletin*, no. 36 (July 2007), p. 44; and Kalinowski, M. B. et al., 'Environmental sample analysis', eds Avenhaus et al. (note 54), pp. 376–77; International Panel on Fissile Materials (IPFM), *Global Fissile Material Report 2007* (IPFM: Princeton, NJ, Oct. 2008), p. 107; Heim, M., et.al., 'Discovery of the krypton isotopes', *Atomic Data and Nuclear Data Tables*, vol. 96, no. 4 (July 2010), <http://arxiv.org/pdf/0904.2362>, pp. 333–40; and Born, H. J. and Seelmann-Eggebert, W., 'Über die Identifizierung einiger Uranspaltprodukte mit entsprechenden durch (nα)- und (np)-Prozesse erhaltenen Isotopen' [On the identification of some uranium fission products with the related isotopes received through (nα)- und (np)-processes], *Naturwissenschaften*, vol. 31, nos 7–8 (Feb. 1943).

ating today or in the foreseeable future. If the UF_6 production and inventory is verified at all uranium conversion plants, and all imports are monitored, then undeclared enrichment activities would not be possible without producing the UF_6 covertly. Published research suggests that remote detection of facilities producing UF_6 through detection in the atmosphere of UF_6 degradation products, namely UO_2F_2 aerosols, is possible in principle, but will be problematic at longer distances and in case of an installation of a high-efficiency particulate air filters at the clandestine facility.[92]

III. Analysis after a nuclear explosion

The nuclear tests in North Korea in 2006, 2009 and 2013[93]

North Korea has conducted three nuclear tests to date: on 9 October 2006, 25 May 2009 and 12 February 2013. In each case, North Korea's official news agency, the Korean Central News Agency (KCNA), issued statements announcing the tests and making claims about their magnitude and technical details. These claims had to be verified using available technologies. On each occasion, numerous measurements and studies using nuclear forensic techniques were carried out by international, governmental and independent experts outside North Korea to determine whether there had been an explosion and, if so, its nature, location and actual yield (see table 9.2).

The technologies used for verification of underground nuclear tests include seismology, radionuclide monitoring and satellite imagery analysis.[94]

The recording of a seismic event is usually the first indication of an underground nuclear explosion. Seismic monitoring networks record the various seismic waves propagating from a source through the earth's deep interior and surface. Analysis of these records often allows the azimuth (direction) and the distance of the event to be calculated. In order to discriminate between seismic signals from an earthquake and those from an

[92] Kemp, R. S., 'Initial analysis of the detectability of UO_2F_2 aerosols produced by UF_6 released from uranium conversion plants', *Science and Global Security*, vol. 16, no. 3 (2008); Kemp, R. S., 'Source terms for routine UF6 emissions', *Science and Global Security*, vol. 18, no. 2 (2010); and Kemp, R. S., 'The non-proliferation emperor has no clothes', *International Security*, vol. 38, no. 4 (2014), p. 49.

[93] These accounts of the 3 North Korean tests are based on Fedchenko, V. and Ferm Hellgren, R., 'Nuclear explosions, 1945–2006', *SIPRI Yearbook 2007: Armaments, Disarmament and International Security* (Oxford University Press: Oxford, 2007); Fedchenko, V., 'Nuclear explosions, 1945–2009', *SIPRI Yearbook 2010: Armaments, Disarmament and International Security* (Oxford University Press: Oxford, 2010); and Fedchenko, V., 'Nuclear explosions, 1945–2013', *SIPRI Yearbook 2014: Armaments, Disarmament and International Security* (Oxford University Press: Oxford, 2014).

[94] US National Academy of Sciences, *Technical Issues Related to the Comprehensive Nuclear Test Ban Treaty* (National Academy Press: Washington, DC, 2002), pp. 39–41; and Dahlman, O. et al., *Detect and Deter: Can Countries Verify the Nuclear Test Ban?* (Springer: Dordrecht, 2011), pp. 29–76.

Table 9.2. Data on North Korea's nuclear explosions, 2006, 2009 and 2013

Source	Origin time (UTC)	Latitude	Longitude	Error margin[a]	Body wave magnitude[b]
9 October 2006					
IDC[c]	01:35.27.6	41.3119° N	129.0189° E	±20.6 km[d]	4.1
CEME (Russia)	01:35.26.0	41.31° N	128.96° E	..	4.0
KIGAM (South Korea)	01:35	40.81° N	129.10° E	..	3.9
NEIC (United States)	01:35.28	41.29° N	129.09° E	±8.1 km[e]	4.3
25 May 2009					
IDC[c]	00:54:42.8	41.3110° N	129.0464° E	±9.6 km[d]	4.52
BJI (China)	00:54:43.10	41.3000° N	129.0000° E	..	4.6
CEME (Russia)	00:54:40.9	41.29° N	129.07° E	..	5.0
NEIC (United States)	00:54:43	41.306° N	129.029° E	±3.8 km[e]	4.7
NORSAR (Norway)	00:54:43	41.28° N	129.07° E	..	4.7
12 February 2013					
IDC[c]	02:57:51	41.3005° N	129.0652° E	±8.1 km[d]	4.9
CEME (Russia)	02:57:49.4	41.31° N	129.1° E	..	5.3
IGGCAS (China)	02:57:51.3	41.2927° N	129.0730° E	..	4.93±0.21
NEIC (United States)	02:57:51	41.308° N	129.076° E	±11.2 km[e]	5.1
NORSAR (Norway)	02:57:51	41.28° N	129.07° E	..	5.0

UTC = Coordinated Universal Time; .. = Data not available.

Note: Because of differences between estimates, particularly regarding the precise site of the explosion, in each case data from a number of sources is given, including an internationally recognized body—the Comprehensive Nuclear-Test-Ban Treaty Organization (CTBTO)—and national bodies: BJI = China Earthquake Administration, Institute of Geophysics, Beijing; CEME = Russian Academy of Sciences, Geophysical Survey, Central Experimental Methodical Expedition, Obninsk, Kaluga oblast; IDC = CTBTO, International Data Centre, Vienna, Austria; IGGCAS = Chinese Academy of Sciences, Institute of Geology and Geophysics, Beijing; KIGAM = Korea Institute of Geoscience and Mineral Resources (South Korea); NEIC = US Geological Survey, National Earthquake Information Center, Denver, CO; and NORSAR = Norwegian Seismic Array, Karasjok.

[a] The error margins are as defined by the data sources.

[b] Body wave magnitude indicates the size of the event. In order to give a reasonably correct estimate of the yield of an underground explosion, detailed information is needed, e.g. on the geological conditions in the area where the explosion took place. Body wave magnitude is therefore an unambiguous way of giving the size of an explosion.

[c] In 2006, the IDC was 'in a test and provisional operation mode only' and only 60% of the monitoring stations in the CTBTO's International Monitoring System (IMS) were contributing data at the time of the event. By 2009, 75% of the IMS monitoring stations were contributing data.

[d] These figures are the length of the semi-major axis of the confidence ellipse.

[e] These figure are the horizontal location error, defined as the 'length of the largest projection of the three principal errors on a horizontal plane'.

Sources: Fedchenko, V. and Ferm Hellgren, R., 'Nuclear explosions, 1945–2006', *SIPRI Yearbook 2007: Armaments, Disarmament and International Security* (Oxford University Press: Oxford, 2007), table 12B.1; Fedchenko, V., 'Nuclear explosions, 1945–2009', *SIPRI Yearbook 2010: Armaments, Disarmament and International Security* (Oxford University Press: Oxford, 2010), table 8B.1; and Fedchenko, V., 'Nuclear explosions, 1945–2013', *SIPRI Yearbook 2014: Armaments, Disarmament and International Security* (Oxford University Press: Oxford, 2014), table 6.15.

explosion, the location, depth and wave patterns associated with the event are normally studied. If the seismic signals originate in an area where earthquakes are rare or unheard of, the event would naturally receive additional scrutiny. If the depth of an event can confidently be estimated as being more than 10 km, then it can normally be screened out as being too deep for a man-made explosion.[95]

The fact that the seismic event in question is an explosion, rather than an earthquake, can also be confirmed by analysis of seismic wave patterns. Two types of seismic wave propagate through the earth's deep interior: compressional (or longitudinal) waves and shear (or transverse) waves. An explosion has a compact symmetrical wave source, which blasts outwards evenly in all directions and excites compressional waves particularly efficiently. Earthquakes, in contrast, are caused by large sections of the earth's crust grinding past each other along a fault line, which produces stronger shear waves. Thus, if the compressional waves—which travel faster than shear waves and arrive first at seismic monitoring stations—are the stronger of the two types associated with an event, then that would suggest that the seismic event was an explosion. Other wave characteristics can also help to distinguish an earthquake from an explosion.[96]

Data on the seismic magnitude of the event is normally used to estimate the yield of an explosion, although the accuracy of such estimates is affected by the amount of information available on the geology of the test site.[97]

Seismic data alone is insufficient to confirm that an underground explosion is nuclear—other evidence needs to be collected. Air sampling—which aims to collect and identify radioactive material, such as particulate and gaseous effluents and debris in the vicinity of an event—can provide the most useful evidence following a nuclear explosion. Air sampling can help to measure: (*a*) the neutron fluence and spectrum (i.e. the volume and energy of neutrons emitted by the explosion); (*b*) the yield of the device; (*c*) the design of the nuclear weapon (using information from *a* and *b*); (*d*) the age of the plutonium used (if plutonium was detonated); (*e*) when the detonation took place; (*f*) the general location of the detonation; (*g*) whether the detonation was atmospheric or submerged; and (*h*) the source of the nuclear materials used.[98]

For air sampling to be successful, there must be an accessible radioactive plume; the meteorology must be favourable; and the background radionuclide concentration must be low relative to the concentration of the

[95] US National Academy of Sciences (note 94), p. 43.

[96] US National Academy of Sciences (note 94), p. 39.

[97] US National Academy of Sciences (note 94), pp. 41–42.

[98] Williams, D. L., 'Characterizing nuclear weapons explosions based upon collected radionuclide effluents', Memorandum, Massachusetts Institute of Technology, Department of Nuclear Science and Engineering, 21 Oct. 2006. See also chapter 6.

nuclides in the plume.[99] With these limitations, a negative result from air sampling cannot be used to conclude that no nuclear test took place: if no radionuclides are detected, the conclusion must be that either no nuclear explosion took place or one of the above limitations influenced the detection process.

The 1996 Comprehensive Nuclear-Test-Ban Treaty is a multilateral treaty that will prohibit the carrying out of any nuclear explosion.[100] The Preparatory Commission for the Comprehensive Nuclear-Test-Ban Treaty Organization (CTBTO) has been established to prepare for the entry into force of the CTBT, including through the establishment of an International Monitoring System (IMS) to detect nuclear explosions. The IMS consists of a global network of 50 primary and 120 auxiliary seismic stations, 60 infrasound stations, 11 hydroacoustic stations, 80 radionuclide stations (including 40 with a noble gas system) and 16 radionuclide laboratories. These stations relay information to the International Data Centre (IDC) at the CTBTO headquarters in Vienna, Austria.

While the CTBT had been ratified by 163 states and signed by another 20 states as of March 2015, it cannot enter into force until it has been ratified by 44 states with certain nuclear facilities. North Korea, which is one of these 44 states, has not signed the treaty and therefore does not participate in the IMS.

The 2006 test

On 9 October 2006 the KCNA reported that North Korea had on that day successfully conducted an underground nuclear test explosion 'under secure conditions'.[101] The Chinese Government was given 20 minutes' prior warning of the test and was informed that the explosion's yield would be 4 kilotons.[102]

A seismic event was recorded by several monitoring networks at 01:35 UTC on 9 October 2006, originating 70 km north of the city of Kimchaek in North Korea's North Hamgyong province. The wave patterns recorded at monitoring stations and the depth of the event (less than 1 km) indicate that it was an explosion rather than an earthquake. Data on the seismic magnitude of the event was used to estimate the yield of the explosion, although the lack of information on the geology of the test site affects

[99] Williams (note 98).

[100] Comprehensive Nuclear-Test-Ban Treaty (CTBT), opened for signature 24 Sep. 1996, not in force, <http://treaties.un.org/Pages/CTCTreaties.aspx?id=26>.

[101] Korean Central News Agency, 'DPRK successfully conducts underground nuclear test', 9 Oct. 2006. The KCNA's statements are available on the website of the Korean News Service in Tokyo, <http://www.kcna.co.jp/>.

[102] CNN, 'North Korea claims nuclear test', 9 Oct. 2006; and Linzer, D., 'Low yield of blast surprises analysts', *Washington Post*, 10 Oct. 2006.

the reliability of such estimates.[103] Based on the seismic data, the governments of France, South Korea and the USA, along with independent researchers, concluded with a high degree of certainty that there had been an explosion and that its yield was well below 1 kt.[104]

North Korea had announced on 3 October its intention to conduct a nuclear test.[105] Immediately after that announcement, the USA had deployed its WC-135W Constant Phoenix atmospheric collection aircraft, which is normally used for collection of particulate and gaseous effluents and debris in support of the 1963 Partial Test-Ban Treaty.[106] Based on analysis of atmospheric radioactive debris collected by the aircraft, the US Government announced on 16 October that the event had been a nuclear explosion.[107] This was corroborated by the findings of South Korea and Sweden and, later, the CTBTO.[108] Analysis of the debris also indicated that the test used plutonium, which was confirmed by North Korean officials.[109]

The extent to which the 2006 nuclear test was successful is uncertain. The discrepancy between the pre-announced yield of 4 kt and the estimated actual yield of less than 1 kt made some experts speculate that the test ended in a 'fizzle'—that is, an inefficient detonation releasing less explosive energy than expected. As reported in the South Korean press, a North Korean diplomat acknowledged that the test was 'smaller in scale than expected'.[110]

[103] US National Academy of Sciences (note 94), pp. 41–42.

[104] Garwin, R. L. and von Hippel, F. N., 'A technical analysis: deconstructing North Korea's October 9 nuclear test', *Arms Control Today*, vol. 36, no. 9 (Nov. 2006). An early official Russian estimate that the yield was 5–15 kt was dismissed by US Government officials as inaccurate. It was identical to an estimate reportedly given to Russia by North Korea in the run-up to the test. Chanlett-Avery, E. and Squassoni, S., *North Korea's Nuclear Test: Motivations, Implications, and U.S. Options*, US Library of Congress, Congressional Research Service (CRS) Report for Congress RL33709 (CRS: Washington, DC, 4 Oct. 2006); and Linzer (note 102).

[105] Korean Central News Agency, 'DPRK foreign ministry clarifies stand on new measure to bolster war deterrent', 3 Oct. 2006.

[106] Chin, T., 'Seoul's intelligence capabilities "a total failure"', *Korea Herald*, 18 Oct. 2006; US Department of the Air Force, 'WC-135 Constant Phoenix', Fact sheet, Oct. 2005, <https://web.archive.org/web/20051021200300/http://www.af.mil/factsheets/factsheet.asp?fsID=192>; and Treaty Banning Nuclear Weapon Tests in the Atmosphere, in Outer Space and Under Water (Partial Test-Ban Treaty), opened for signature 5 Aug. 1963, entered into force 10 Oct. 1963, *United Nations Treaty Series*, vol. 480 (1963).

[107] Office of the Director of National Intelligence (ODNI), 'Statement by the Office of the Director of National Intelligence on the North Korea nuclear test', ODNI News Release no. 19-06, 16 Oct. 2006, <https://web.archive.org/web/20061104015730/http://www.dni.gov/announcements/20061016_release.pdf>.

[108] 'S. Korean gov't officially confirms N. Korea's nuclear test', Yonhap News, 25 Oct. 2006; 'ROK confirms radiation level normal following Pyongyang's nuke test', Yonhap News, 25 Oct. 2006; and US Defense Treaty Readiness Inspection Program, 'CTBTO observatory detects radioactive materials from DPRK nuclear test', *Weekly Treaty Review*, 5–11 Jan. 2007, p. 19.

[109] Shanker, T. and Sanger, D. E., 'North Korean fuel identified as plutonium', *New York Times*, 17 Oct. 2006; and Hecker, S. S., *Report on North Korean Nuclear Program* (Stanford University, Center for International Security and Cooperation: Stanford, CA, 15 Nov. 2006).

[110] MacAskill, E., 'Diplomat says test was smaller than expected', *The Guardian*, 11 Oct. 2006.

The 2009 test

On 29 April 2009 the KCNA issued a statement warning that the country was prepared to conduct a nuclear test explosion.[111] On 25 May 2009, the Chinese and US governments were reportedly given less than one hour's notice that North Korea would conduct a nuclear test.[112] The explosion itself took place at 00:54 UTC. At 02:24 UTC the CTBTO's IMS issued the first report to CTBTO member states on the time, location and magnitude of the event.[113] A few hours later the KCNA announced that North Korea had conducted 'one more successful underground nuclear test' that was 'on a new higher level in terms of its explosive power and technology'.[114]

A seismic event was recorded at 00:54 UTC on 25 May 2009 by several seismic monitoring networks, which calculated its origin as being no more than a few kilometres from the location of the 2006 nuclear test (see table 9.2).[115] The wave patterns recorded at monitoring stations, the depth of the event (less than 1 km) and the fact that it occurred so close to the site of the 2006 nuclear test indicated that it was an explosion rather than an earthquake.[116]

Based on the seismic data, most estimates of the yield of the May 2009 explosion vary between 2 and 7 kt, which is 'about 5 times stronger' than the 2006 test.[117] In June 2009 the US Government estimated the yield as 'approximately a few kilotons'.[118] Non-governmental scientists tend to agree with this assessment.[119] The Russian Ministry of Defence estimated that 'The nuclear device had a yield of between 10 and 20 kilotons'—this estimate was the highest of all announced and was not independently confirmed.[120] The South Korean Institute of Geoscience and Mineral

[111] Korean Central News Agency (KCNA), 'UNSC urged to retract anti-DPRK steps', 29 Apr. 2009.

[112] Agence France-Presse, 'NKorea informed US of nuclear test: official', 25 May 2009.

[113] CTBTO, 'CTBTO's initial findings on the DPRK's 2009 announced nuclear test', Press release, 25 May 2009, <http://www.ctbto.org/press-centre/press-releases/2009/ctbtos-initial-findings-on-the-dprks-2009-announced-nuclear-test/>.

[114] Korean Central News Agency (KCNA), 'KCNA report on one more successful underground nuclear test', 25 May 2009, <http://www.kcna.co.jp/item/2009/200905/news25/20090525-12ee.html>.

[115] Pearce, R. G. et al., 'The announced nuclear test in the DPRK on 25 May 2009', *CTBTO Spectrum*, no. 13 (Sep. 2009), p. 27.

[116] Pearce et al. (note 115), p. 27.

[117] MacKenzie, D., 'North Korea's nuke test could have positive outcome', *New Scientist*, 26 May 2009.

[118] US Office of the Director of National Intelligence (ODNI), 'Statement by the Office of the Director of National Intelligence on North Korea's declared nuclear test on May 25, 2009', ODNI News Release no. 23-09, 15 June 2009, <http://www.dni.gov/index.php/newsroom/press-releases/170-press-releases-2009>.

[119] Kalinowski, M. B., 'Second nuclear test conducted by North Korea on 25 May 2009', Fact sheet, University of Hamburg, Carl Friedrich von Weizsäcker Centre for Science and Peace Research (ZNF), 27 May 2009.

[120] 'Russia confirms N.Korea nuclear test, voices concern—2', 25 May 2009, RIA Novosti, <http://en.rian.ru/russia/20090525/155081541.html>.

Resources estimated the yield to have been 5.2 kt.[121] Won-Young Kim and Paul Richards of Columbia University estimated a yield of about 2.2 kt, assuming that the explosion took place in hard rock. This estimate matched the results of Jungmin Kang of Stanford University, who estimated the yield as being 2.2–2.8 kt for an underground nuclear test in hard rock.[122]

Unlike in 2006, after the 2009 event no trace of radioxenon or other debris was reported to have been found.[123] The failure to find radioactive effluents in 2009 might be attributable to two reasons: the test was buried deeper than the 2006 event; or a broad region of high atmospheric pressure suppressed the exhalation of xenon isotopes when they might have been detected before their decay to background levels.[124]

Despite this, there is consensus among scientists and CTBTO officials that the explosion was most probably nuclear. The alternative to a nuclear explosion would have been the highly synchronized detonation of thousands of tonnes of explosive material. Chemical explosions of this size are not unheard of. For example, the Soviet Union was reported to have set off 'several immense explosions' in the last half of 1956, with charges of 1640–9200 tons.[125] However, the preparation by North Korea of such an explosion would have been a massive undertaking, easily detectable by satellite imagery.[126]

In order to establish the nuclear nature of the event with absolute certainty, on-site inspection would have been needed. As the CTBTO pointed out, had the CTBT been in force, it would have been possible to conduct such an inspection because the location of the explosion was determined 'precisely enough to stay within the 1000 km^2 to which on-site inspections are limited'.[127] Indeed, the size of the error ellipse determined by the CTBTO was about four times smaller than the area of 1000 km^2 allowed by the treaty (see table 9.2).

Due to the absence of detected radioactive effluents from the explosion, it is not possible to establish whether the North Korean test in 2009 used uranium or plutonium. It is widely assumed that it used plutonium.[128] The extent to which the North Korean nuclear test was successful is uncertain

[121] Yoo, J., 'Test threatens regional stability', *JoongAng Daily* (Seoul), 26 May 2009.

[122] Kang, J., 'The North Korean nuclear test: Seoul goes on the defensive', *Bulletin of the Atomic Scientists*, 12 June 2009.

[123] Pearce et al. (note 115), pp. 28–29.

[124] Pearce et al. (note 115), p. 29.

[125] Kramish, A., *Atomic Energy in the Soviet Union* (Stanford University Press: Palo Alto, CA, 1960), p. 137.

[126] CTBTO, 'Experts sure about nature of the DPRK event', Press release, 12 June 2009, <http://www.ctbto.org/press-centre/highlights/2009/experts-sure-about-nature-of-the-dprk-event/>; and Clery, D., 'Verification experts puzzled over North Korea's nuclear test', *Science*, 19 June 2009.

[127] CTBTO, 'Homing in on the event', Press release, 29 May 2009, <http://www.ctbto.org/press-centre/highlights/2009/homing-in-on-the-event/>.

[128] Kile, S. N., 'Nuclear arms control and non-proliferation', *SIPRI Yearbook 2010* (note 93).

because, unlike in 2006, North Korea did not pre-announce the expected yield of the explosion. Some experts have questioned the success of the test, because the several-kiloton yield of the North Korean device is still a few times smaller than the yield that the initial nuclear tests by nuclear weapon states have historically produced.[129]

The 2013 test

On 24 January 2013 the KCNA issued a statement announcing that the country would conduct 'a nuclear test of higher level'.[130] An explosion took place at 02:57 UTC on 12 February. A few hours later the KCNA announced that the event was North Korea's third successful underground nuclear test that was 'conducted in a safe and perfect way on a high level with the use of a smaller and light A-bomb unlike the previous ones, yet with great explosive power'.[131] The announcement added that 'the test did not give any adverse effect to the surrounding ecological environment'.

The IMS issued its first report to CTBTO member states less than two hours after the explosion.[132] The recorded seismic wave patterns, the depth of the event (less than 1 km) and the fact that it occurred so close to the site of both the 2006 and 2009 nuclear tests indicated that the 2013 event was an explosion rather than an earthquake (see table 9.2).[133]

Based on seismic data, satellite imagery and information from the previous two tests, most estimates of the yield of the explosion varied between about 5 and 16 kt, which is 'about 2.5 to 3 times larger in yield' than the 2009 test (and therefore 12.5 to 15 times larger than the 2006 test).[134]

On 23 April 2013 the CTBTO announced that two of its stations, in Takasaki, Japan, and Ussuriisk, Russia, had earlier in the month detected two radioactive isotopes of xenon—xenon-131m and xenon-133—which act as a 'chronometer', determining the time of the event in which these isotopes were created.[135] The ratio of concentrations of these isotopes in air samples was consistent with a nuclear fission event that would have taken place

[129] Park, J., 'The North Korean nuclear test: what the seismic data says', *Bulletin of the Atomic Scientists*, 26 May 2009.

[130] Korean Central News Agency (KCNA), 'DPRK NDC vows to launch all-out action to defend sovereignty of country', 24 Jan. 2013.

[131] Korean Central News Agency (KCNA), 'KCNA report on successful 3rd underground nuclear test', 12 Feb. 2013.

[132] CTBTO, 'On the CTBTO's detection in North Korea', Press release, 12 Feb. 2013, <http://www.ctbto.org/press-centre/press-releases/2013/on-the-ctbtos-detection-in-north-korea/>.

[133] Richards, P., 'Seismic detective work: CTBTO monitoring system "very effective" in detecting North Korea's third nuclear test', *CTBTO Spectrum*, no. 20 (July 2013), p. 22.

[134] Richards (note 133); and Zhang, M. and Wen, L., 'High-precision location and yield of North Korea's 2013 nuclear test', *Geophysical Research Letters*, vol. 40, no. 12 (28 June 2013).

[135] CTBTO, 'CTBTO detects radioactivity consistent with 12 February announced North Korean nuclear test', Press release, 23 Apr. 2013, <http://www.ctbto.org/press-centre/press-releases/2013/ctbto-detects-radioactivity-consistent-with-12-february-announced-north-korean-nuclear-test/>.

more than 50 days before detection. (The xenon detection in Japan occurred 55 days after the explosion.) Atmospheric transfer modelling, conducted by the CTBTO, identified the site of the previous two nuclear tests as a possible source of the xenon emission. Since the radioxenon was detected so late after the event, it was not possible to determine if it had been produced by fission of uranium or plutonium.[136] This, in turn, meant that it was not possible to assess whether North Korea may have used highly enriched uranium in the explosive device.

Analysis of the debris from the 1954 Castle Bravo thermonuclear test

The nuclear weapon test Bravo conducted by the USA in the Pacific on 1 March 1954 as part of the Operation Castle test series has proven to be more informative on more levels and to more people and organizations than originally expected. For the USA, it proved the existence of unforeseen explosive properties of thermonuclear fuel. For Japanese and British scientists, it served as a confirmation of previous research and an indicator of one of the possible features of the thermonuclear weapon design. For the Soviet Union, it was the first foreign nuclear explosion systematically studied with nuclear forensic methods, allowing these methods to be tested and calibrated.

Background: thermonuclear weapon development before Castle Bravo

After a few years of researching less fruitful designs, in 1951 the USA began to pursue the line of research that ultimately led to successful deployment of thermonuclear weapons. In March and April 1951, Edward Teller and Stanisław Ulam proposed a set of ideas for the design of thermonuclear weapons, which has become known as the Teller–Ulam design.[137] This design included the following features. First, it employed the idea of 'separate stages', where the detonation of the fission explosive device at the first stage provides the necessary energy to compress the thermonuclear materials in the separate, adjacent second stage. Second, Teller and Ulam introduced the idea of implosion of the second stage by radiation. Third, the area of thermonuclear burn was surrounded with a layer of natural or depleted uranium, which would fission after irradiation with high-energy neutrons. The thermonuclear weapon would thus have three stages: fission, fusion and fission again. A key advantage of the Teller–Ulam design was the possibility of producing weapons with practically unlimited yield.

[136] 'Detection of radioactive gases consistent with North Korean test underlines strength of CTBTO monitoring system', *CTBTO Spectrum*, no. 20 (July 2013), p. 26.

[137] Rhodes, R., *Dark Sun: The Making of the Hydrogen Bomb* (Simon & Schuster: New York, 1995), pp. 466–75.

These ideas were put forward just a few weeks before the Operation Greenhouse weapon test series was to be conducted at Eniwetok Atoll, in the US territory of the Marshall Islands.[138] The first test of the series, named George, was successfully conducted on 9 May 1951. It was planned as a 'physics experiment' in the hope of advancing the previous concept of the Classical Super. The design of the explosive device to be used in the test, known as Cylinder, was frozen in October 1950, months before the Teller–Ulam design was proposed.[139] The original purpose of Cylinder was to test the possibility of fusion of deuterium and tritium by placing 'less than an ounce' of that material next to the 200-kt fission explosion, but it turned out to be fortuitously designed to provide data on radiation implosion as well. The device included complex cryogenic equipment to keep the hydrogen isotopes in liquid form. It produced a yield of 225 kt, 25 kt of which came from the fusion reaction.[140]

A full-scale test of the Teller–Ulam design was done by the famous Mike shot during the Operation Ivy test series at Eniwetok on 1 November 1952. The explosive device used in the test, called Sausage, was a cylinder almost 6.2 metres tall with a diameter of about 2 metres, weighing 82 tonnes. At that time, it was the largest experimental cryogenic device ever constructed, and it contained 'dozens of litres' of liquid deuterium.[141] The USA was aware at the time that the technological extremes associated with liquefied gas can be avoided by using the stable powder of lithium-6 deuteride (^{6}LiD), which can both provide deuterium and 'breed' tritium directly in the weapon by neutron activation of ^{6}Li. However, ^{6}LiD was not used in Ivy Mike, partially because its behaviour in the thermonuclear 'burn' was more difficult to calculate, and partially because it was not yet available in the necessary quantities.[142] Ivy Mike's yield was 10.4 megatons from fission and fusion.[143] Later, various specific aspects of thermonuclear weapon design were tested in shots named Nancy and Simon during Operation Upshot-Knothole in March and April 1953.[144]

Castle Bravo

The first lithium deuteride-fuelled US thermonuclear device incorporating the Teller–Ulam design, named Shrimp, was scheduled to be tested on

[138] Rhodes (note 137), pp. 467, 472.

[139] Hansen, C., *Swords of Armageddon*, vol. 2, *Thermonuclear Weapons Development, 1942–1952* (Chukelea Publications: Sunnyvale, CA, 2007), p. 157.

[140] Hansen, C., *US Nuclear Weapons: The Secret History* (Orion Books: New York, 1988), p. 93; and Rhodes (note 137), p. 474.

[141] Hansen, C., *Swords of Armageddon*, vol. 3, *Thermonuclear Weapons Development, 1952–1954* (Chukelea Publications: Sunnyvale, CA, 2007), pp. 38, 42, 43.

[142] Hansen (note 139), p. 271.

[143] Hansen (note 140), p. 60.

[144] Hansen, C., *Swords of Armageddon*, vol. 7, *Arming & Fuzing: Technologies & Equipment* (Chukelea Publications: Sunnyvale, CA, 2007), pp. 96–99.

1 March 1954 in Bikini Atoll, Marshall Islands. This test, named Bravo, was the first in the Operation Castle series.[145] The US weapon designers wanted the isotope ^{6}Li as a relatively cheap and convenient room-temperature precursor of tritium. Lithium found in nature contains only about 7.6 per cent of ^{6}Li, the rest being ^{7}Li.[146] The lithium used in Shrimp was enriched to 40 per cent ^{6}Li.[147]

The US Atomic Energy Commission estimated that the yield of the Castle Bravo shot would be approximately 6 Mt 'as the most likely figure'.[148] The actual yield of 15 Mt was completely unexpected.[149] The US weapon designers overlooked the fact that ^{7}Li undergoes a (n, 2n) reaction with fast neutrons (i.e. a neutron 'entering' the nucleus 'knocks' two neutrons out). This reaction was essentially multiplying neutrons and simultaneously making more ^{6}Li available for tritium production in the explosion and was thus considerably increasing the yield.[150]

The unexpectedly high yield, exacerbated by the inadequate meteorological forecasting, caused a bigger than expected volume of radioactive fallout to be distributed over a wider area than predicted. It resulted in radiological contamination of personnel at the test site, inhabitants of the Marshall Islands and at least three Japanese fishing vessels.[151]

Sample collection

One of the fishing vessels, the *Fukuryu Maru No. 5*, happened to be a few kilometres outside the 'danger zone' designated by the US Government. The crew witnessed the explosion. They speculated that it was nuclear and, by roughly measuring the time between seeing its flash and hearing its sound, were able to calculate its distance to be about 140 kilometres. A few hours later, clouds of precipitating white dust overtook the boat.[152]

The crew collected this dust on several occasions. On 1 March a crewman collected a small sample of the dust, which he gave to Dr Takanobu Shiokawa of Shizuoka University on 16 March.[153] On 3 March the 'fishing master', Yoshio Misaki, collected a sizeable quantity of ash into a vinyl bag,

[145] Rhodes (note 137), p. 541.

[146] De Laeter, J. R. et al., International Union of Pure and Applied Chemistry (IUPAC), 'Atomic weights of the elements: review 2000 (IUPAC Technical Report)', *Pure and Applied Chemistry*, vol. 75, no.6 (June 2003), p. 735.

[147] Hansen (note 144), p. 100.

[148] Hewlett, R. G. and Holl, J. M., *A History of the United States Atomic Energy Commission*, vol. 3, *1952-1960*, DOE/NBM-7010972 (Department of Energy: Washington, DC, 1987), p. VI-23.

[149] Hansen (note 144), p. 101.

[150] Barth, K.-H., Interview with Dr Marshall Rosenbluth, Oral History Transcript, American Institute of Physics, Niels Bohr Library and Archives, 11 Aug. 2003, <http://www.aip.org/history/ohilist/28636_1.html>; Rhodes (note 137), p. 541.

[151] Hansen (note 141), pp. 292–303.

[152] Lapp, R., *The Voyage of the Lucky Dragon* (Frederick Mueller Ltd: London, 1958), pp. 31–33. Fukuryu maru translates as 'lucky dragon'.

[153] Lapp (note 152), pp. 34, 81.

Table 9.3. Results of Soviet radiochemical analysis of debris from the Castle Bravo explosion

Isotope	Concentration ratio to Ba-140
U-237	0.94 ± 0.2
Ag-111	0.073 ± 0.010
Sr-89	0.58 ± 0.06
Ru-103	1.15

Source: Lobikov, E. A. et al., [Development in the USSR of physical methods of long-range detection of nuclear explosions], Proceedings of the Second International Symposium on the History of Atomic Projects (HISAP'99), International Institute for Applied Systems Analysis (IIASA), Laxenburg, 4–8 Oct. 1999.

which was later delivered to Dr Kenjiro Kimura of the University of Tokyo.[154] On 17 March Dr Yasushi Nishiwaki of Osaka City University collected some radioactive dust from the *Fukuryu Maru*, and later rainwater, seawater, fish and contamination from other boats.[155]

It is not clear if the Soviet Union had direct access to the *Fukuryu Maru* or the ash collected from it. By 1954, however, it had deployed 120 'gauze trays' across its territory and began to conduct regular flights of an aircraft equipped with specialized particle-collection equipment between Leningrad and Odessa and from bases in China.[156] In April or May 1954 those measures allowed collection of radioactive debris from the Castle Bravo test. This was reportedly the first instance of intentional debris collection by the Soviet Union from a foreign nuclear weapon test.[157]

The British Government was informed in advance by the USA of the forthcoming Castle test series. The USA also gave the UK an opportunity to monitor the tests by providing 'facilities' at Kwajalein, the largest of the Marshall Islands.[158] The British effort to monitor the Castle test series, which included collection of radioactive debris, was given the code name Operation Aconite.

[154] Lapp (note 152), pp. 44, 99.

[155] Lapp (note 152), p. 89; and Nishiwaki, Y., 'Bikini ash', *Atomic Scientists Journal*, vol. 4, no. 2 (Nov. 1954), p. 98.

[156] Vasil'ev, A. P., *Rozhdennaya Atomnym Vekom* [Created by the nuclear age], vol. 1 (Self published, Moscow, 2002), p. 8 (in Russian); Lobikov, et al., [Development in the USSR of physical methods of long-range detection of nuclear explosions], and Vasil'ev, A. P., [The initial stage of development of the nuclear explosions detection system in the USSR], Proceedings of the Second International Symposium on the History of Atomic Projects (HISAP'99), International Institute for Applied Systems Analysis (IIASA), Laxenburg, 4–8 Oct. 1999, p. 2. The proceedings of HISAP'99 were prepared but never published. Referenced texts were provided to SIPRI by the Kurchatov Institute, Moscow. Information on the symposium is available at <http://webarchive.iiasa.ac.at/Admin/INF/PR/PR-99-10-08.html>.

[157] Lobikov et al. (note 156), p. 1.

[158] Goodman, M. S., *Spying on the Nuclear Bear: Anglo-American Intelligence and the Soviet Bomb* (Stanford University Press: Stanford, CA, 2007), pp. 99–100, 111.

Sample characterization

According to one account of fallout analysis by Japanese scientists, the examination of dust by optical microscopy revealed that most of the dust particles were about 0.2–0.3 millimetres in diameter, had a porous structure and resembled volcanic ash.[159] Further physical characterization with 'electron micro diffraction and X-ray diffraction' showed that it was calcite, probably recrystallized after the explosion from aragonite from Bikini Atoll.[160] A group of scientists at Kyoto University used electron microscopy to confirm that the dust contained granules with diameter of about 0.1–0.5 mm (0.3 mm on average), which, in turn, consisted of 'finer unit particles of the size 0.1–3 [μm] with cubic or spindle shapes'.[161]

Chemical characterization revealed that it was 55.2 per cent calcium oxide, 7.0 per cent magnesium oxide, 11.8 per cent carbon dioxide and 26.0 per cent water.[162] Isotopic characterization was done by all four universities—Kyoto, Osaka City, Shizuoka and Tokyo—using radiochemical methods ('with the ordinary method of chemical analysis with carrier, as well as with the ion exchange method').[163] This yielded the same results: the presence of the fission products (^{89}Sr, ^{90}Sr, ^{90}Y, ^{91}Y, ^{95}Zr, ^{95m}Nb, ^{95}Nb, ^{103}Ru, ^{106}Ru, ^{106}Rh, ^{111}Ag, ^{125}Sb, ^{127}Sb, ^{127}Te, ^{129m}Te, ^{129}Te, ^{132}Te, ^{131}I, ^{132}I, ^{140}Ba, ^{140}La, ^{141}Ce, ^{143}Ce, ^{144}Ce, ^{143}Pr, ^{144}Pr, ^{147}Nd), activation products of calcium and chlorine from the environment (^{45}Ca and ^{35}S, respectively), some ^{239}Pu from the weapon and, unexpectedly, the rare isotope ^{237}U.

The Soviet radiochemical characterization of Castle Bravo was more interested in calculating ratios of specific isotopes than in their detection per se (see table 9.3). These ratios are more informative for discerning the design of the explosive device, which was the Soviet goal.[164] In contrast, the Japanese scientists approached this study from the radiation protection and general interest points of view.

Nuclear forensic interpretation

Kimura was in the group of five distinguished Japanese scientists who discovered the isotope ^{237}U in 1940.[165] He knew that ^{237}U is produced from ^{238}U in a (n, 2n) reaction with fast neutrons. This reaction is only possible

[159] Lapp (note 152), pp. 144, 148.
[160] Nishiwaki, Y., 'Effects of H-bomb tests in 1954', *Atomic Scientists Journal*, vol. 4, no. 5 (May 1955), p. 282.
[161] Nishiwaki (note 160), p. 282.
[162] Lapp (note 152), pp. 144, 148.
[163] Nishiwaki (note 160), p. 283.
[164] On early Soviet methodologies of debris analysis see chapter 7 in this volume.
[165] Nishina, Y. et al., 'Induced β-activity of uranium by fast neutrons', *Physical Review*, vol. 57, no. 12 (June 1940), p. 1182.

with neutrons with energies higher than 6.6 megaelectronvolts (MeV).[166] He also knew that a fission bomb does not produce fast neutrons with energies greater than 6.6 MeV unless they are boosted or have thermonuclear stages. In fact, even the most common thermonuclear reaction of fusion of two atoms of deuterium (a D–D reaction) only produces neutrons with energies of 2.45 MeV. The presence of ^{237}U therefore indicated that fusion of deuterium and tritium (a D–T reaction) was taking place, producing neutrons with energies of 14 MeV.

Kimura could therefore interpret the characterization results to conclude that the explosion was partially thermonuclear and that the device contained a sizeable amount of ^{238}U. He consulted Dr Mituo Taketani of St Paul University (Rikkyo University), Tokyo, and together they concluded that the device that exploded on 1 March was thermonuclear and had three 'stages', as described above. Taketani estimated that 'a few hundred kilograms' of natural uranium had fissioned at the third stage of the detonation, and understood that this technology allowed for detonations 'without limit in power'.[167] The British scientist Joseph Rotblat reached similar conclusions on the basis of measurements provided to him by Nishiwaki.[168]

The British Government's effort to develop thermonuclear weapons reportedly benefited from debris samples from the Castle test series, most likely including Bravo, collected as part of Operation Aconite.[169] Information obtained from radiochemical analysis of those samples reportedly 'confirmed ideas' that already existed in the British nuclear weapon establishment.[170]

Since the date of the test was well known due to the press coverage of the *Fukuryu Maru* incident, Soviet scientists were able to test and calibrate their theoretical models for calculation of explosion time from the decay rate of fission products.[171] The first methods for extraction of debris from nuclear explosions from air filters, trays with filter material and soil samples and their further radiochemical analysis were developed in the Soviet Union between 1952 and 1954.[172] These methods were capable in principle of determining a weapon's type (i.e. whether nuclear or thermonuclear), the type of nuclear explosive (i.e. HEU or plutonium) and the presence of

[166] Lapp, R. E., 'Local fallout radioactivity', *Bulletin of the Atomic Scientists*, vol. 15, no. 5 (May 1959), p. 182; and Knight, J. D., Smith, R. K. and Warren, B., '$U^{238}(n,2n)U^{237}$ cross section from 6 to 10 MeV', *Physical Review*, vol. 112, no. 1 (Oct. 1958), p. 261.

[167] Lapp (note 152), pp. 148, 149.

[168] Rotblat, J., 'The hydrogen-uranium bomb', *Bulletin of the Atomic Scientists*, vol. 11, no. 5 (May 1955).

[169] Goodman (note 158), p.111.

[170] Arnold, L., *Britain and the H-Bomb* (Palgrave: Basingstoke, 2001), p. 91.

[171] Lobikov et al. (note 156), p. 1.

[172] Vasil'ev, HISAP'99 (note 156), p. 2.

^{238}U.[173] The first known discussion by Soviet weapon designers of the presence of ^{237}U in foreign weapon debris as an indicator of thermonuclear explosion occurred in a memorandum written on 23 August 1952.[174]

In all the cases discussed above, the analysis of fallout allowed the fission–fusion–fission nature of the device to be deduced by determining the thermonuclear nature of the explosion and the presence of ^{238}U fissioning by fast neutrons. As shown in an article published decades after the Castle Bravo test, information on another crucial feature of the Teller–Ulam design—the physically separated primary and secondary—can in principle be derived from the weapon debris.[175] However, there is no indication in available sources that any of the scientists involved was able to deduce at the time that fission and fusion stages were spatially separate, or that the implosion of a thermonuclear secondary by radiation has occurred.

IV. Combating nuclear trafficking

The examples described above essentially involve the behaviour of a state being investigated by other states or international organizations. In the case of trafficking of nuclear or other radioactive materials, the focus is on the behaviour of individuals, and often requires investigations of domestic crimes by domestic law-enforcement agencies (albeit in the context of international agreements and in cooperation with other states). In addition to forensic analysis of nuclear material, trafficking investigations may draw on all sources of information about the circumstances surrounding the sample, such as the movements and motivations of the traffickers. Another difference with the above examples is that there is no overarching multilateral legal framework for combatting nuclear trafficking. Instead, nuclear security is primarily the responsibility of individual states, and international cooperation in this area is voluntary and subject to goodwill.

Nuclear smuggling by individuals is a relatively new phenomenon, not known before the early 1990s.[176] For this reason, a lot of relevant information has not yet been published, although since 1995 the IAEA has maintained the Incident and Trafficking Database (ITDB), which records inci-

[173] Lobikov et al. (note 156), p. 2. See also chapter 6 in this volume.

[174] Vasil'ev, A. P., *Rozhdennaya Atomnym Vekom* [Created by the nuclear age], vol. 3 (Self published, Moscow, 2002), p. 224.

[175] De Geer, L.-E., 'The radioactive signature of the hydrogen bomb', *Science and Global Security*, vol. 2, no. 4 (1991).

[176] Fissile materials have allegedly been stolen before 1990s. E.g. there are allegations that Israel stole c. 100 kg of weapon-grade uranium from a US plant in the early 1960s. Gilinsky, V. and Mattson, R. J., 'Did Israel steal bomb-grade uranium from the United States?', *Bulletin of the Atomic Scientists*, 17 Apr. 2014, <http://thebulletin.org/did-israel-steal-bomb-grade-uranium-united-states7056>. However, trafficking of fissile materials by non-state actors was not an issue before the break-up of the Soviet Union.

Table 9.4. Analytical methods used for characterization of uranium pellets confiscated by Hungary

Parameter	Obtainable information	Analytical techniques or instruments
Dimensions	Reactor type, intended use	Micrometer, measuring gauge
Uranium content	Chemical composition	Titration, hybrid K-edge densitometry, isotope dilution mass spectrometry
Isotopic composition	Reactor type, intended use	High-resolution gamma spectrometry, thermal ionization mass spectrometry (TIMS)
Impurities	Production process or facility	Inductively coupled plasma mass spectrometry (ICP-MS)
Age	Production date	Alpha spectrometry, TIMS and multi-collector ICP-MS
Microstructure	Production process	Scanning electron microscopy, transmission electron microscopy

Source: Stefanka, Z. et al., 'Hungarian joint analysis: report on investigation of uranium pellets', Technical Note JRC-ITU-TN-2007/44, Institute for Transuranium Elements, 11 Sep. 2007, p. 5.

dents of nuclear trafficking and other events involving radioactive material beyond regulatory control.[177]

In Europe during the 1990s nuclear fuel pellets and other fuel elements constituted a significant part of intercepted nuclear material.[178] Below is the description of an investigation of fuel pellets intercepted in Hungary on three separate occasions.

Fuel pellets intercepted in Hungary: background

Hungarian officials detected 10 nuclear trafficking incidents during the 1990s.[179] On at least three occasions the nuclear material was intercepted and identified as uranium oxide fuel pellets: once in 1992 and twice in 1995. The material was studied at the time by Hungarian authorities. Additionally, in 2006 the Institute of Isotopes (Izotópkutató Intézet, IKI), of the Hungarian Academy of Sciences and the Institute for Transuranium Elements (ITU), Karlsruhe, of the Joint Research Centre of the European

[177] On the ITDB see chapter 2 in this volume.

[178] Koch, L. et al., 'International cooperation in combating illicit trafficking of nuclear materials by technical means', European Safeguards Research and Development Association (ESARDA), *21st ESARDA Annual Meeting*, Seville, 4–6 May 1999 (ESARDA: Ispra, 1999), p. 809.

[179] Hungarian Atomic Energy Authority (Országos Atomenergia Hivatal, OAH), *Nuclear Non-Proliferation Activities in Hungary 1999-2009* (OAH: Budapest, [2010]), p. 30.

Table 9.5. Dimensions and weights of pellets confiscated by Hungary

Pellet no.	Weight (g)	Diameter (mm)	Height (mm)
590-1	16.0061	12.33	13.11
590-2	15.6353	12.35	12.69
590-5	17.1390	12.31	14.01
642-1	15.3515	11.43	14.30
642-2	15.9104	11.42	14.88
642-5	15.5325	11.42	14.70
643-1	2.7391	5.81	10.12
643-2	2.7110	5.82	10.08
643-5	2.1489	5.81	8.3 (chopped)

Source: Stefanka, Z. et al., 'Hungarian joint analysis: report on investigation of uranium pellets', Technical Note JRC-ITU-TN-2007/44, Institute for Transuranium Elements, 11 Sep. 2007, p. 7.

Commission began a joint analysis of the intercepted pellets in order to both validate existing methods of analysis, train in collaborative analysis procedures and provide more information in the context of the national nuclear forensic library effort.

From each of the three intercepted batches (numbered 590, 642 and 643), the IKI selected five pellets for measurement (numbered 590-1, 590-2, 590-3 etc.). The IKI conducted some measurements itself and then sent three pellets from each group of five to the ITU for joint analysis.[180]

Characterization of the fuel pellets

The analysis was done in accordance with an established sequence (see table 9.4). First, the pellets were inspected visually, their shapes were described, and their dimensions and weights taken (see table 9.5). It was noted that pellets showed signs of damage, which might have been an indication of 'inappropriate storage and transport conditions', as well as that 'the pellets were rejected during the production process and considered as scrap (intended to be recycled)'.[181]

Second, the ITU chose one pellet of each group of three to be kept as an archive, one as a sample for electron microscopy examination and one for further destructive measurements (pellet numbers 590-2, 642-1 and 643-2). It determined the uranium content of these three pellets to be very close to

[180] Stefanka, Z. et al., 'Hungarian joint analysis: report on investigation of uranium pellets', Technical Note JRC-ITU-TN-2007/44, Institute for Transuranium Elements, 11 Sep. 2007, pp. 5, 13; Mayer, K. et al., 'Recent advances in nuclear forensic science', European Safeguards Research and Development Association (note 36); and Koch et al. (note 178), p. 809.

[181] Mayer et al. (note 180), p. 4.

Table 9.6. Isotopic composition of uranium in pellets confiscated by Hungary, measured by thermal ionization mass spectrometry

Figures are in weight per cent, with confidence interval (2μ) given in brackets.

Isotope	Pellet 590-2	Pellet 642-1	Pellet 643-2
U-232	..	0.000000032(9)[a]	..
U-234	0.00494(31)	0.0347(21)	0.00128(80)
U-235	0.71121(41)	2.5121(14)	0.25501(15)
U-236	..	0.47(44)	0.0061(57)
U-238	99.2839(20)	96.9823(20)	99.7376(20)

[a] The IKI determined this figure by low-background gamma-spectrometry using another pellet from the same batch (642).

Source: Stefanka, Z. et al., 'Hungarian joint analysis: report on investigation of uranium pellets', Technical Note JRC-ITU-TN-2007/44, Institute for Transuranium Elements, 11 Sep. 2007, p. 8.

88 per cent—a theoretical value for UO_2, the most common nuclear fuel material.

Third, in order to compare and validate measurement methods, the IKI and the ITU both measured the isotopic composition of uranium in the three pellets 590-2, 642-1 and 643-2 using high-resolution gamma spectrometry and a number of mass spectrometry techniques (see e.g. table 9.6).[182] A measurement with just one or two methods would probably have been enough for most normal nuclear forensic investigations.

Fourth, the age of the uranium in the pellets (i.e. time passed since previous chemical purification) was determined using the uranium-234 : thorium-230 ratio, where the amount of Th-230 was measured by alpha spectrometry and the amount of U-234 was known from previous measurements (see table 9.7).[183]

Fifth, the microstructure of the pellets was investigated by scanning electron microscopy (SEM). The SEM images of pellet surfaces demonstrated that the pellets from batches 590 and 642 had similar microstructures, consisting of homogeneous crystals of similar size and shape. This suggests a similar production technology for these two batches. The pellet from batch 643 had a very different surface microstructure, with much smaller crystals or grains, suggesting a different production technology for this batch.[184] Finally, the impurities in the samples (27 chemical elements) were determined by sector-field ICP-MS.[185]

[182] Stefanka et al. (note 180), p. 8.
[183] For detailed discussion of age determination of nuclear materials see chapter 5 in this volume.
[184] Stefanka et al. (note 180), p. 12.
[185] Stefanka et al. (note 180), p. 9.

Table 9.7. Results of the uranium age determination in pellets confiscated by Hungary

Results	Pellet 590-2	Pellet 642-1	Pellet 643-2
Age (years)[a]	16.85 ± 0.3	13.6 ± 0.2	16.8 ± 0.3
Year of production	1989	1993	1990

[a] The age was determined in March 2007.

Source: Stefanka, Z. et al., 'Hungarian joint analysis: report on investigation of uranium pellets', Technical Note JRC-ITU-TN-2007/44, Institute for Transuranium Elements, 11 Sep. 2007, p. 10.

Interpretation of the characterization results

The characterization results are summarized in table 9.8.

Batch 590

The pellet 590-2 (and, by extension, the whole batch 590) turned out to be made of natural uranium. The literature search for nuclear fuel pellets made of uranium with nearly natural content of U-235 and diameter and height close to that of the pellet 590-2 would turn up two possible candidates: fuel for CANDU-type reactors or pellets from the breeding blankets of two Soviet-era fast reactors, BN-350 at Aktau, Kazakhstan, and BN-600 at Zarechny, Sverdlovsk Oblast, Russia.[186] As described in chapter 2, the ITU and the High-Technology Scientific Research Institute for Inorganic Materials (VNIINM), Moscow, operate a database for use in identification of trafficked nuclear fuels. Detailed information provided to that database by the VNIINM allowed the ITU and IKI researchers to exclude fuel pellets for BN-350 and BN-600 from further analysis, leaving CANDU-type reactor fuel as the only match.[187]

There are only two CANDU-type reactors in Europe, both located at Cernavodă in Romania, close to Hungary. The first began operation in 1996, and the second in 2007.[188] Nuclear fuel for Cernavodă is produced at Pitești, Romania, where CANDU-type fuel has been manufactured since 1980s. Thus, pellets in batch 590, produced in 1989, might have been manufactured there. Information available to the ITU and IKI researchers from literature allowed them to conclude that pellets from Pitești and from batch 590 are a possible match, even though their heights did not coincide

[186] CANDU is an acronym for 'Canadian deuterium-uranium. See also chapter 5, table 5.1, in this volume.

[187] Mayer et al. (note 180), p. 5.

[188] IAEA, Power Reactor Information System, 'Romania', 23 Dec. 2014, <http://www.iaea.org/PRIS/CountryStatistics/CountryDetails.aspx?current=RO>.

Table 9.8. Summary of characterization results of pellets confiscated by Hungary

Results	Pellet 590-2	Pellet 642-1	Pellet 643-2	Pellet 'Fund 21'
Year of seizure	1992	1995	1995	1994
Year of production	1989	1993	1990	. .
Mass (g)	16.54(68)	16.03(22)	3.18(9)	14.964
Diameter (mm)	12.40(5)	11.50(5)	6.0(1)	12.17
Height (mm)	13.2(5)	14.8(3)	10.5(2)	13.01
U-235 content (weight %)	0.71121(41)	2.5121(14)	0.25501(15)	0.7113

Source: Stefanka, Z. et al., 'Hungarian joint analysis: report on investigation of uranium pellets', Technical Note JRC-ITU-TN-2007/44, Institute for Transuranium Elements, 11 Sep. 2007, 11 Sep. 2007, p. 13; and European Commission, Institute for Transuranium Elements, 'Fund 21', BE/VBM/021/9, 20 Dec. 1994.

perfectly.[189] The ITU researchers noted that fuel pellets with similar characteristics, probably also for CANDU reactors, had been seized in Germany in 1992 (at Windsbach), 1993 (at Munich) and 1994 (at Pforzheim-Ost) and investigated at the ITU.[190] For example, the pellet investigated in 1994 (designated by the ITU as 'Fund 21') had very similar characteristics to that of pellet 590-2 (see table 9.8).

Batch 642

Pellet 642-1 had diameter and height similar to those of the pellets in the fuel of RBMK-1000 reactors built at Kursk and Smolensk, Russia, and at Chernobyl, Ukraine (both then in the Soviet Union).[191] The only two fuel manufacturers for such pellets are Elemash (Mashinostroitel'niy Zavod, MSZ) in Elektrostal, Moscow Oblast, Russia, and Ulba Metallurgical Plant (UMP) in Ust-Kamenogorsk, Kazakhstan.

Reprocessed uranium is uranium recovered from nuclear fuel that was irradiated in the reactor. It contains uranium isotopes not found in natural uranium in quantities relevant in this context, in particular U-236 and U-232, as well as elevated amounts of U-234. The isotopic composition of the fuel pellets in batch 642 (see table 9.6) shows that they are indeed made of reprocessed uranium. Of the two possible fuel pellet producers, only UMP used reprocessed uranium in the manufacture of RBMK-1000 fuel pellets.

189 Stefanka et al. (note 180), p. 13; and Mayer, K., et.al. (note 180), p. 5.

190 Mayer et al. (note 180), p. 5; Mayer, K., Institute for Transuranium Elements, Personal communication with the author; European Commission, Institute for Transuranium Elements, 'Fund 21', BE/VBM/021/9, 20 Dec. 1994, p. 3; and Nuclear Threat Initiative, 'Information from the Federal Ministry for Environment, Nature Protection and Reactor Safety, Germany', 19 Aug. 1996, <http://www.nti.org/analysis/articles/information-federal-ministry-environment-nature-protection-and-reactor-safety-germany/>.

191 See chapter 5, table 5.1, in this volume.

Moreover, at the time that the intercepted pellets were manufactured, in 1993, MSZ had not yet begun to produce such pellets.[192]

Batch 643

Pellet 643-2 was made of depleted, but previously unirradiated, uranium, with a diameter smaller than usually seen in fuel for power reactors that use water as coolant (which accounts for almost all power reactors in the world).

Once nuclear fission heats up nuclear fuel, that heat is transferred by coolant to turbines or steam generators. Water has relatively limited cooling capabilities, which in practice means that it can remove only a limited amount of energy every second from a given surface of the fuel. In other words, there is a maximum allowable surface heat flux for water-cooled fuels (since higher heat flux would lead to burnout of fuel cladding). Because the surface heat flux is inversely proportional to the fuel diameter, there is a practical minimum diameter of fuel pellets in water-cooled reactors.

This problem does not exist in fast reactors that are cooled with liquid sodium, which removes heat much more efficiently.[193] For this reason fuel pellet diameters smaller than about 7 millimeters are more characteristic of fast reactors. Of those, there were probably fewer than 12 operating at the time the pellets of batch 643 were manufactured, in the early 1990s.[194]

The ITU and IKI researchers concluded that pellets in batch 643 were characteristic of the fast breeder reactors BN-350 and BN-600, and that 'this material was probably intended as a fertile material for a breeder reactor' (i.e. the pellets were intended for the part of the reactor core designed to produce plutonium from uranium-238 in the pellets).[195] The pellets for BN-350 and BN-600 were produced by MSZ.[196]

Although the material had been studied at the time of its seizure, it did not lead to any conviction for nuclear trafficking and the full history of the material was not revealed at the time. The subsequent study by the IKI and the ITU was able to determine possible sources and probable intended uses of the uranium pellets. The detailed conclusions reached in this case illustrate what can be achieved by nuclear forensic analysis of radioactive material intercepted in a trafficking case, using only the information inherent to the material itself.

[192] Bibilashvili, Yu. K. and Reshetnikov, F. G., 'Russia's nuclear fuel cycle: an industrial perspective', *IAEA Bulletin*, vol. 35, no. 3 (July/Sep. 1993), pp. 28–31; and Stefanka et al. (note 180), p. 13.

[193] Waltar, A. E. and Reynolds, A. B., *Fast Breeder Reactors* (Pergamon Press: NY, 1981), pp. 47–48.

[194] World Nuclear Association, 'Fast neutron reactors', Dec. 2014, <http://www.world-nuclear.org/info/Current-and-Future-Generation/Fast-Neutron-Reactors/>.

[195] Stefanka et al. (note 180), p. 13.

[196] Bibilashvili and Reshetnikov (note 192), p. 31.

About the authors

Lars-Erik De Geer (Sweden) is a retired nuclear physicist and Associate Professor at the Royal Institute of Technology, Stockholm. He has served as consultant to the United Nations Scientific Committee on the Effects of Atomic Radiation (UNSCEAR) and chaired the international working group that assessed the radiological situation at Fangataufa and Mururoa atolls. In 1997 he was appointed head of the Scientific Methods and Data Fusion Section at the Comprehensive Nuclear-Test-Ban Treaty Organization (CTBTO), Vienna. Until 2012 he worked at the Swedish Defence Research Agency (FOI), studying nuclear debris from nuclear weapon tests and large nuclear power accidents.

Nicole Erdmann (Germany) studied physics and obtained her PhD in 1998 from the nuclear chemistry department of the Johannes Gutenberg University Mainz, Germany. She subsequently worked as a postdoctoral researcher at the European Commission's Joint Research Centre, Institute for Transuranium Elements (JRC-ITU) for two years. From 2000–2003 she worked at the JRC Institute for Environment and Sustainability (Ispra, Italy), from 2003–2006 again at the nuclear chemistry department of the Johannes Gutenberg University Mainz. In August 2006 she returned to JRC-ITU, where she has been involved in the development of analytical techniques for nuclear Safeguards purposes. Since late 2011 she has been the research programme manager of JRC-ITU.

Vitaly Fedchenko (Russia) is a Senior Researcher with the SIPRI European Security Programme, responsible for nuclear security issues and the political, technological and educational dimensions of nuclear arms control and non-proliferation. Previously he was a visiting researcher at SIPRI and worked at the Center for Policy Studies in Russia at the Institute for Applied International Research in Moscow. He is the author or co-author of several publications on international non-proliferation and disarmament assistance, the international nuclear fuel cycle and Russian nuclear exports.

Sophie Grape (Sweden) received her PhD in fundamental nuclear physics in 2009. She currently works as a researcher in nuclear safeguards at Uppsala University in Sweden. Her field of expertise is instrumentation for non-destructive assay of spent nuclear fuel for nuclear safeguards purposes. She supervises PhD students, teaches at the university, is the chair of the European Safeguards Research and Development Association

(ESARDA) working group for Training and Knowledge management and is a member of the Swedish National Council for Nuclear Waste.

Magnus Hedberg (Sweden) has worked at the Nuclear Safeguards and Forensics unit of the European Commission's Joint Research Centre, Institute for Transuranium Elements (JRC-ITU) in Karlsruhe, Germany, since 2006. He heads the LG-SIMS laboratory and his special interest is particle analysis for nuclear safeguards purposes. He received an MSc in electrical engineering at Lund University, Lund, Sweden, in 1987. Previous employments include a period of seven years as a mass spectrometry unit head at the International Atomic Energy Agency (IAEA) in Seibersdorf Austria, two and a half years at St Jude Medical in Järfälla, Sweden, and ten years at the Laboratory for Isotope Geology (LIG) in Stockholm, Sweden.

Robert Kelley (United States) is a veteran of over 35 years in the US Department of Energy (US DOE) nuclear weapons complex, most recently at Los Alamos. He worked in research and engineering before turning to information analysis in the 1980s. He managed the centrifuge and plutonium metallurgy programs at Lawrence Livermore National Laboratory and later was Director of the Department of Energy Remote Sensing Laboratory, the premier US nuclear emergency response organization. He was also seconded by the US DOE to the IAEA where he served twice as a Director of the nuclear inspections in Iraq in 1992 and again in 2001.

Klaus Mayer (Germany) obtained a PhD in radiochemistry and analytical chemistry from the University of Karlsruhe, Germany, in 1987 and subsequently worked at the European Commission's Joint Research Centre, Institute for Transuranium Elements (JRC-ITU) as a post-doctoral researcher. In 1990–96 he worked at the Institute for Reference Materials and Measurement (JRC-IRMM), in Geel, Belgium. In 1996 he returned to JRC-ITU and took responsibility for the nuclear safeguards analytical activities. From 1997 to 2010 he chaired the European Safeguards Research and Development Association Working Group on Destructive Analysis. Since 2004 he has served as co-chairman of the Nuclear Forensics International Technical Working Group (ITWG). Currently, he is in charge of JRC-ITU's activities on combating illicit trafficking and nuclear forensics.

Zsolt Varga (Hungary) received his PhD in chemistry in 2007 from the Institute of Isotopes of the Hungarian Academy of Sciences. After working there as a research fellow until 2008, he joined the nuclear forensic group at the European Commission's Joint Research Centre, Institute for Transuranium Elements (JRC-ITU) as a post-doctorate researcher, then from 2013 as a scientific researcher. His current research interests include the elemental and isotopic analysis of illicit nuclear materials by mass spectrometric and radioanalytical methods and the provenance assessment of uranium samples of unknown origin.

Maria Wallenius (Finland) received an MSc in radiochemistry at the University of Helsinki, Finland, where she worked as a research scientist in the safeguards project, focusing on analysis of uranium fuel pellets by potentiometric titration and thermal ionization mass spectrometry. From 1996 she conducted doctoral studies at the European Commission's Joint Research Centre, Institute for Transuranium Elements (JRC-ITU), in Karlsruhe, Germany, in nuclear forensics, obtaining her PhD in 2001. She continued at ITU as a research scientist and developed new methods using mass spectrometry techniques (especially TIMS and ICP-MS) in the safeguards and nuclear forensics fields. Currently, she coordinates nuclear forensics casework and cooperative nuclear forensic projects at ITU.

Index

Note: page numbers in *italics* refer to figures; page numbers in **bold** refer to tables.

The manufacturer's authorised representative in the EU for product safety is Oxford University Press España S.A. of El Parque Empresarial San Fernando de Henares, Avenida de Castilla, 2 - 28830 Madrid (www.oup.es/en or product.safety@oup.com). OUP España S.A. also acts as importer into Spain of products made by the manufacturer.
Printed and bound by CPI Group (UK) Ltd, Croydon, CR0 4YY
14/09/2026
02216307-0010